MONOGRAPHS ON
STATISTICS AND APPLIED PROBABILITY

General Editors

D. R. Cox, D. V. Hinkley, D. Rubin and B. W. Silverman

The Statistical Analysis of Compositional Data
J. Aitchison

Probability, Statistics and Time
M. S. Bartlett

The Statistical Analysis of Spatial Pattern
M. S. Bartlett

Stochastic Population Models in Ecology and Epidemiology
M. S. Bartlett

Risk Theory
R. E. Beard, T. Pentikäinen and E. Pesonen

Residuals and Influence in Regression
R. D. Cook and S. Weisberg

Point Processes
D. R. Cox and V. Isham

Analysis of Binary Data
D. R. Cox

The Statistical Analysis of Series of Events
D. R. Cox and P. A. W. Lewis

Analysis of Survival Data
D. R. Cox and D. Oakes

Queues
D. R. Cox and W. L. Smith

Stochastic Modelling and Control
M. H. A. Davis and R. Vinter

Stochastic Abundance Models
S. Engen

The Analysis of Contingency Tables
B. S. Everitt

An Introduction to Latent Variable Models
B. S. Everitt

Finite Mixture Distributions
B. S. Everitt and D. J. Hand

Population Genetics
W. J. Ewens

Classification
A. D. Gordon

Monte Carlo Methods
J. M. Hammersley and D. C. Handscomb

Identification of Outliers
D. M. Hawkins

Generalized Linear Models
P. McCullagh and J. A. Nelder

Distribution-free Statistical Methods
J. S. Maritz

Multivariate Analysis in Behavioural Research
A. E. Maxwell

Applications of Queueing Theory
G. F. Newell

Some Basic Theory for Statistical Inference
E. J. G. Pitman

Density Estimation for Statistics and Data Analysis
B. W. Silverman

Statistical Inference
S. D. Silvey

Models in Regression and Related Topics
P. Sprent

Sequential Methods in Statistics
G. B. Wetherill and K. D. Glazebrook

An Introduction to Latent Variable Models
B. S. Everitt

Bandit Problems
D. A. Berry and B. Fristedt

(Full details concerning this series are available from the Publishers)

Regression Analysis with Applications

G. BARRIE WETHERILL

Department of Statistics
The University of
Newcastle upon Tyne
UK

P. DUNCOMBE

Applied Statistics Research Unit
University of Kent at Canterbury
UK

M. KENWARD
and the late
J. KÖLLERSTRÖM

Mathematical Institute
University of Kent at Canterbury
UK

S. R. PAUL

Department of Mathematics and Statistics
University of Windsor
Canada

B. J. VOWDEN

Mathematical Institute
University of Kent at Canterbury
UK

LONDON NEW YORK
CHAPMAN AND HALL

First published in 1986 by
Chapman and Hall Ltd
11 New Fetter Lane, London EC4P 4EE
Published in the USA by
Chapman and Hall
29 West 35th Street, New York NY 10001

Printed in Great Britain at the
University Printing House, Cambridge

ISBN 0 412 27490 6

British Library Cataloguing in Publication Data

Regression analysis with applications.——
(Monographs on statistics and applied probability)
1. Regression analysis
I. Wetherill, G. Barrie II. Series
519.5′36 QA278.2

ISBN 0–412–27490–6

Library of Congress Cataloging in Publication Data

Regression analysis with applications.
(Monographs on statistics and applied probability)
Includes bibliographies and index.
1. Regression analysis. I. Wetherill, G. Barrie.
II. Series.
QA278.2.R44 1986 519.5′36 86–9751

ISBN 0–412–27490–6

Contents

Sections with an asterisk contain proofs and other more mathematical material and may be omitted by readers interested only in results.

Preface

Multiple regression analysis is one of the most used – and misused – statistical techniques, but elementary texts go little further than 'least squares theory' and residuals plotting.

This volume arose out of work carried out for ICI (Mond) to write a user-friendly program which contained validity checks and diagnostics. A group was formed to carry out the work and many people were consulted during the research. We owe a particular debt to Dr Al Baines, whose vision set us going and whose constant advice was a help in many ways.

We assume that the reader has had a first course in regression covering 'least squares theory', practical use of multiple regression and residuals. Chapters 3–5 cover some algebraic and numerical problems involved in regression analysis, with special emphasis on the problem of multicollinearity and on the use of generalized inverses. Chapter 3 is preliminary to this section. It contains a discussion of 'sweep' methods of obtaining regression models, methods for updating regressions, and a treatment of eigenvalues and eigenvectors to facilitate a thorough understanding of later chapters.

Chapters 6–9 contain a detailed study of diagnostic checks and plots with an emphasis on those which can be routinely incorporated into statistical packages. Many diagnostic plots, such as the normal plot, are subjective and although they are extremely valuable for presentation and interpretation, they did not satisfy our requirements. We wanted techniques which could be used for routine diagnosis and output messages saying that there was a problem. As we shall see, this objective can only be partly met and further work needs be done.

Chapters 10–12 cover model selection and prediction, including the fitting of response surface models. The final chapter contains an introduction to models in which the usual assumptions about errors are relaxed in various ways.

I am indebted to my co-authors, who have all undertaken a major responsibility for part of the work. The sudden death of my colleague, Dr Julian Köllerström, on December 31, 1983, left us a problem with Chapters 7, 8 and 9. I have considerably revised his original drafts and Professor S. C. Pearce helped with Chapter 7. The main responsibility for other chapters was as follows: Chapters 2 and 4, P. Duncombe; Chapters 3 and 5, B. J. Vowden and G. B. Wetherill; Chapter 12, M. Kenward; Chapter 6, S. R. Paul. Sections 4 and 5 of Chapter 13, and the exercises, were contributed by Dr C. M. Theobald, based on some joint work by him and myself. The whole project was a team effort, and we hope that the result shows the value of combining different people's expertise in a project like this.

A number of people have commented on drafts of the book, including P. Kelly, R. E. Kempson, G. Thackray, G. Wetherill, and M. K. Williams.

Finally, I am indebted to ICI (Mond) for permission to publish the data set used in Chapter 12, and to Professor D. Ruppert for permission to use the final data set in the Appendix.

Most people realize that for safe application, a somewhat deeper understanding of multiple regression is necessary than is usual in the standard first course. We hope that this book will help to provide this. We also hope that practitioners will be able to select and use some of the techniques.

G. Barrie Wetherill
December 1985

CHAPTER 1

A review of multiple regression by least squares

1.1 Introduction

For many years multiple regression analysis has been one of the most heavily-used techniques in statistics, and there are applications of it in many different areas. For example, there are applications to control charts (Mandel, 1969), calibration (Mendenhall and Ott, 1971), biology and medicine (Armitage, 1971), survey analysis (Holt *et al.*, 1980) and to time series data (Coen *et al.*, 1969). Unfortunately, the technique is often abused and two factors contribute to this:

(1) The basic mathematics is simple, leading to the idea that all that is necessary to use multiple regression is to program some standard formulae.
(2) There are many computer programs enabling the unwary to proceed with the regression analysis calculations, whether or not the analysis fits.

Even some applications by experts have been criticized (see Coen *et al.* (1969), and the discussion on that paper). It was back in 1966 that Professor G. E. P. Box wrote a paper (Box, 1966) on the use and abuse of regression.

One simple, but unfortunately real, example will suffice here. One user found that at the end of each regression analysis the multiple correlation coefficient was printed out. He also found that by taking the option of 'regression through the origin', this multiple correlation coefficient could be dramatically increased. Thereafter, this user always used regression through the origin!! A blunder of this kind is easily spotted by those with more knowledge of statistics. However, there are other difficulties and dangers which are not so easy to spot.

The work in this volume arose out of a request by ICI (Mond

Division) to the University of Kent at Canterbury to write a user-friendly regression analysis program which incorporated checks of assumptions, etc., and which would guide people round difficulties. A research group was set up and the work was divided between us. Eventually a program (U-REG) came out of the project. This volume does not set out to discuss U-REG directly, but instead it offers a coverage of some of the more important points related to validation of models, checks of assumptions, etc.

It is our opinion that some of these checks are so readily performed, and the costs of abuse so great, that programs lacking the kind of facilities and checks given here should be under very restricted use.

In the remainder of this chapter our purposes are:

(1) To introduce two numerical examples which will be used in the text.
(2) Review some well-known results in least squares regression, and by so doing introduce some notation.
(3) To comment on assumptions which need checking.

1.2 Two examples

It will be helpful to have some examples ready to hand and we shall use the following data sets.

EXAMPLE 1.1

There are many applications of regression to the fitting of quadratic response surfaces and this is illustrated by the data given in Table 1.1. The variables are

x_1: reactor temperature in °C;
x_2: ratio of H_2 to n-heptane (mole ratio);
x_3: contact time (in seconds);
x_4: conversion of n-heptane to acetylene (in %).

It is anticipated that an equation of the following form would fit the data

$$E(X_4) = \beta_0 + \sum \beta_i x_i + \sum \beta_{ii} x_i^2 + \sum_{i<j}\sum \beta_{ij} x_i x_j,$$

$$V(X_4) = \sigma^2.$$

This model is linear in the unknown parameters β_i, β_{ij} and β_{ii}, and can be fitted by the ordinary least squares (OLS) procedure. The objective in the analysis of such data is to determine values of the explanatory variables which give an optimum value of X_4, the response variable, and to determine the shape of the response surface near the optimum. This particular data set has

Table 1.1 *Data for Example 1.1.*

x_1 *Reactor temperature* (°C)	x_2 *Ratio of* H_2 *to n-heptane (mole ratio)*	x_3 *Contact time (sec.)*	x_4 *Conversion of n-heptane to acetylene (%)*
1300	7.5	0.0120	49.0
1300	9.0	0.0120	50.2
1300	11.0	0.0115	50.5
1300	13.5	0.0130	48.5
1300	17.0	0.0135	47.5
1300	23.0	0.0120	44.5
1200	5.3	0.0400	28.0
1200	7.5	0.0380	31.5
1200	11.0	0.0320	34.5
1200	13.5	0.0260	35.0
1200	17.0	0.0340	38.0
1200	23.0	0.0410	38.5
1100	5.3	0.0840	15.0
1100	7.5	0.0980	17.0
1100	11.0	0.0920	20.5
1100	17.0	0.0860	29.5

been used frequently in discussions on regression problems. The data are as shown in Table 1.1 (from Marquardt and Snee, 1975).

An analysis of Example 1.1 looks fairly straightforward and all that appears to be necessary is to fit the given quadratic model using least squares. In fact, such a straightforward fitting approach with this data set should be very likely to produce misleading or wrong results. The data set has two problems within it. Firstly, the explanatory variables vary in size by more than 10^4 so that the sums of squares vary by more than 10^8. This leads to a rather ill-conditioned matrix in the least squares analysis. Secondly, the data set has a condition referred to as a multicollinearity; a multicollinearity exists in a data set if there is a near linear relationship among the explanatory variables. The presence of a multicollinearity in a data set leads to the estimated regression coefficients being 'unstable'. The effect of these problems is that rounding errors in the fitting procedure can cause changes in significant places of the estimated regression coefficients.

Some years ago the Royal Statistical Society circulated some regression data sets and asked for people to try fitting regressions using the computer packages available to them. When the results were collated the regression coefficients fitted by different programs (on different computers) did not always even have the same sign! While the data set of Example 1.1 is nowhere near as extreme as those circulated, if does illustrate the sort of problem which can arise.

EXAMPLE 1.2 (Prater, 1956).

In the petroleum refining industry scheduling and planning plays a very important role. One phase of the process is the production of petroleum spirit

Table 1.2 *Data on percentage yields of petroleum spirit.*

x_1	x_2	x_3	x_4	y
38.4	6.1	220	235	6.9
40.3	4.8	231	307	14.4
40.0	6.1	217	212	7.4
31.8	0.2	316	365	8.5
40.8	3.5	210	218	8.0
41.3	1.8	267	235	2.8
38.1	1.2	274	285	5.0
50.8	8.6	190	205	12.2
32.2	5.2	236	267	10.0
38.4	6.1	220	300	15.2
40.3	4.8	231	367	26.8
32.2	2.4	284	351	14.0
31.8	0.2	316	379	14.7
41.3	1.8	267	275	6.4
38.1	1.2	274	365	17.6
50.8	8.6	190	275	22.3
32.2	5.2	236	360	24.8
38.4	6.1	220	365	26.0
40.3	4.8	231	395	34.9
40.0	6.1	217	272	18.2
32.2	2.4	284	424	23.2
31.8	0.2	316	428	18.0
40.8	3.5	210	273	13.1
41.3	1.8	267	358	16.1
38.1	1.2	274	444	32.1
50.8	8.6	190	345	34.7
32.2	5.2	236	402	31.7
38.4	6.1	220	410	33.6
40.0	6.1	217	340	30.4
40.8	3.5	210	347	26.6
41.3	1.8	267	416	27.8
50.8	8.6	190	407	45.7

from crude oil and its subsequent processing. To enable the plant to be scheduled in an optimum manner, an estimate is required of the percentage yield of petroleum spirit from the crude oil, based upon certain rough laboratory determinations of properties of the crude oil. Table 1.2 shows values of actual per cent yields of petroleum spirit (y) and four explanatory variables as follows

x_1: specific gravity of the crude, a function of the API measurement;
x_2: crude oil vapour pressure, measured in pounds per square inch absolute;
x_3: the ASTM 10% distillation point, in °F;
x_4: the petroleum fraction endpoint, in °F.

Variables x_1 to x_4 are properties of the crudes. Variables x_3 and x_4 are measurement of volatility; they measure the temperatures at which a given amount of the liquid has been distilled. It is required to use this data to provide an equation for predicting the per cent yield of petroleum spirit from measurements of these four variables.

Again an analysis of Example 1.2 data looks very straightforward. All we seem to have to do is to fit a linear regression of y on the four explanatory variables and then test to see if any of them can be deleted. However, if the data is reordered as in Table 1.3, we see that there is a very definite structure in the data set and a straightforward multiple regression analysis is not appropriate.

There are seen to be only ten crudes present, not 32. In addition, there are regressions of x_4 on y nested within the ten crudes. The analysis should therefore check whether or not a common regression of y on x_4 is acceptable. There are also other features which arise here, due to the variability due to the crudes, which affect variances, confidence intervals, etc., see Daniel and Wood (1980, Chapter 8) for details.

The point here is that a straightforward regression analysis would be likely to have missed the special structure in the data, and this emphasizes the importance of the data exploration techniques discussed in Chapter 2.

Another common problem with regression data is the presence of outliers, and two examples are introduced in Chapter 6 to illustrate methods for detecting them. The presence of outliers can easily be 'masked' (see Chapter 6) and a few outliers can easily lead to a totally wrong interpretation.

At this point we suggest that the reader should study the data sets given in the Appendix, and to attempt an analysis of them before

Table 1.3 *Reordered data for Example 1.2.*

x_1	x_2	x_3	x_4	y
31.8	0.2	316	365	8.5
			379	14.7
			428	18.0
32.2	2.4	284	351	14.0
			424	23.2
	5.2	236	267	10.0
			360	24.8
			402	31.7
38.1	1.2	274	285	5.0
			365	17.6
			444	32.1
38.4	6.1	220	235	6.9
			300	15.2
			365	26.0
			410	33.6
40.0	6.1	217	212	7.4
			272	18.2
			340	30.4
40.3	4.8	231	307	14.4
			367	26.8
			395	34.9
40.8	3.5	210	218	8.0
			273	13.1
			347	26.6
41.3	1.8	267	235	2.8
			275	6.4
			358	16.1
			416	27.8
50.8	8.6	190	205	12.2
			275	22.3
			345	34.7
			407	45.7

consulting the references given. It will soon become apparent that multiple regression analysis is not as straightforward as it might seem, and that an almost endless variety of problems can occur. This volume sets out to examine some of the common problems, and to propose techniques which can be used routinely so as to warn users of the presence of outliers, special data structures, etc., which need closer examination.

Before proceeding, we first give a brief review of least squares. Proofs will not be given in the brief review which follows but page references will be given to Wetherill (1981) where adequate proofs may be found.

1.3 Basic theory

A regression model is one in which the expected value of one variable, called the response variable, is related to the actual values of other variables, called explanatory variables. Thus in Example 1.1 the objective is to fit an equation relating the expected value of X_4 to the actual values of X_1, X_2 and X_3. In linear regression, the models used are limited to those which are linear in the unknown parameters. A general form for such a model is to consider an $n \times 1$ vector $\mathbf{Y}$ (representing the response variable), and $n \times m$ matrix $\boldsymbol{a}$ (representing the explanatory variables), and an $m \times 1$ vector $\boldsymbol{\theta}$ of unknown parameters. We assume that the rank of $\boldsymbol{a}$ is $k \leq m$ and that the model is written

$$\begin{aligned} E(\mathbf{Y}) &= \boldsymbol{a}\boldsymbol{\theta}, \\ V(\mathbf{Y}) &= \boldsymbol{I}\sigma^2, \end{aligned} \tag{1.1}$$

where $\boldsymbol{I}$ is the identity matrix. In the discussion here we shall assume that $m = k$, but in Chapter 5 we deal with the more general case.

The sum of squares is

$$S = (\mathbf{Y} - \boldsymbol{a}\boldsymbol{\theta})'(\mathbf{Y} - \boldsymbol{a}\boldsymbol{\theta}), \tag{1.2}$$

and by differentiating with respect to $\boldsymbol{\theta}$ we get the 'normal equations' for the least squares estimators

$$(\boldsymbol{a}'\boldsymbol{a})\hat{\boldsymbol{\theta}} = \boldsymbol{a}'\mathbf{Y} \tag{1.3}$$

leading to

$$\hat{\boldsymbol{\theta}} = (\boldsymbol{a}'\boldsymbol{a})^{-1}\boldsymbol{a}'\mathbf{Y} \tag{1.4}$$

if $(\boldsymbol{a}'\boldsymbol{a})$ is non-singular.

It is then readily seen (Wetherill, 1981, p. 110) that the covariance matrix of the parameter estimators is

$$V(\hat{\boldsymbol{\theta}}) = (\boldsymbol{a}'\boldsymbol{a})^{-1}\sigma^2. \tag{1.5}$$

If we define the fitted values $\hat{\mathbf{Y}}$, then we have

$$\hat{\mathbf{Y}} = \boldsymbol{a}\hat{\boldsymbol{\theta}}$$

and the residuals are

$$\begin{aligned} \mathbf{R} &= \mathbf{Y} - \hat{\mathbf{Y}} = \mathbf{Y} - \boldsymbol{a}(\boldsymbol{a}'\boldsymbol{a})^{-1}\boldsymbol{a}'\mathbf{Y} \\ &= (\boldsymbol{I} - \boldsymbol{H})\mathbf{Y}, \end{aligned} \tag{1.6}$$

where $\boldsymbol{H}$ is the projection matrix, sometimes called the 'hat' matrix

$$\boldsymbol{H} = \boldsymbol{a}(\boldsymbol{a}'\boldsymbol{a})^{-1}\boldsymbol{a}'.$$

It now follows that the covariance matrix of the residuals is (Wetherill, 1981, p. 114)

$$V(\mathbf{R}) = (\boldsymbol{I} - \boldsymbol{H})\sigma^2. \tag{1.7}$$

It should be noted that this is an $n \times n$ matrix.

1.4 Sums of squares

The minimized sum of squares, or sum of squared residuals is given by inserting (1.4) into (1.2) leading to

$$S_{\text{min}} = \mathbf{Y}'\mathbf{Y} - \hat{\boldsymbol{\theta}}'\boldsymbol{a}'\mathbf{Y}. \tag{1.8}$$

Clearly, another expression for S_{min} is

$$S_{\text{min}} = \mathbf{R}'\mathbf{R}.$$

Now it can be shown that (Wetherill, 1981, p. 117)

$$E(S_{\text{min}}) = (n - k)\sigma^2, \tag{1.9}$$

so that the mean square residual

$$S_{\text{min}}/(n - k) \tag{1.10}$$

always provides an unbiased estimate of σ^2.

The sum of squares of the fitted values is (Wetherill, 1981, p. 118)

$$S_{\text{par}} = \hat{\mathbf{Y}}'\hat{\mathbf{Y}} = \hat{\boldsymbol{\theta}}'\boldsymbol{a}'\mathbf{Y}, \tag{1.11}$$

which is denoted S_{par} because it is a sum of squares due to the fitted parameters $\hat{\boldsymbol{\theta}}$. It can be shown that the expectation of S_{par} is (Wetherill, 1981, p. 119)

$$E(S_{\text{par}}) = k\sigma^2 + \boldsymbol{\theta}'\boldsymbol{a}'\boldsymbol{a}\boldsymbol{\theta}. \tag{1.12}$$

We can now present our results in an analysis of variance table and we shall use the 'centred' form of the model. In a 'centred' model we assume that there is a constant term and that all (other) explanatory variables have their means subtracted. We write

$$\boldsymbol{\theta}' = (\alpha_1 \boldsymbol{\beta}),$$

where α is a scalar and the first column of the corresponding **a** matrix is a column of 1's (see Section 1.1). That is, the model has the form

$$\begin{aligned} E(\mathbf{Y}) &= \boldsymbol{a\theta} \\ &= \alpha \mathbf{1} + X\boldsymbol{\beta}, \end{aligned} \tag{1.13}$$

and the matrix X is the matrix of centred explanatory variables (without the column of 1's), such that

$$\mathbf{1}'X = \mathbf{0}. \tag{1.14}$$

Thus X is a matrix of size $n \times (k-1)$. It will be helpful to use the notation

$$p = k - 1,$$

as the number of explanatory variables, so as to avoid the use of $(k-1)$ appearing in many formulae. With this change it can be shown that

$$\hat{\alpha} = \bar{Y} = \sum Y_i/n, \tag{1.15}$$

$$\hat{\boldsymbol{\beta}} = (X'X)^{-1}X'\mathbf{Y}, \tag{1.16}$$

$$S_{\text{par}} = \mathbf{Y}'\mathbf{Y} - n\bar{Y}^2 - \hat{\boldsymbol{\beta}}X'\mathbf{Y}, \tag{1.17}$$

and the analysis of variance is given in Table 1.4. In this table $\mathrm{CS}(y, x_j)$ is the corrected sum of products of y and the jth explanatory variable,

$$\mathrm{CS}(y, x_j) = \sum_{i=1}^{n} (y_i - \bar{y})(x_{ji} - \bar{x}_{j.}),$$

Table 1.4 *ANOVA for regression.*

Source	*CSS*	*d.f.*	*E(mean square)*
Due to the constant	$n\hat{\alpha}^2$	1	$\sigma^2 + n\alpha^2$
Due to regression	$\sum \hat{\beta}_j \mathrm{CS}(y, x_j)$	p	$\sigma^2 + \boldsymbol{\beta}'\boldsymbol{a}'\boldsymbol{a}\boldsymbol{\beta}$
Residual	$S_{\min}$	$n-p-1$	σ^2
Total	$y'y$	n	

Table 1.5 *ANOVA for regression.*

Source	*CSS*	*d.f.*	*E(mean square)*
Due to regression	$\sum \hat{\beta}_j \text{CS}(y, x_j)$	p	$\sigma^2 + \boldsymbol{\beta}' x' x \boldsymbol{\beta}$
Residual	$S_{\min}$	$n - p - 1$	σ^2
Total	$\text{CS}(y, y)$	$n - 1$	

where

$$\bar{y} = \sum_{i=1}^{n} y_i/n$$

and

$$\bar{x}_{j.} = \sum_{i=1}^{n} x_{ji}/n.$$

The mean squares are simply the corrected sums of squares divided by the appropriate degrees of freedom. The more usual form of Table 1.4 is to subtract the sum of squares due to the constant from the total, leading to the corrected sum of squares. This leads to Table 1.5.

From this point the analysis proceeds by assuming normality, carrying out F-tests, etc. See standard texts for details.

1.5 The extra sum of squares principle

The final step in this review of least squares is the method of testing the contribution of certain variables to the regression. The method used is called the extra sum of squares principle and is illustrated below. We simply calculate the sum of squares due to regression on a set of p variables, and then also calculate the sum of squares due to regression on a subset $s(<p)$ of the p explanatory variables, completely ignoring the $(p-s)$ other variables. The results are presented in Table 1.6.

Table 1.6 *Calculation of adjusted sum of squares.*

Source	*d.f.*
Due to regression on $X_1, \ldots, X_p$	p
Due to regression on $X_1, \ldots, X_s$ (ignoring $X_{s+1}, \ldots, X_p$)	s
Due to $X_{s+1}, \ldots, X_p$ (adjusting for $X_1, \ldots, X_s$)	$(p-s)$

The sum of squares in the final line is the 'extra' sum of squares obtained by incorporating $X_{s+1}, \ldots, X_p$ into the equation, given that $X_1, \ldots, X_s$ are already included. It is this sum of squares which is used in F-tests. See Wetherill (1981) for illustrations and examples.

As a final note here, it is vital to check that the denominator of all F-tests is a valid estimate of σ^2, uncontaminated by small but possibly real effects.

1.6 Scaled and centred variables

Although the mathematical theory of Sections 1.3 and 1.4 holds whatever the range of the variables, severe numerical problems can arise if some of the variables differ greatly in magnitude and range, as in Example 1.1. In practice, therefore, regression is carried out using scaled and centred variables. We start here by assuming a model of the form (1.13) and (1.14), which we restate for convenience,

$$\begin{aligned} E(\mathbf{Y}) &= \boldsymbol{a}\boldsymbol{\theta} \\ &= \alpha\mathbf{1} + \boldsymbol{X}\boldsymbol{\beta}, \end{aligned} \tag{1.18}$$

where $\boldsymbol{X}$ is an $n \times p$ matrix representing n observations and p variables such that

$$\mathbf{1}'\boldsymbol{X} = \mathbf{0}.$$

We now form the vector $\mathbf{S}$ of square roots of the sums of squares of the explanatory variables,

$$\mathbf{S} = \left\{ \sqrt{\left(\sum_i X_{ji}^2 \right)} \right\}. \tag{1.19}$$

Let $\boldsymbol{D}$ be a matrix with diagonal elements $\mathbf{S}$ and all off-diagonal elements zero which we call a scaling matrix

$$\boldsymbol{D} = \text{Diag}\{\mathbf{S}\}. \tag{1.20}$$

Then we scale the $\boldsymbol{X}$ matrix by writing

$$z = \boldsymbol{X}\boldsymbol{D}^{-1} \tag{1.21}$$

so that

$$z'z = \boldsymbol{D}^{-1}\boldsymbol{X}'\boldsymbol{X}\boldsymbol{D}^{-1} \tag{1.22}$$

is the correlation matrix of the explanatory variables, and has unity down the leading diagonal.

The variables z are scaled and centred, and it is usually better to work with these rather than with the original variables in any computer program. There are simple relationships between results obtained on scaled and centred variables and results obtained on the original variables. We now obtain these relationships.

The model (1.18) can be put in the form

$$E(\mathbf{Y}) = \alpha\mathbf{1} + z\boldsymbol{\phi}, \tag{1.23}$$

where the vector $\boldsymbol{\phi}$ is given by

$$\boldsymbol{\phi} = \boldsymbol{D}\boldsymbol{\beta}. \tag{1.24}$$

The least squares estimators are therefore

$$\hat{\boldsymbol{\beta}} = (X'X)^{-1}X'\mathbf{Y},$$

or

$$\begin{aligned}\hat{\boldsymbol{\phi}} &= (z'z)^{-1}z'\mathbf{Y}\\ &= \boldsymbol{D}(X'X)^{-1}X'\mathbf{Y},\end{aligned} \tag{1.25}$$

and

$$\hat{\boldsymbol{\beta}} = \boldsymbol{D}^{-1}\hat{\boldsymbol{\phi}}. \tag{1.26}$$

Also we have

$$V(\hat{\boldsymbol{\beta}}) = (X'X)^{-1}\sigma^2,$$

and

$$V(\hat{\boldsymbol{\phi}}) = \boldsymbol{D}(X'X)^{-1}\boldsymbol{D}\sigma^2. \tag{1.27}$$

Finally, we notice that the regression sum of squares is unaltered by scaling,

$$\begin{aligned}S_{\text{par}} &= \hat{\boldsymbol{\beta}}'X'\mathbf{Y} = \hat{\boldsymbol{\phi}}'\boldsymbol{D}^{-1}\boldsymbol{D}z'\mathbf{Y}\\ &= \hat{\boldsymbol{\phi}}z'\mathbf{Y}.\end{aligned} \tag{1.28}$$

1.7 Regression through the origin

As an illustration of the extra sum of squares principle, we shall consider the question of whether or not the regression should be through the origin.

Suppose we have the model

$$E(\mathbf{Y}) = \boldsymbol{a}\boldsymbol{\theta},$$

where $\boldsymbol{a} = (\mathbf{1}\,\boldsymbol{W})$ and where $\boldsymbol{W}$ is a $p \times p$ matrix of raw explanatory variables, neither centred nor scaled. We also write the centred variables as $\boldsymbol{X}$, where

$$\boldsymbol{X} = \boldsymbol{W} - \mathbf{1}\bar{\boldsymbol{W}} \tag{1.29}$$

and where $\bar{\boldsymbol{W}}$ is a $p \times 1$ vector of column means of $\boldsymbol{W}$. The ordinary regression model, including the constant term, is, from (1.13)

$$E(\mathbf{Y}) = \alpha\mathbf{1} + \boldsymbol{X}\boldsymbol{\beta}$$

and the residual sum of squares for this model is from (1.17)

$$\mathrm{SS_c} = \mathbf{Y}'\mathbf{Y} - n\bar{Y}^2 - \hat{\boldsymbol{\beta}}\boldsymbol{X}'\mathbf{Y} \tag{1.30}$$

on $(n-k)$ d.f., where

$$\hat{\boldsymbol{\beta}} = (\boldsymbol{X}'\boldsymbol{X})^{-1}\boldsymbol{X}'\mathbf{Y}.$$

If we insist on the regression passing through the origin the model is

$$E(\mathbf{Y}) = \boldsymbol{W}\boldsymbol{\beta}^0$$

and we obtain the least squares estimator

$$\hat{\boldsymbol{\beta}}^0 = (\boldsymbol{W}'\boldsymbol{W})^{-1}\boldsymbol{W}'\mathbf{Y}.$$

The residual sum of squares is

$$\mathrm{SS_0} = \mathbf{Y}'\mathbf{Y} - \hat{\boldsymbol{\beta}}^0\boldsymbol{W}'\mathbf{Y} \tag{1.31}$$

on $(n-p)$ d.f., where $p = k - 1$. The 'extra' sum of squares for testing the significance of the constant term is therefore (1.31) minus (1.30) which will have $(n-p)-(n-k) = 1$ d.f. This sum of squares should be tested against the mean square error obtained from (1.30).

In actual calculations, it is preferable, for two reasons, to invert the matrix of sums of squares of centred variables $(\boldsymbol{X}'\boldsymbol{X})$ rather than $\boldsymbol{W}'\boldsymbol{W}$. Firstly, the matrix $(\boldsymbol{X}'\boldsymbol{X})$ has a lower condition number (see Chapter 4). Secondly, as we see from the argument above, the inverse of $(\boldsymbol{X}'\boldsymbol{X})$ is needed as well. In doing these calculations, use can be made of the following identity

$$(\boldsymbol{W}'\boldsymbol{W})^{-1} = (\boldsymbol{X}'\boldsymbol{X})^{-1} - \frac{n(\boldsymbol{X}'\boldsymbol{X})^{-1}\boldsymbol{W}'\boldsymbol{W}(\boldsymbol{X}'\boldsymbol{X})^{-1}}{1 + n\bar{\boldsymbol{W}}(\boldsymbol{X}'\boldsymbol{X})^{-1}\bar{\boldsymbol{W}}'}. \tag{1.32}$$

1.8 Problems and pitfalls

The previous sections of this chapter review the mathematical theory behind the applications of least squares. The examples in the earlier part of the chapter, and in the Appendix, show that in practice many difficulties arise and in this section these points are simply listed. Some of them will be discussed in detail in succeeding chapters. There are two main classes of problems:

(1) Problems due to the assumptions.
(2) Problems arising out of the form of the data.

1.8.1 Problems due to the assumptions

(1) The assumptions of normality, homoscedasticity and independence of the error may not be valid. Three questions arise for each of these. How severe does each have to be before it matters? How do we detect discrepancies from these assumptions? What action do we take when the assumptions are found to be false?

(2) The assumed functional form (linear) may be false. We may need a transformation of the response variable, or of the explanatory variables, or we may need non-linear terms in the model. Again three questions arise. When and for what purposes does an error in the function form matter? How is a wrong functional form to be detected? What action do we take?

1.8.2 Problems arising out of the form of the data

(1) One chief difficulty arises out of multicollinearity, which is when there are (near or exact) linear relationships among the explanatory variables. Multicollinearity causes problems with the numerical routines used to fit the model, and it also causes problems of interpretation, if deciding upon a final equation. Sometimes multicollinearity is not severe enough to render the matrix $(x'x)$ singular, but severe enough to render the 'inverses' calculated by most programs as nonsense. Again we need to study when multicollinearity causes problems, how to detect it, and what action to take.

(2) Sometimes there are a large number of explanatory variables, and a numerical difficulty arises of exploring the many possible forms of model in an efficient way.

(3) Often data contain outliers and the presence of one or more

outliers can lead us to detect non-normality, heteroscedasticity, and the need for a transformation, etc. We need to discuss when outliers matter, how to detect them, and what action to take.

(4) Sometimes data has a special structure, such as replications of the response variable at certain points, or some orthogonality. It is quite wrong in such cases to proceed with a standard analysis, but sometimes it is difficult to detect these features in a large data set.

(5) Sometimes the data has not been recorded with sufficient accuracy and this can render the whole data set useless.

Discussion

Satisfactory answers do not exist to all of the questions raised above. Some of the questions involve discussions of numerical algorithms, while others can take us into deep theoretical discussions.

In some cases, a simple solution such as omitting an outlier, or taking logarithms of the response variable, will simultaneously solve all of the problems. However, there is no 'automated' way of replacing the professional statistician in many cases, and what is needed is a program which will indicate the presence of some of the problems, as they arise, and point towards a solution of the simpler ones.

For references and reading around the problems raised in this section the reader should consult Seber (1977), Draper and Smith (1981), Daniel and Wood (1980) and Weisberg (1980). For access to some of the large literature on linear regression methodology, see Hocking (1983) or Hocking and Pendleton (1983).

Attention is drawn to the data sets given in the Appendix which should be used throughout the book to try the techniques suggested.

Exercises 1

1. You are given some data (y_i, x_i), $i = 1, 2, \ldots, n$, and it is desired to fit a simple linear regression model. In order to test whether the regression line goes through the origin the following procedure is suggested. Fit the model

$$\left.\begin{aligned} E(y_i) &= \alpha + \beta x_i, \\ V(y_i) &= \sigma^2, \end{aligned}\right\} \qquad (1)$$

and test the significance of $\hat{\alpha}$ from a hypothesis value of zero. An alternative procedure is to fit a second model

$$\left.\begin{aligned} E(y_i) &= \beta x_i, \\ V(y_i) &= \sigma^2. \end{aligned}\right\} \qquad (2)$$

A test is then performed by differencing the residual sum of squares for model 1 using an $F(1, n-2)$-distribution. Are these procedures equivalent, or not?

2. Suppose that you are given data (y_{ij}, x_{ij}), $i = 1, 2$; $j = 1, 2, \ldots, n_i$, and that you wish to fit simple linear regression models,

$$E(Y_{ij}) = \alpha_i + \beta_i x_{ij} \quad \text{for } i = 1, 2.$$

Show how to test whether or not the regressions are parallel for the two data sets $i = 1, 2$.

References

Armitage, P. (1971) *Statistical Methods in Medical Research*. Blackwell, Oxford.

Box, G. E. P. (1966) Use and abuse of regression. *Technometrics*, **8**, 625–629.

Coen, P. J., Gomme, E. D. and Kendall, M. G. (1969) Lagged relationships in economic forecasting, *J. Roy. Statist. Soc.* A, **132**, 133–163.

Daniel, C. and Wood, F. S. (1980) *Fitting Equations to Data*. Wiley, New York.

Draper, N. R. and Smith, H. (1981) *Applied Regression Analysis*. Wiley, New York.

Hocking, R. R. (1983) Developments in linear regression methodology. *Technometrics*, **25**, 219–230 (with discussion).

Hocking, R. R. and Pendleton, J. (1983) The regression dilemma. *Commun. Statist. Theor. Methods*, **12** (5), 497–527.

Holt, D., Smith, T. M. F. and Winter, P. D. (1980) Regression analysis of data from complex surveys. *J. Roy Statist. Soc.* A, **143**, 474–487.

Mandel, B. J. (1969) The regression control chart. *J. Qual. Tech.*, **1**, 1–9.

Marquardt, D. W. and Snee, R. D. (1975) Ridge regression in practice. *Amer. Statist.*, **29**, 3–19.

Mendenhall, W. and Ott, L. (1971) A method for the calibration of an on-line density meter. *J. Qual. Tech.*, **3**, 80–86.

Prater, N. H. (1956) Estimate gasoline yields from crudes. *Petrol. Refiner*, **35** (5).

Seber, G. A. F. (1977) *Linear Regression Analysis*. Wiley, New York.

Weisberg, S. (1980) *Applied Linear Regression*. Wiley, New York.

Wetherill, G. B. (1981) *Intermediate Statistical Methods*. Chapman and Hall, London.

CHAPTER 2

Data exploration

2.1 Introduction

The rôle of exploratory data analysis is to reveal the features of the data set under study. All statistical analyses should include a thorough exploration of the data. For some sets of data this may just show a complete lack of unusual or interesting aspects. For others interesting observations, or groups of observations, unexpected structure or relationships may be found in the data. Failure to explore the data before embarking on a formal statistical analysis may cause much time, effort and resources to be wasted, or even worse, incorrect conclusions to be drawn.

However, it must also be noted that some features of the data may be accidental or coincidental. For example, if a set of data on people who drink a certain brand of cola contained mainly females, this may either be accidental or symptomatic of the section of public that uses the brand in question. It is the purpose of exploratory data analysis to uncover all such features, whether they be accidental or important, so that the evidence can be weighed and investigated.

The techniques presented in this chapter provide basic tools for the exploratory data analyst. Many of them can be done fairly easily by hand. Most are fairly generally available in statistical packages. It is the job of the data analyst to become conversant with techniques like the ones given here and to become expert in interpreting the figures and values produced. Data exploration is very important, but those who feel confident about this material should pass straight to Section 2.6.2 or to Chapter 3.

2.2 Objectives

The main objectives of data exploration can be divided into two distinct areas:

(1) *Detecting errors*
Many raw data sets contain errors. These may be introduced, for example, at the point of data collection or coding, when entering the data into the computer, when editing the data, or caused by a mechanical problem such as 'dirt' when transmitting data down a telephone line. Whenever the data are altered, in any way, errors may be introduced into the data. This is particularly true if humans are involved in the altering of the data.

(2) *Exploring the features of the data*
It is important to investigate the structure of the data as this can completely alter the type of statistical analysis that is necessary. It may also indicate the type of model that is appropriate or suggest that the proposed model may be inadequate. Certain shortcomings in the data, or peculiar points or groups, may be highlighted and these should be taken into consideration when performing an analysis and interpreting the results.

We shall now expand on these two areas of interest. However, it is not possible to give a definitive coverage of the two topics, but only to sketch the scenery so that the reader may obtain an overall impression of the areas.

2.3 Detecting errors

(a) *Digit transposition (etc.)*

We define an observation to be a set of measurements on a number of variates, all related by the fact that they are taken on the same individual, sample, item, or that they are measurements on a given system taken at a specified time point. A common error is then for values of two successive variates to be transposed in a number so that, for example 2982 becomes 9282. Obviously, this can be catastrophic and the example cited may well be spotted easily. However, even a transposition such as 2982 to 2928, which is not so easy to spot, can have a marked effect on the analysis, particularly if the data set in question is ill-conditioned (see Chapter 4). Other similar errors to transposition include missing out a digit, or a decimal point, or adding an extra digit.

(b) *Incorrect values for a variate*

In a similar fashion to (a) above, values for two successive variates can be transposed or the value of one variate repeated and another missed out completely. A similar error is when all the values of the variate are incorrect (due possibly to incorrect editing of the data), use of the wrong variate or even use of the wrong data set!

(c) *Incorrect values for an observation*

Similarly to (b), it is possible to have the correct values for a variate but in the wrong order due to a transposition. Also values can be duplicated by mistake. This sort of error is likely to happen when the data have been entered by variate, rather than observation, at some point in its history and is usually more difficult than (b) to detect.

(d) *Experimental errors*

A further wealth of errors can be introduced at the time of performing the experiment. Possibly the process under investigation went out of control, a new operator was using the machinery, the wrong fertilizer or weed killer was used, secondary infection set in, etc. Whatever the subject area of the data there are usually numerous things that can go wrong during the experiment.

Should some data points appear discordant these must be thoroughly investigated. All possible sources of error should be checked, if possible, and corrective action taken when necessary. It cannot be stressed strongly enough that even one data point in error can completely ruin the analysis of your data. However, just because a data point looks 'odd' this is no reason to delete the data point from the analysis. To do so, if the observation is good, is in itself introducing an error into the data set! In some cases it is the unusual points that are the most important aspects of the data. For example, if one patient dies whilst using a drug under investigation it may be caused, either directly or indirectly (due possibly to a general lowering of resistance levels), by the drug itself. Clearly, in this case it would not do just to ignore the results on this patient!

There is no foolproof way of finding all of the errors in a data set. There is no easy way either. Double entry of the data set by independent persons will catch many of the more basic forms of error. Logical checks performed automatically on the data by a computer program can, when appropriate, reduce the amount of bad data allowed into a data set. However, use of the techniques listed below can help to find many other errors. Unfortunately, exploratory data analysis is not, in many ways, as simple as confirmatory analysis (Tukey, 1980), where a statistical method is used and the resulting test statistic examined. With exploratory data analysis techniques it is necessary for the investigator to examine each plot or statistic carefully, and to decide whether it is entirely as expected or whether it suggests or indicates an error has occurred.

2.4 Exploring the features of the data

(a) *Linear relationships*

A frequent problem with data sets in regression is caused by near linear relationships between two or more of the explanatory variables. If the relationships are exact then the regression estimates cannot be formed, although an analysis of variance table and the residuals can be. This is discussed in Chapter 5. When the linear relationships are not exact the variance of the regression estimates can be grossly inflated by their presence. Chapter 4 deals with this problem of so-called 'multicollinearities' amongst the explanatory variables.

(b) *Replication*

Sometimes the observations on the response variable in regression analysis will have been taken repeatedly at the various points in the explanatory variable space. In this case the replications can be used to obtain an estimate of the underlying error σ^2 and so an entirely different form of analysis of variance is appropriate (see Wetherill, 1982, Chapter 11) that takes into account the replications.

(c) *Crossed or nested data*

An example of a nested design is given by the data of Example 1.2, shown by the structure of Table 1.3. Nesting may be caused by the explanatory variable data having a crossed or nested design. In an extreme form of crossed structure the design of the data points may be such that the variables are orthogonal to each other and so a different form of analysis is appropriate.

(d) *Time trends*

If there is a natural order to the observations then the possibility exists that the model may have to be modified. Sometimes this merely means that a linear, periodic or seasonal effect due to time has to be added to the model. However, it frequently arises that successive observations are correlated and this conflicts with one of the basic assumptions of least squares. In this latter case, therefore, a different form of model should be fitted to the data; see Chapter 13.

(e) *Boundary points*

The data may show evidence of boundaries beyond which observations do not exist. Such boundaries should be thoroughly explored and an attempt made to justify their existence. The model to be fitted may need altering to account for any boundaries. If the model form is not changed then the fit may not be applicable to data close to the boundaries, and so it may be necessary to ignore data in certain regions to prevent the fit from being poor in the region of interest. This may occur, for example, in a response surface analysis and so cause poor predictions of the response.

(f) *System change points*

Somewhere within the space spanned by the explanatory variables the underlying system relating these variables to the response may change. In a similar fashion to (e) above such points need to be identified and if possible a reason for their existence formulated. It is also necessary to attempt to quantify the change observed and to alter the fitted model accordingly. An example of this is when bankruptcy of a competitor causes demands for one's own produce to increase by some constant factor. In this case we should consider fitting equations separately to the pre- and post-bankruptcy data.

(g) *Outliers*

Frequently, data sets contain one or more outliers, that is, points that appear separated in some way from the remainder of the data. Outliers may be individual observations or groups of two or more points. Such outliers need to be identified and investigated thoroughly, for the same reasons as noted in Section 2.3. It is worth reiterating here, however, that such observations must not be removed from all further analyses purely because they seem discordant.

(h) *Clumping*

Sometimes the data set can be separated into two or more sets of points or 'clumps'. These clumps may pull the fitted equation around through themselves. Also the existence of clumps in the data set implies that there are regions with sparse density of points in which the fitted model cannot be validated with any confidence.

As well as the items given above there are many other points to consider when exploring the data. For example, thought should be given to the possibility of important variables missing from those under study. Of course, such considerations should be entertained, when appropriate, prior to the experiment being performed and all variables that may be of importance included in the analysis. However, this is not always feasible and there is also the possiblity of the unexpected occurring. Omitted variables will sometimes account for unexpected features of the data (for example, outliers and system change points).

A change of scale for one or more variables is also worth considering. Many basically non-linear systems can be fitted quite adequately by taking an appropriate transformation. Also, sometimes, two or more variables can be combined in some manner and the resulting transformed variable used with more success than the original ones. (See Chapter 7 for more transformations.)

If the data come from a designed experiment then it should be checked that adequate randomization has actually been performed.

The use required of the model should also be considered when exploring your data. For example, if the use is to predict the response at some points then you should check that the prediction points are not in regions of sparse density, or that you will not be extrapolating too far beyond the region covered by the experiment. (Computer packages like the one written by Meeker *et al.* (1975, 1977) can help here.)

2.5 Basic statistics

Some of the features mentioned in the previous sections can be detected by examining various basic statistics for all of the variables in the data. It should be checked that the maxima and minima are not larger or smaller than expected. The mean and/or median (a more robust estimator of location than the mean), as well as the variance or standard deviation, should be as expected.

The correlation matrix of the variables should be examined and any high correlations explored. It must be noted, however, that the simple correlation coefficient measures the linear interdependence between the two variables only and as such will not help detect anything but the most simple relationships. That is, relationships of the form

$$\mathbf{v}_i \simeq \alpha + \beta \mathbf{v}_j,$$

where $\mathbf{v}_i, \mathbf{v}_j$ are variables and α, β are constants. The problem of

Table 2.1 *Basic statistics 1 for Example 1.1.*

				Quartiles	
Variable	*Minimum*	*Maximum*	*Range*	*Lower*	*Upper*
RESP	15.0000	50.5000	35.5000	28.75000	48.0000
X1	1100.0000	1300.0000	200.0000	1150.0000	1300.0000
X2	5.3000	23.0000	17.7000	7.5000	17.0000
X3	0.0115	0.0980	0.0865	0.0125	0.0625

Variable	*Inter-quartile range*	*Median*	*Mean*	*Variance*	*Standard deviation*
RESP	19.2500	36.5000	36.1063	141.5806	11.8988
X1	150.0000	1200.0000	1212.5000	6500.0000	80.6226
X2	9.5000	11.0000	12.4438	32.0586	5.6620
X3	0.0500	0.0330	0.0403	1.001E–03	0.0316

Correlation matrix

	RESP	*X1*	*X2*	*X3*
RESP	1.0000			
X1	0.9450	1.0000		
X2	0.3700	0.2236	1.0000	
X3	−0.9140	−0.9582	−0.2402	1.0000

The following pairs of variables are highly correlated:
X1 with X3

identifying more complex linear relationships is tackled in Chapter 4.

Table 2.1 shows the output from one regression program of some basic statistics for the data from Example 1.1. As well as the statistics mentioned above it also lists the quartiles, the range and interquartile range. It also aids the user by listing any high correlations (defined as correlations where absolute value is greater than or equal to 0.95). The same regression program also produced Table 2.2 for this data set. If the minimum non-zero difference (mnzd) between values of a variable given in Table 2.2 is larger than expected, then this should be investigated. It may be that rounding of the data has not been performed as desired. This value, expressed as a percentage of the range, indicates how relatively close the two nearest (but non-coincidental) points are and, if this is large, it should be checked that the points were supposed to be well separated. Table 2.2 also contains an estimate of the number of digits of accuracy that the data were measured to. For various reasons this is a fairly difficult statistic for a computer to estimate. The actual estimate produced for a vector of data $\mathbf{x} = (x_1, x_2, \ldots, x_n)$ is calculated as

$$\begin{cases} U\left[\log_{10}\left\{\operatorname*{Max}_{i=1,n} |x_i|/\text{mnzd}\right\}\right], & \text{mnzd} \neq 0 \\ 0 \quad , & \text{mnzd} = 0, \end{cases}$$

where mnzd is the minimum non-zero difference given in column 2 of Table 2.2 and U is the function that rounds up to the nearest whole

Table 2.2 *Basic statistics 2 for Example 1.1.*

Variable	*Minimum non-zero difference* *Absolute value*	*% of range*	*Number of significant digits?*	*Number of distinct values*
RESP	0.300 000	0.845	3	16
X1	100.000 000	50.000	2	3
X2	1.500 000	8.475	2	7
X3	0.000 500	0.578	3	14

The following variable may be qualitative: X1.
X1 is replicated four times.
The values of X1 are in sequential order.
The values of X2 are nearly sequentially ordered.

number. The reasoning behind this estimate is that, if as is hoped, the observed mnzd is close in size to the rounding error in the data, then dividing this into the largest absolute data value will indicate the magnitude of the data relative to the mnzd. Taking logs to base 10 and rounding up converts this into an estimate of the number of digits of accuracy. If mnzd is not close to the rounding error in the data this method produces an underestimate of the accuracy.

The number of digits' accuracy in the data is important in assessing the effects of rounding and ill-conditioning in the data on the usefulness of any results produced. Chapter 4 discusses this in more detail. It is worth noting in passing that the estimated accuracy for variate X1 from Example 1.1 is only two digits. This is since these four figure numbers all end with two zeros. The investigator should check how accurate these data really are. Obviously, we would have more confidence in any results produced if it were known that the data are rounded to the nearest integer rather than to the nearest hundred.

The final column in Table 2.2 indicates the number of distinct points at which each variate was observed. It is noted here that the variable X1 is only measured at three distinct points and the variable X2 at seven. Beneath the table any variable that contains complete replications of its values, together with details of the number of replications, is listed. Also, a message suggests that X1 may be qualitative (e.g. machine number) rather than quantitative.

Finally, it is pointed out that the values of X1 and X2 are in, and nearly in, sequential order, respectively, as we go through the observations in order of rows. This may be important if there is a time sequence inherent in the order in which the rows were presented.

2.6 Graphical methods

Formal numerical statistics are too often designed to give specific answers to specific questions. Graphical techniques do not suffer from this rigidity and consequently are a major tool of the exploratory data analyst (see also Feder, 1974; Mosteller and Tukey, 1977; Tukey, 1977; and Velleman and Hoaglin, 1981). Presented in this section are various methods for plotting data that will help us to gain an insight into the structure of the data at hand. If the plots for individual variates are produced on similar scales then these can be used to compare the distributions of these variables.

2.6.1 Univariate plots

Such plots are useful for finding clumping, outliers and hard boundaries that occur in individual variates. We can also observe the general distribution of points within each variate and spot unusual features here.

(a) *Histograms and bar charts*

Histograms and bar charts are a very easy way to illustrate tabular data graphically and can be readily drawn by hand. Also, most statistical packages contain facilities for producing such plots. As a start to exploring our data it is well worth producing histograms or bar charts of all our data. Histograms are given in Fig. 2.1 for the data

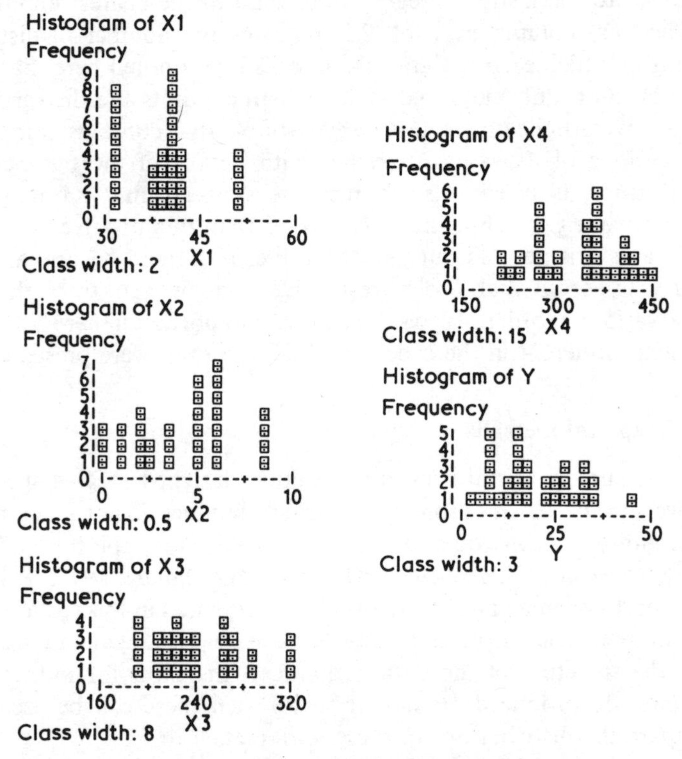

Fig. 2.1 *Histograms for Example 1.2.*

Stem and leaf plot of X1

```
3 |111122222
3 |8888888
4 |0000000001111
4 |
5 |0000
```

Leaf unit:1 - Stem and leaf of 1|2 represents: 12.

Stem and leaf plot X2

```
0 |2222228888
2 |44555
4 |888222
6 |1111111
8 |6666
```

Leaf unit: 0.1 - Stem and leaf of 1|2 represents: 1. 2.

Stem and leaf plot X3

```
1 |9999
2 |111111
2 |2222333333
2 |
2 |6666777
2 |88
3 |111
```

Leaf unit: 10 - Stem and leaf of 1|2 represents: 120.

Stem and leaf plot of X4

```
2 |01133
2 |677778
3 |00444
3 |556666679
4 |0011224
```

Leaf unit: 10 - Stem and leaf of 1|2 represents: 120

Stem and leaf plot Y

```
0 |2566788
1 |02344456788
2 |2346667
3 |012344
4 |5
```

Leaf unit: 1 - Stem and leaf of 1|2 represents: 12.

Fig. 2.2 *Stem and leaf plots for Example 1.2.*

from Example 1.2. From these we can see, for example, that the data for variate X1 appears to divide into three groups.

(b) *Stem and leaf plots*

A stem and leaf plot is somewhat like a histogram on its side, but with more information recorded about the actual values of the observations falling within each class.

The range of the variable is divided into classes called stems. Each stem indicates the leading digit (or digits) of the numbers that fall within the range of that stem. The stems are listed on the left-hand side of the vertical axis. In the basic plot the next most significant digit in the data, after those given in the stem, are printed on the right-hand side of the axis. These are called the leaves of the plot. The number of leaves after each stem then represents the number of observations in the class. The data from Example 1.2 are plotted in this way in Fig. 2.2 so that a comparison can be made with the histograms of the same data given in Fig. 2.1.

The leaves of stem and leaf plots need not be restricted to single digits from each data point. Figure 2.3 shows a stem and leaf plot where the next two digits have been listed for each leaf with successive leaves separated by commas.

(c) *Box and whisker plots*

Box and whisker plots, or box plots for short, give a simplified view of the data so that only the general distribution and extreme 'outliers' are discernible.

A box is drawn between the upper and lower quartiles of the data. The median is then marked by some suitable symbol such as an asterisk. Arms, called whiskers, are drawn from the sides of the box

```
Stem and leaf plot of RESP

1 |50,70
2 |05,80,95
3 |15,45,50,80,85
4 |45,75,85,90
5 |02,05

Leaf unit: 1 - Stem and leaf of 1| 22 represents: 12.2.
```

Fig. 2.3 *Stem and leaf plot of RESP from Example 1.1 with two-digit leaves.*

out to the furthest points either side of the quartiles to within one and a half times the length of the box. Any points outside this are indicated separately. Those within two and a half times the length of the box are marked with an '0' and those beyond that are marked by ϕ.

Tukey (1977) uses values he calls 'hinges' to define a box plot. These values sometimes coincide with the quartiles although sometimes they fall slightly closer to the median. This difference is of no practical significance.

Some example box plots are given in Fig. 2.4. The figures given on the left and right above each box plot are the extreme values of the variate being plotted. From these plots we can see, for example, that X3 and RESP are skewed to the left and right respectively.

McGill *et al.* (1978) suggested an extension to box plots that enables us to compare whether or not the medians of two variables are significantly different at approximately the 5% level. This is done by placing 'notches' at the

$$\text{median} \pm 1.58 \times (\text{interquartile range})/\sqrt{n},$$

where n is the number of data points in the box plot. Significance is determined by whether or not the notches in these so-called notched

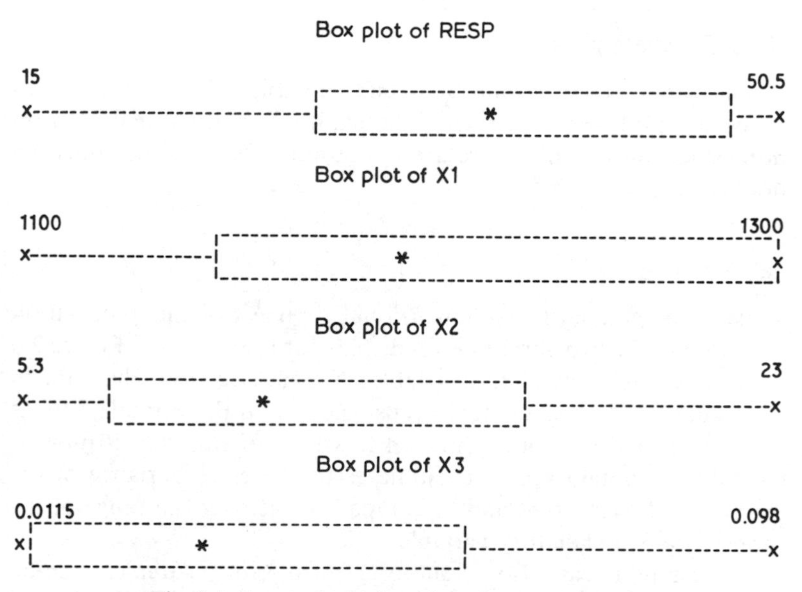

Fig. 2.4 *Box and whisker plots for Example 1.1.*

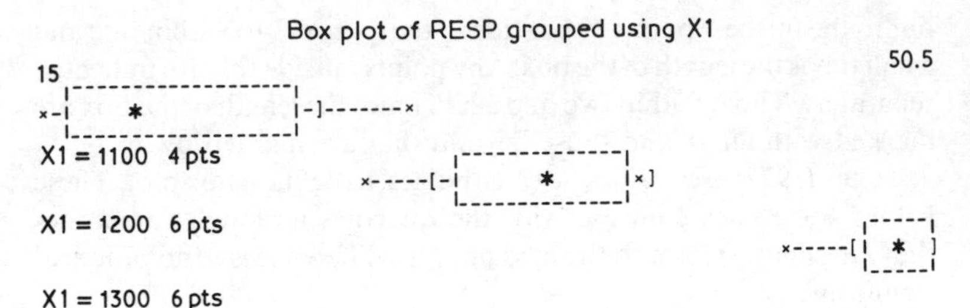

Fig. 2.5 *Notched box and whisker plots of RESP grouped using X1 for Example 1.1.*

box plots overlap. Figure 2.5 gives some examples of notched box plots where the notches are marked using square brackets. In the same paper it was also proposed that the width of box plots be made proportional to the square root of the number of points being plotted. This draws attention to any box plots that are based on a comparatively large or small number of points, and this can affect box plots for regression data if some variates have replicated observations.

2.6.2 Bivariate plots

To investigate relationships between two variates further techniques must be employed. The rôle of the exploratory data analyst is no longer as simple since there are many more features that must be looked for.

(a) *Scatter plots*

This is the standard two-dimensional plot. All of the information inherent in the two variates is on display for examination. Figure 2.6 gives a scatter plot of the variables X1 and X2 from the data in Example 1.1. This diagram clearly suggests that the variable X1 was measured at three points only. In this type of situation it may be possible to obtain separate estimates of the error variance of the response at each point and to compare these over the range of the appropriate explanatory variable.

Scatter plots can also be smoothed using, for example, running means or medians, to obtain less confused pictures of the behaviour of

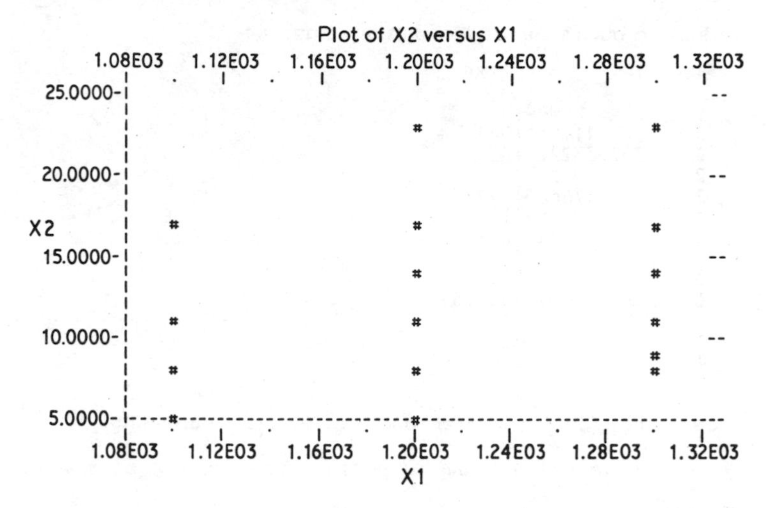

Fig. 2.6 *Scatter plot of X1 and X2 for Example 1.1.*

the data. This is a fairly large topic; see Cleveland (1982) for an introduction. However, when examining such plots we must remember that the picture is a simplification, not a reality.

(b) *Grouped box plots*

As well as comparing box plots of two variates it can also be useful to compare separate box plots for observations selected from a variable using some criterion, usually involving another variate. As an example, Figure 2.5 gives notched box plots of RESP from Example 1.1, in which the observations were selected according to the value of the variable X1. This gives three box plots in total, one for X1 at each of the values 1100, 1200 and 1300. This plot clearly shows how the observations on RESP change as X1 increases.

(c) *Back-to-back stem and leaf plots*

Comparisons of two stem and leaf plots can be made easier by drawing them back-to-back as in Fig. 2.7. Variates can also be split into two sections using some criterion (in a similar way to box plots) and a back-to-back stem and leaf plot formed using these two sections.

```
Back-to-back stem and leaf plot of X3 and X4

Stem          X3       X4

 1          9999|
 2        1111111|011
 2    33333322222|33
 2               |
 2      777666666|67777
 2            888|8
 2            111|00
 3               |
 3               |44455
 3               |666667
 3               |9
 4               |0011
 4               |22
 4               |4

Leaf unit:  10   -   Stem and leaf of 1|2 represents: 120
```

Fig. 2.7 *Back-to-back stem and leaf plots for X3 and X4 of Example 1.2.*

2.6.3 The third dimension and beyond

Even with the technology available today it is still very difficult to produce true three-dimensional plots. The author at the time of writing knows only of one successful attempt to introduce true perspective into three-dimensional plots and this requires much expensive equipment and is not readily available (see Stover, 1981). However, using suitable graphical devices we can obtain reasonable approximations to three-dimensional surfaces and indeed many packages now contain the required software to produce such graphs.

Once we decide to go beyond the third dimension it is necessary to produce plots which indicate reality in some approximate way that people can visually interpret.

(a) *Pseudo three-dimensional plots*

Basic scatter plots produce only two-dimensional representations. However, a third dimension can be introduced into the plot by using various symbols in the plots with each symbol representing a different depth, or range of depth, in the third dimension. Interpretation is made easier if the visual effect of the symbols used is matched, in some way, with the range of values they represent. For example, when plotting a variate that takes both negative and positive values we could use the symbols:

O: Large negative;
o: medium negative;
·: small modulus;
x or +: medium positive;
X or #: large positive.

Single digits can also be used to represent ranges of values, but this notation is most natural if the values are either purely positive or purely negative.

An example of a pseudo three-dimensional plot is given in Fig. 2.8. This graph is of the three explanatory variables from Example 1.1 and clearly shows the structure of the data. The third dimension coming out from the 'graph' indicates the three distinct values of X1.

(b) *Principal component plots*

Principal component plots are a method of plotting three or more variates on a scatter diagram in such a way as to account (hopefully) for a high proportion of the variation amongst the variates. The

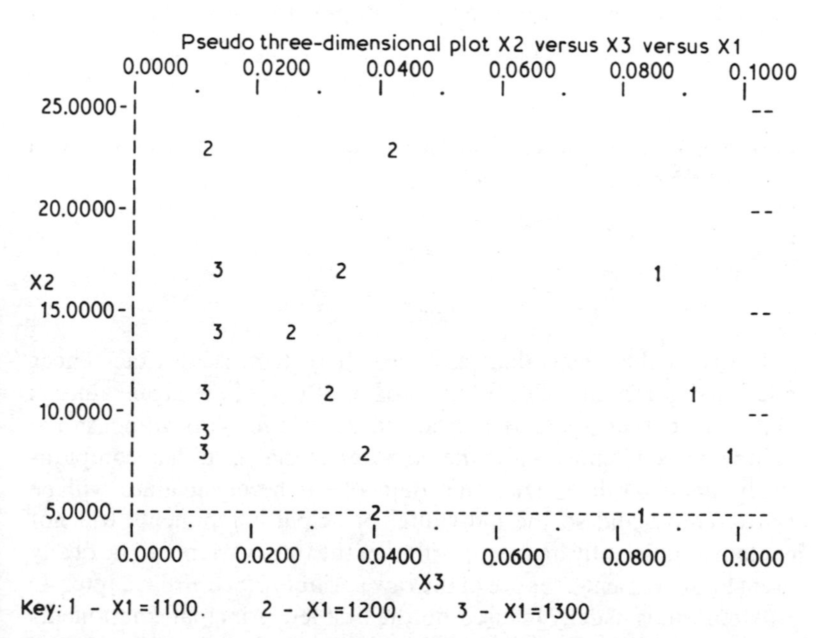

Fig. 2.8 *Pseudo three-dimensional plot of X1, X2 and X3 for Example 1.1.*

justification for the formation of the principal components is as follows:

Let the covariance matrix of the variates $\mathbf{X}$ be C. Then the variance of a linear combination $\mathbf{Xv} \to \mathbf{a} = \mathbf{Xv}$ is $\mathbf{v}'C\mathbf{v}$. The initial problem is to find the linear combination $\mathbf{Xv}_1$ that has maximum variance subject to the normalizing constraint $\mathbf{v}'_1\mathbf{v}_1 = 1$. Following this we then find a set of linear combinations $\mathbf{Xv}_i, i = 2, \ldots, p$, where p is the number of variates, each with maximum variance, subject to $\mathbf{v}'_1\mathbf{v}_i = 1$ and $\mathbf{v}'_i\mathbf{v}_j = 0, j = 1, \ldots, i-1$.

The solution to this problem is given by the eigenvectors (see Chapter 3) of the matrix C. If V is the matrix of eigenvectors of C ordered in accordance with decreasing size of eigenvalues λ_i then

$$V = [\mathbf{v}_1, \ldots, \mathbf{v}_p]$$

and the principal components are formed by the transformation

$$A = XV, \qquad \text{with } X \text{ the design matrix.}$$

If we define the total variance to be the sum of the variances of the estimated regression coefficients, then the proportion of variation accounted for by the ith principal component is

$$\lambda_i \Big/ \sum_{i=1}^{p} \lambda_i,$$

and the proportion of variation accounted for by a set Ω of principal components is

$$\frac{\sum_{\mathbf{a}_j \in \Omega} \lambda_j}{\sum_{i=1}^{p} \lambda_i}.$$

If any of the eigenvalues are zero then there is an exact linear relationship amongst the columns of X. If one of the eigenvalues is nearly zero then there is a near linear relationship amongst the columns (see Chapter 4). If the covariance matrix C has comparatively small off-diagonal terms then often the eigenvalues will be nearly equal, and so the individual principal components will not explain a sufficiently high proportion of the total variance to be of any great improvement over use of the raw variables. (See also Chapter 4.)

Attention is usually focused on the first few principal components since these will, in general, explain a high proportion of the total variance amongst the variables. As a rule-of-thumb we would normally consider producing histograms, scatter diagrams and pseudo three-dimensional plots of the set of principal components that explain about 70% of the total variance. However, the final few principal components may also be of interest since this may indicate that the matrix X is singular apart from a few isolated points (ones with large scores in the final few components).

The discussion above has been in terms of the covariance matrix of the variables. We could also use the correlation matrix. This is equivalent to performing a principal component analysis on the variables scaled to have unit length (by dividing by their standard deviation). Clearly, if all of the variables have equal variance then the two methods will be equivalent. Unfortunately, if this is not the case, and in general it is not, the two methods will usually produce different components and eigenvalues. Furthermore, there will be no simple correspondence between the two alternative results. A principal component analysis on the covariance matrix will tend to identify the early components with those variables that have a high variance whereas an analysis on the correlation matrix treats each variable equally. Thus if the variances of the variables indicate their relative importance we should use the covariance matrix. If this is not the case and there is no meaningful transformation that will achieve this desirable situation then we should use the correlation matrix. An example where we might wish to use the correlation matrix would be in the analysis of the human shape with one variable measuring height in millimetres and another measuring weight in kilograms. The variable height would have a much larger variance and so would dominate the early component(s).

The first two and last two principal components for the data from Example 1.2, based on the correlation matrix, are plotted together in Fig. 2.9. The first of these plots shows that the four observations 8, 16, 26 and 32 appear to be in a line slightly away from the remainder of the data. The second plot shows points 30, 23 and 5 to be fairly distant from the remainder of the data in the direction of the final component.

(c) *Other multivariate plots*

Probably the most simple way to represent multivariate data is by representing each observation by a series of bars whose heights each

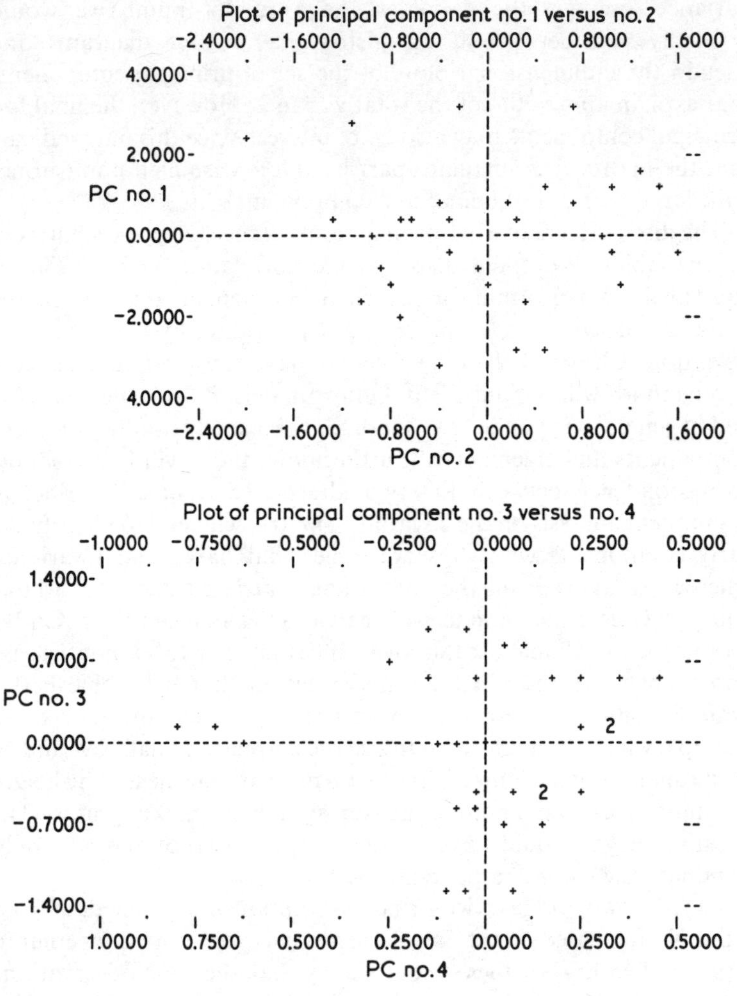

Fig. 2.9 *Principal component plots of X1 versus X2 and X3 versus X4 for Example 1.2.*

indicate the value of a variate. To form a satisfactory plot it will probably be necessary to standardize the variables in some manner. Figure 2.10 gives an example of such a plot for the four variables of Example 1.1 standardized into the range [0, 1]. The standardized data are rounded to the nearest 0.1 and then the symbols ∩ and ⌹ used to represent 0.1 and 0.2 respectively. This plot clearly indicates

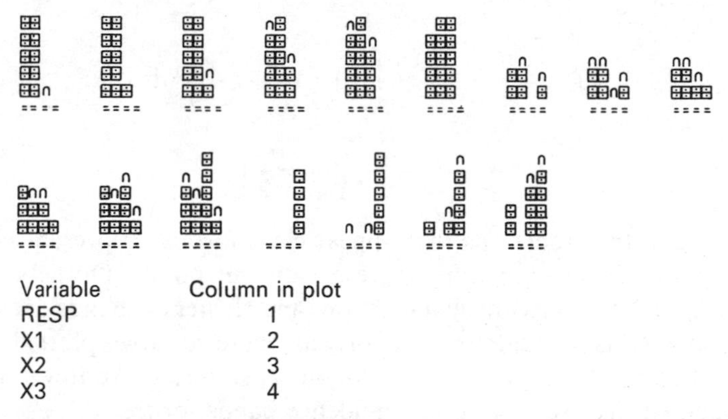

Fig. 2.10 *Multidimensional plot of RESP, X1, X2 and X3 for Example 1.1 standardized to lie in [0,1].*

how the data, including the response, fall into three fairly distinct clusters corresponding to observations 1 to 6, 7 to 12 and finally 13 to 16. Variable X2 varies fairly consistently within each group.

Anderson (1960) suggested using glyphs (circular objects with lines radiating at various angles) with the lengths of the lines representing the values of the variables. Picket and White (1966) used triangles to represent four-dimensional data. The values of three of the variables defined the side lengths with the final variable defining the orientation of the triangles. An extension of this to p dimensions was used by Siegel *et al.* (1971) where p-sided polygons were formed by plotting p points around a centroid and then joining adjacent points together. (See also Fienberg (1979).) The dimensionality of these plots can be increased by one or two by locating the figures produced on a line or scatter diagram at positions defined by the extra variable(s).

Andrews (1972) proposed plotting p-dimensional data as a finite Fourier series. If $\mathbf{x} = (x_1, x_2, \ldots x_p)$ is a data point then its Andrews curve is obtained by plotting

$$f_{\mathbf{x}}(t) = x_1\sqrt{2} + x_2 \sin t + x_3 \cos t + x_4 \sin 2t + x_5 \cos 2t + \cdots$$

for $-\pi \leq t \leq \pi$. This plot has the desirable properties that f preserves means and distances. That is

$$f_{\bar{\mathbf{x}}}(t) = \frac{1}{n}\sum_{i=1}^{n} f_{\mathbf{x}_i}(t)$$

and

$$\| f_{\mathrm{x}}(t) - f_{\mathrm{y}}(t) \|_{L_2} = \int_{-\pi}^{\pi} [f_{\mathrm{x}}(t) - f_{\mathrm{y}}(t)]^2 \, dt$$

$$= \pi \sum_{i=1}^{p} (x_i - y_i)^2.$$

Because of this, functions that appear close for all t represent near points and distant functions represent distant points. Outliers will appear on the plot as unusual functions and clusters as bands of close functions. Clusters that do not involve all of the variables plotted will have values of f that are close for at least one t. Andrews also considered the production of confidence bands for f.

Chernoff (1973) proposed the use of faces, something we are all used to comparing, to represent multidimensional data. Each variate in the data affects a particular characteristic of the drawn face (such as length of nose, shape of head, position, slant and size of eyebrows, curvature and size of mouth, etc.). Up to 18 variables can be plotted using this method. However, visual interpretation of faces is affected more by slight changes in some characteristics than in others and so it is advisable to try several variate to characteristic correspondences in order to obtain the most satisfactory plot. An example of the use of Chernoff faces and Andrews curves can be seen in Jones (1979).

An elaborate constellation graphical method was suggested by Wakimoto and Taguri (1978). This involved transforming the values of all the variables to lie between 0 and π and then plotting on a half-unit circle in the complex space C^1 the weighted sum

$$z_k = \sum_{j=1}^{p} w_j(\cos \xi_{jk} + i \sin \xi_{jk}), \quad \text{where } i = \sqrt{(-1)}, \tag{2.1}$$

$$\sum_{j=1}^{p} w_j = 1; \qquad w_j \geq 0 \quad (j = 1, 2, \ldots, p)$$

are the weights and ξ_{jk} is the transformed jth observation on the kth variate. The proximity of the points, z_k, in the half-circle represents the closeness of points in $\mathbb{R}^k$ when equal weighting is used. The path through the p partial sums of (2.1) indicates the structure of the kth observation. However, it is probably desirable to plot these paths separately to avoid cluttering the plot.

All of the above plots may be drawn using either the raw data or some constructed variables such as the principal components (or a subset thereof). However, comparisons often become difficult where there are a large number of observations. Most of the graphical methods require special equipment for them to be drawn using a computer, unlike the plots of preceding sections, all of which can be adequately drawn on standard teletype output devices. Also, some of the more advanced methods are rather difficult to produce without the computational aid of a computer.

2.6.4 Graphical considerations

All graphs should be drawn so as to exhibit the data in as clear a manner as possible. Care should be taken not to present plots in a way that is misleading. Wainer (1984) presents an amusing, although serious, illustration of some bad practices taken from various published literature. (See also Huff, 1973.) A list of considerations when drawing graphs is given in Cox (1978), although clearly the appropriateness of the various considerations depends upon the use made of the graph.

Colour is finding use in an increasing number of statistical graphs and is now also available on computers by using colour monitors. Apart from aesthetic uses, colour can also be used to aid the visual interpretation and to record more information on the graph. For example, colour can be used to indicate the values of the third dimension on one of the pseudo three-dimensional plots of Section 2.6.3 by using varying shades or colours. However, extreme care must be taken with the use of colour. Cleveland and McGill (1983) report an experiment that indicated a colour caused optical illusion. Colour blindness adds a further dimension to the problems of colour interpretation.

2.7 Further exploratory data analysis techniques

Investigating clumping and outlying groups in a data set can be done using clustering techniques. In order to use cluster analysis we must initially choose an appropriate distance or similarity measure to represent distances between points. Frequently the standard Euclidean distance will be appropriate, particularly in the context of multiple regression; however, one is not confined to using this metric. Points are joined, or fused, together to form clusters starting with the individuals or groups of individuals that are closest together and continuing until all of the observations are combined into one group. A diagram called a dendrogram can be drawn to represent the results from using a clustering technique (see Everitt, 1974). A decision must then somehow be made, possibly using the dendrogram, as to how many clusters the data actually divides into.

Different cluster analysis methods occur due to different ways of defining the distance between an individual and a group, or between two groups. Single-linkage cluster analysis defines the distance between two clusters to be the minimum distance between any two individuals in the clusters and is particularly easy to operate. Unfortunately, this method tends to create long 'straggly' clusters and is possibly not ideal for the intended purpose. Complete linkage cluster analysis is the opposite of single linkage in the sense that the distances between two groups is defined to be the maximum distance between any two individuals from the groups. Complete linkage cluster analysis does not tend to create long, straggly clusters. Various other methods of defining the intergroup distances, together with a more complete description of these hierarchical cluster analysis methods, can be found in Cormack (1971) and Everitt (1974). Cluster analysis can be performed using the original variables or using a subset of the principal components in order to simplify the situation.

When the observations have an inherent time-ordering the possibility of a time trend must be entertained. This can be examined by plotting the autocorrelation function for varying time lags. Box and Jenkins (1976) and Chatfield (1975) contain a description of this together with a general discussion of time series problems.

Further exploratory data analysis can be performed on the residuals resulting from an analysis (be it fitting a regression equation or some other statistical model-fitting technique). In particular, the residuals are extremely useful in finding discordant observations where the model does not fit well, and for examining the possible need for transformations or extra variables to improve the model. Later chapters deal with this in more detail, but see also Belsley *et al.* (1980), Barnett and Lewis (1978) and Atkinson (1982).

Exercises 2

1. Use the techniques set out in this chapter to explore the data sets given in Chapter 1 and in the Appendix.

References

Anderson, E. (1960) A semigraphical method for the analysis of complex problems. *Technometrics*, **2**, 387–391.

Andrews, D. F. (1972) Plots of high-dimensional data. *Biometrics*, **28**, 125–136.

Atkinson, A. C. (1982) Regression diagnostics, transformations and constructed variables. *J. Roy. Statist. Soc.* B, **42**, 1–36.

Barnett, V. and Lewis, T. (1978) *Outliers in Statistical Data.* Wiley, Chichester.

Belsley, D. A. Kuh, E. and Welsch, R. E. (1980) *Regression Diagnostics: Identifying Influential Data and Sources of Collinearity.* Wiley, New York.

Box, G. E. P. and Jenkins, G. M. (1976) *Time Series Analysis, Forecasting and Control*, 2nd edn. Holden-Day, San Francisco.

Chatfield, C. (1975) *The Analysis of Time Series: Theory and Practice.* Chapman and Hall, London.

Chernoff, H. (1973) Using faces to represent points in *K*-dimensional space graphically. *J.A.S.A.*, **68**, 361–368.

Cleveland, W. S. (1982) A reader's guide to smoothing scatterplots and graphical methods for regression, in *Modern Data Analysis* (eds R. L. Launer and A. F. Siegel). Academic Press, New York, pp. 37–43.

Cleveland, W. S. and McGill, R. (1983) A colour-caused optical illusion on a statistical graph. *Amer. Statist.*, **37**, 101–105.

Cormack, R. M. (1971) A review of classification. *J. Roy. Statist. Soc.* A, **134**, 321–367.

Cox, D. R. (1978) Some remarks on the roll in statistics of graphical methods. *Appl. Statist.*, **27**, 4–9.

Everitt, B. S. (1974) *Cluster Analysis.* Heineman, London.

Feder, P. I. (1974) Graphical techniques in statistical data analysis – Tools for extracting information from data. *Technometrics*, **16**, 287–299.

Fienberg, S. E. (1979) Graphical methods in statistics, *Amer. Statist.*, **33**, 165–178.

Huff, D. (1973) *How of Lie with Statistics.* Penguin, Hardmondsworth.

Jones, B. (1979) Cluster analysis of some social survey data. *Bull. Appl. Statist.*, **6**, 25–56.

McGill, R., Tukey, J. W. and Larsen, W. A. (1978) Variations of box plots. *Amer. Statist.*, **32**, 12–16.

Meeker, W. Q., Hahn, G. J. and Feder, P. I. (1975) A computer program for evaluating and comparing experimental designs and some applications. *Amer. Statist.*, **29**, 60–64.

Meeker, W. Q., Hahn, G. J. and Feder, P. I. (1977) New bias evaluation features of EXPLOR – A program for assessing experimental design properties. *Amer. Statist.*, **31**, 95–96.

Mosteller, F. and Tukey, J. W. (1977) *Data Analysis and Regression.* Addison-Wesley, Reading, MA.

Pickett, R. and White, B. W. (1966) Constructing data pictures. *Proceedings of the Seventh National Symposium of the Society for Information Display*, pp. 75–81.

Siegel, J. H., Goldwyn, R. M. and Friedman, H. P. (1971) Pattern and process of the evolution of human septic shock. *Surgery*, **70**, 232–245.

Stover, H. S. (1981) Terminal puts three-dimensional graphics on solid ground. *Electronics*, July, 150–155.

Tukey, J. W. (1977) *Exploratory Data Analysis.* Addison-Wesley, Reading, MA.

Tukey, J. W. (1980) We need both exploratory and confirmatory. *Amer. Statist.*, **34**, 23–25.

Velleman, P. F. and Hoaglin, D. C. (1981) *Applications, Basics and Computing of Exploratory Data Analysis.* Duxbury Press, Boston.

Wainer, H. (1984) How to display data badly. *Amer. Statist.*, **38**, 137–147.

Wakimoto, K. and Taguri, M. (1978) Constellation graphical method for representing multi-dimensional data. *Ann. Inst. Statist. Math.*, **30**, 97–104.

Wetherill, G. B. (1982) *Intermediate Statistical Methods* Chapman and Hall, London.

Further reading

Anderson, T. W. (1958) *Introduction to Multivariate Statistical Analysis.* Wiley, New York.

Anscombe, F. J. (1973) Graphs in statistical analysis. *Amer. Statist.*, **27**, 17–21.

Kendall, M. G. (1957) A Course in Multivariate Analysis. Charles Griffin, London.

Morrison, D. F. (1976) Multivariate Statistical Methods, 2nd edn McGraw-Hill, New York.

CHAPTER 3

Obtaining least squares solutions

3.1 Introduction

Our summary of least squares given in Sections 1.3–1.5 shows that the core problem is one of solving the set of linear equations (1.3). In some accounts of regression, therefore, we are led directly to this algebraic problem.

However, it is really most important that the 'data exploration' phase be thoroughly worked through before turning attention to this core problem, which is one of numerical linear algebra.

In this chapter we assume the 'data exporation' phase to be completed, and we turn our attention to this problem of solving equations (1.3). In fact, many computer programs exist which do all of the calculations for us, so it is tempting to think that we need not go further into the methods of solution. Unfortunately this is not so, for several reasons. Firstly, a number of multiple regression programs in common use will produce erroneous output, without warning, for certain data sets. It is important to understand something of the underlying numerical methods, and to verify that those built into the programs we use are sound. A point linked with this is that very large data sets are frequently analysed, and then it is of vital importance to use computationally efficient algorithms.

A second major reason for the material in this chapter is that sometimes we come across data sets which have near linear relationships between the explanatory variables. This condition is called multicollinearity and was referred to briefly in Chapters 1 and 2. In order to give an adequate account of multicollinearity (in Chapter 4) we must be familiar with certain aspects of linear algebra, including eigenvalues and eigenvectors, and some matrix decompositions. We shall cover these topics in this chapter, in a manner suitable for our purposes.

We start by describing Gaussian elimination, which is one of the simplest, but effective, methods of solving linear equations, and is at the heart of most of the computer routines. This leads us to a discussion of modern 'sweep' methods of obtaining regression estimates. A further application of algebra, covering orthogonality and the QR factorization, leads us to a discussion of methods of 'updating' regressions for further data. An application of the final sections on eigenvalues and eigenvectors has already been mentioned. Those who feel confident about their knowledge of these matters may care to pass to Chapter 4, with possible study of Sections 3.5, 3.8 and 3.10. For those wishing to pursue the subject further, a list of references is given at the end of the chapter.

3.2 Gaussian elimination

Suppose that we wish to solve the following equations

$$\left.\begin{aligned} 2x_1 + 4x_2 - x_3 &= 13, \\ 4x_1 + 6x_2 + 5x_3 &= 17, \\ 2x_1 + 2x_2 + 3x_3 &= 7, \end{aligned}\right\} \tag{3.1}$$

then a natural approach is to use the first equation to eliminate x_1 from the second and third equations, and then to eliminate x_2 from

Table 3.1 *Gaussian elimination.*

	Coefficient					
	x_1	x_2	x_3	*Constant*	*Operation*	*Pivot*
(1)	2	4	−1	13		2
(2)	4	6	5	17	$-2 \times (1)$	
(3)	2	2	3	7	$-1 \times (1)$	
(4)	2	4	−1	13		
(5)	0	−2	7	−9		−2
(6)	0	−2	4	−6	$-1 \times (5)$	
(7)	2	4	−1	13		
(8)	0	−2	7	−9		
(9)	0	0	−3	3		

Solutions by back substitution:

from (9) $x_3 = -1,$

from (8) (using $x_3 = -1$) $x_2 = 1,$

from (7) (using $x_2 = 1$, $x_3 = -1$) $x_1 = 4.$

the new third equation by using the new second equation. This is all set out in Table 3.1. The operations listed in Table 3.1 take us from the set of equations (1) to (3) via the set (4) to (6), to the set (7) to (9). Once this final set has been reached, the solutions are readily obtained by 'back substitution'.

In Table 3.1 using equation (1) to eliminate x_1 from equations (2) and (3) we have depended on the fact that the coefficient of x_1 in equation (1) is not zero (in our case it is 2). Likewise, in equation (5) we have used the non-zero coefficient of x_2 to eliminate that variable from equation (6). These critical coefficients, which must be non-zero if the method is to proceed, are called 'pivots'.

The situation arising when one of the pivots is zero will interrupt the method of Gaussian elimination. The entries beneath the zero pivot cannot be eliminated. For example, consider the set of equations

$$\left.\begin{aligned} 2x_1 + 4x_2 - x_3 &= 13, \\ 4x_1 + 8x_2 + 5x_3 &= 17, \\ 2x_1 + 2x_2 + 3x_3 &= 7. \end{aligned}\right\} \tag{3.2}$$

When we eliminate x_1 from the second equation, x_2 disappears as well – we have a zero pivot!

The following two possibilities can arise when we arrive at a zero pivot:

(1) At least one entry beneath the zero pivot is non-zero. In this case we can get round the trouble by permuting the rows of the original coefficient matrix to bring a non-zero coefficient into the pivot position.

(2) All the entries beneath the zero pivot are also zero. This means that the original coefficient matrix is singular. There are special features about the solution of a linear equation system whose coefficient matrix is singular and we defer consideration of this case until Chapter 5.

Some illustrations of these possibilities are given in the exercises at the end of this section.

In the discussion above we found that a zero pivot caused serious problems. In fact, in computational work a non-zero (but very small) pivot may be expected to cause problems. A near-zero pivot leads to large multiplying factors. In this situation the elimination algorithm coupled with round-off errors can remove significant digits from the original data and give rise to substantial errors.

In practice, therefore, it is best to permute the rows of the coefficient matrix at each stage so that we work with the largest possible pivots. This is the important method known as 'Gaussian elimination with partial pivoting'. We could employ 'complete pivoting' by permuting the columns also, but this is usually not necessary.

We have now arrived at a satisfactory algorithm for solving normal equations such as (1.3). If all we need are the estimated regression coefficients, then this kind of algorithm solves our problem. However, we often need the inverse matrix as well, either to provide variances of the estimated parameters, or else to use as a starting point for updating the solution for further observations (see Chapter 6). A method for determining a matrix inverse is discussed in Section 3.4. Before passing to this we pause to explain how the method of Gaussian elimination is related to an important decomposition of matrices, the 'LDU factorization'.

Exercises 3.2

1. Show that when applied to the linear equation system

$$\begin{aligned}
3x_1 + 2x_2 - x_3 + 4x_4 &= 8,\\
6x_1 + 4x_2 + 2x_3 - 3x_4 &= 9,\\
15x_1 + 10x_2 - 9x_3 - 6x_4 &= 10,\\
-3x_1 + x_2 + 4x_3 + 9x_4 &= 11,
\end{aligned}$$

the method of Gaussian elimination leads to a zero pivot, but if the original equations are permuted so that the fourth equation is promoted to second place, then Gaussian elimination may proceed in the usual way.

2. Replace the left-hand side of the final equation in Exercise 3.2.1 by

$$-3x_1 - 2x_2 + 4x_3 + 9x_4.$$

Show that Gaussian elimination now breaks down because no permutation of the equations will produce a non-zero pivot. Prove that the new linear equation system has no solution. What value must replace 11 on the right-hand side of the fourth equation in order for there to be a solution? Show that the solution of the equations is then no longer unique by finding two solutions, for one of which $x_1 = 1$ whilst for the other $x_1 = 3$.

3.3 The LDU factorization of a matrix

Consider the matrix of coefficients of the linear equation system of the last section, namely

$$A = \begin{bmatrix} 2 & 4 & -1 \\ 4 & 6 & 5 \\ 2 & 2 & 3 \end{bmatrix}.$$

We may view the first stage of our elimination process as effecting the matrix factorization

$$A = L_1 A_1,$$

where

$$L_1 = \begin{bmatrix} 1 & 0 & 0 \\ 2 & 1 & 0 \\ 1 & 0 & 1 \end{bmatrix} \quad \text{and} \quad A_1 = \begin{bmatrix} 2 & 4 & -1 \\ 0 & -2 & 7 \\ 0 & -2 & 4 \end{bmatrix}.$$

Indeed, if the matrix representation of our original linear equation system is

$$A\mathbf{x} = \mathbf{b}$$

that of equations (4)–(6) in Table 3.1 is

$$A_1\mathbf{x} = \mathbf{b}_1,$$

where $\mathbf{b}_1 = L_1^{-1}\mathbf{b}$ and $\mathbf{b}_1$ may be determined, element by element, from the equation $L_1\mathbf{b}_1 = \mathbf{b}$.

Likewise, the second stage in the elimination factorizes A_1 as

$$A_1 = L_2 A_2,$$

where

$$L_2 = \begin{bmatrix} 1 & 0 & 0 \\ 0 & 1 & 0 \\ 0 & 1 & 1 \end{bmatrix} \quad \text{and} \quad A_2 = \begin{bmatrix} 2 & 4 & -1 \\ 0 & -2 & 7 \\ 0 & 0 & -3 \end{bmatrix}.$$

Combining these two factorizations

$$\begin{aligned} A &= L_1 L_2 A_2 \\ &= L A_2, \end{aligned}$$

where

$$L = L_1 L_2 = \begin{bmatrix} 1 & 0 & 0 \\ 2 & 1 & 0 \\ 1 & 1 & 1 \end{bmatrix}.$$

Note that the first column of $\boldsymbol{L}$ may be taken (unchanged) from $\boldsymbol{L}_1$ and the second column from $\boldsymbol{L}_2$. The matrix $\boldsymbol{L}$ is lower triangular and has 1's on its diagonal. The matrix $\boldsymbol{A}_2$ is upper triangular, and we may express it as the product of a diagonal matrix $\boldsymbol{D}$ and an upper triangular matrix $\boldsymbol{U}$ with 1's on its diagonal,

$$A_2 = DU,$$

where

$$D = \begin{bmatrix} 2 & 0 & 0 \\ 0 & -2 & 0 \\ 0 & 0 & -3 \end{bmatrix} \quad \text{and} \quad U = \begin{bmatrix} 1 & 2 & -0.5 \\ 0 & 1 & -3.5 \\ 0 & 0 & 1 \end{bmatrix}.$$

Thus we see that:

Result 3.1
Provided there are no zero pivots a matrix $\boldsymbol{A}$ can be factorized as the product of

$\boldsymbol{L}$, a lower triangular matrix with 1's on its diagonal;
$\boldsymbol{D}$, a diagonal matrix; and
$\boldsymbol{U}$, an upper triangular matrix with 1's on its diagonal;

$$A = LDU. \tag{3.3}$$

Moreover, the matrices $\boldsymbol{L}$, $\boldsymbol{D}$ and $\boldsymbol{U}$ are uniquely determined by the matrix $\boldsymbol{A}$. (A sketch proof of this uniqueness is indicated in Exercise 3.3.3.)

This form of factorization of a matrix is important and has a number of applications.

Exercises 3.3

1. Calculate the LDU factorization of the 4×4 matrix

$$\begin{bmatrix} 2 & 2 & 4 & 0 \\ 2 & 5 & 1 & 9 \\ 4 & 1 & 16 & -29 \\ 0 & 9 & -29 & 108 \end{bmatrix}$$

Do you notice any relation between $\boldsymbol{L}$ and $\boldsymbol{U}$ in this example?

2. Show that there is no LDU factorization for the matrix

$$A = \begin{bmatrix} 1 & 2 & 3 & 4 \\ 2 & 4 & 7 & 9 \\ 1 & 2 & 4 & 6 \\ 1 & 3 & 4 & 5 \end{bmatrix}$$

Let $\boldsymbol{P}$ be the matrix obtained from the 4×4 identity matrix $\boldsymbol{I}$ by permuting its columns so that column 2 becomes the new column 4 and columns 3 and 4 are each shifted one place to the left. Show that $\boldsymbol{P}'\boldsymbol{P} = \boldsymbol{I} = \boldsymbol{P}\boldsymbol{P}'$. Calculate the matrix $\boldsymbol{PA}$ and its LDU factorization. What relation does premultiplication of the matrix $\boldsymbol{A}$ by the 'permutation' matrix $\boldsymbol{P}$ have to the method of partial pivoting employed in solving linear equation systems by Gaussian elimination?

3. We may use induction to prove that the LDU factorization is unique. Let $\boldsymbol{A}$ be an $(n+1) \times (n+1)$ matrix having non-zero pivots. Suppose we have the two factorizations

$$\boldsymbol{A} = \boldsymbol{LDU} = \tilde{\boldsymbol{L}}\tilde{\boldsymbol{D}}\tilde{\boldsymbol{U}}.$$

Let $\boldsymbol{A}_n$ be the matrix obtained from $\boldsymbol{A}$ by deleting its final row and column. Partition $\boldsymbol{L}$, $\boldsymbol{D}$ and $\boldsymbol{U}$ accordingly so that

$$\boldsymbol{L} = \begin{bmatrix} \boldsymbol{L}_n & 0 \\ \mathbf{l} & 1 \end{bmatrix}, \quad \boldsymbol{D} = \begin{bmatrix} \boldsymbol{D}_n & 0 \\ 0 & d \end{bmatrix}, \quad \boldsymbol{U} = \begin{bmatrix} \boldsymbol{U}_n & \mathbf{u} \\ 0 & 1 \end{bmatrix},$$

where $\boldsymbol{L}_n$, $\boldsymbol{D}_n$ and $\boldsymbol{U}_n$ are $n \times n$, $\mathbf{l}$ is $1 \times n$, $\mathbf{u}$ is $n \times 1$ and d is a scalar. Suppose $\tilde{\boldsymbol{L}}_n$, $\tilde{\boldsymbol{D}}_n$, $\tilde{\boldsymbol{U}}_n$, $\tilde{\mathbf{l}}$, $\tilde{\mathbf{u}}$, $\tilde{d}$ are derived similarly from $\tilde{\boldsymbol{L}}$, $\tilde{\boldsymbol{D}}$ and $\tilde{\boldsymbol{U}}$. Show that multiplication of partitioned matrices leads to the equations

$$\begin{aligned} \boldsymbol{A}_n = \boldsymbol{L}_n\boldsymbol{D}_n\boldsymbol{U}_n &= \tilde{\boldsymbol{L}}_n\tilde{\boldsymbol{D}}_n\tilde{\boldsymbol{U}}_n, \\ \mathbf{l}\boldsymbol{D}_n\boldsymbol{U}_n &= \tilde{\mathbf{l}}\tilde{\boldsymbol{D}}_n\tilde{\boldsymbol{U}}_n, \\ \boldsymbol{L}_n\boldsymbol{D}_n\mathbf{u} &= \tilde{\boldsymbol{L}}_n\tilde{\boldsymbol{D}}_n\tilde{\mathbf{u}}, \\ \mathbf{l}\boldsymbol{D}_n\mathbf{u} + d &= \tilde{\mathbf{l}}\tilde{\boldsymbol{D}}_n\tilde{\mathbf{u}} + \tilde{d}. \end{aligned}$$

Argue inductively that, because $\boldsymbol{A}_n$ also has non-zero pivots, $\boldsymbol{L}_n = \tilde{\boldsymbol{L}}_n$, $\boldsymbol{D}_n = \tilde{\boldsymbol{D}}_n$ and $\boldsymbol{U}_n = \tilde{\boldsymbol{U}}_n$. Noting that $\boldsymbol{D}_n$ is invertible, because the diagonal entries of $\boldsymbol{D}_n$ are the pivots of $\boldsymbol{A}$, deduce that $\mathbf{l} = \tilde{\mathbf{l}}$, $\mathbf{u} = \tilde{\mathbf{u}}$ and $d = \tilde{d}$.

4. Suppose $\boldsymbol{A}$ is a symmetric matrix with the factorization

$$\boldsymbol{A} = \boldsymbol{LDU}.$$

Use Exercise 3.3.3 to show that $\boldsymbol{L} = \boldsymbol{U}'$.

5. Write a computer program to carry out the LDU factorization of a matrix.

3.4 The Gauss–Jordan method for the inverse of a matrix

A refinement of the method of Gaussian elimination may be used for obtaining matrix inverses. We illustrate this by continuing with our previous example.

The inverse of a matrix $\boldsymbol{A}$ is a second matrix, written $\boldsymbol{A}^{-1}$, which satisfies

$$\boldsymbol{A}\boldsymbol{A}^{-1} = \boldsymbol{I} = \begin{bmatrix} 1 & 0 & 0 \\ 0 & 1 & 0 \\ 0 & 0 & 1 \end{bmatrix}$$

If, when written in terms of its entries,

$$\boldsymbol{A}^{-1} = \begin{bmatrix} x_1 & y_1 & z_1 \\ x_2 & y_2 & z_2 \\ x_3 & y_3 & z_3 \end{bmatrix}$$

then the matrix relation $\boldsymbol{A}\boldsymbol{A}^{-1} = \boldsymbol{I}$ is equivalent to the three systems of linear equations

$$\boldsymbol{A}\begin{bmatrix} x_1 \\ x_2 \\ x_3 \end{bmatrix} = \begin{bmatrix} 1 \\ 0 \\ 0 \end{bmatrix}, \quad \boldsymbol{A}\begin{bmatrix} y_1 \\ y_2 \\ y_3 \end{bmatrix} = \begin{bmatrix} 0 \\ 1 \\ 0 \end{bmatrix}, \quad \boldsymbol{A}\begin{bmatrix} z_1 \\ z_2 \\ z_3 \end{bmatrix} = \begin{bmatrix} 0 \\ 0 \\ 1 \end{bmatrix}.$$

So the elements of each column of $\boldsymbol{A}^{-1}$ can be found by the method outlined in Section 3.2. However, there is a modification of Gaussian elimination, the Gauss–Jordan method, which simplifies the working. At each stage of the elimination use the pivot to reduce to zero not only all entries in the column coming below the pivot but also the entries above the pivot. Because the coefficient matrix $\boldsymbol{A}$ is always the same the three sets of linear equations may be solved simultaneously, so generating the complete inverse. This is shown in Table 3.2.

At each stage we may use any non-zero element to remove the other entries in its column. The resulting left-hand matrix is then adjusted to be $\boldsymbol{I}$ by rearranging its rows. The corresponding rearrangement of the matrix on the right is the inverse matrix required.

This way of calculating a matrix inverse in known as the Gauss–Jordan method. There are a number of important algorithms for

Table 3.2 *The Gauss–Jordan method.*

							Operation	*Pivot*
(1)	2	4	−1	1	0	0		2
(2)	4	6	5	0	1	0	$-2 \times (1)$	
(3)	2	2	3	0	0	1	$-1 \times (1)$	
(4)	2	4	−1	1	0	0	$-(-2) \times (5)$	
(5)	0	−2	7	−2	1	0		−2
(6)	0	−2	4	−1	0	1	$-1 \times (5)$	
(7)	2	0	13	−3	2	0	$-(-\frac{13}{3}) \times (9)$	
(8)	0	−2	7	−2	1	0	$-(-\frac{7}{3}) \times (9)$	
(9)	0	0	−3	1	−1	1		−3
(10)	2	0	0	$\frac{4}{3}$	$-\frac{7}{3}$	$\frac{13}{3}$	$\frac{1}{2} \times (10)$	
(11)	0	−2	0	$\frac{1}{3}$	$-\frac{4}{3}$	$\frac{7}{3}$	$-\frac{1}{2} \times (11)$	
(12)	0	0	−3	1	−1	1	$-\frac{1}{3} \times (12)$	
(13)	1	0	0	$\frac{2}{3}$	$-\frac{7}{6}$	$\frac{13}{6}$		
(14)	0	1	0	$-\frac{1}{6}$	$\frac{2}{3}$	$-\frac{7}{6}$		
(15)	0	0	1	$-\frac{1}{3}$	$\frac{1}{3}$	$-\frac{1}{3}$		

$$\text{Inverse matrix} = \begin{bmatrix} \frac{2}{3} & -\frac{7}{6} & \frac{13}{6} \\ -\frac{1}{6} & \frac{2}{3} & -\frac{7}{6} \\ -\frac{1}{3} & \frac{1}{3} & \frac{1}{3} \end{bmatrix}$$

determining numerically matrix inverses, and access to the literature may be obtained through Chambers (1977). Many of these algorithms involve operations similar to those employed with the Gauss–Jordan method, or some form of matrix decomposition, such as the LDU factorization. We have looked in detail at this particular technique because a slightly more general form of it, which we now describe, will find an important application in the next section. Suppose we operate on an $n \times n$ matrix $\boldsymbol{A}$, to which is adjoined a further $n \times p$ matrix $\boldsymbol{C}$. The row operations performed above to obtain $\boldsymbol{A}^{-1}$ are equivalent to matrix multiplication on the left by $\boldsymbol{A}^{-1}$, so we get

$$[\boldsymbol{A} \mid \boldsymbol{C}] \rightarrow \boxed{\text{row operations}} \rightarrow [\boldsymbol{I} \mid \boldsymbol{A}^{-1} \quad \boldsymbol{C}]. \tag{3.4}$$

The right-hand part of the resulting matrix presents solutions for the equations

$$A\mathbf{x} = \mathbf{c}_i$$

for each column $\mathbf{c}_i$ of C. This result leads directly to the use of the row operation technique in regression.

Exercises 3.4

1. Use the Gauss–Jordan method to show that

$$\text{(a)} \begin{bmatrix} 2 & 2 & 4 & 0 \\ 2 & 5 & 1 & 9 \\ 4 & 1 & 16 & -29 \\ 0 & 9 & -29 & 108 \end{bmatrix}^{-1}$$

$$= \frac{1}{30}\begin{bmatrix} 2509 & -298 & -1098 & -270 \\ -298 & 46 & 126 & 30 \\ -1098 & 126 & 486 & 120 \\ -270 & 30 & 120 & 30 \end{bmatrix}$$

$$\text{(b)} \begin{bmatrix} 1 & 2 & 3 & 4 \\ 2 & 4 & 7 & 9 \\ 1 & 2 & 4 & 6 \\ 1 & 3 & 4 & 5 \end{bmatrix}^{-1} = \begin{bmatrix} 4 & 0 & -1 & -2 \\ 1 & -1 & 0 & 1 \\ -3 & 2 & -1 & 0 \\ 1 & -1 & 1 & 0 \end{bmatrix}$$

3.5 Application to regression

In applying these techniques to multiple regression we find that a slightly more extended version is useful. We wish to solve equation (1.3) and also obtain the sum of squares due to regression (1.11). To do both we apply row operations to the matrix of augmented normal equations

$$(\boldsymbol{a} \mid \mathbf{Y})'(\boldsymbol{a} \mid \mathbf{Y}) = \left[\begin{array}{c|c} \boldsymbol{a}'\boldsymbol{a} & \boldsymbol{a}'\mathbf{Y} \\ \hline \underbrace{\mathbf{Y}'\boldsymbol{a}}_{m} & \underbrace{\mathbf{Y}'\mathbf{Y}}_{1} \end{array}\right] \begin{matrix} \}m \\ \}1 \end{matrix}. \tag{3.5}$$

For the top half of the matrix we operate as before so that the identity matrix $\boldsymbol{I}$ and the matrix

$$\hat{\boldsymbol{\theta}} = (\boldsymbol{a}'\boldsymbol{a})^{-1}\boldsymbol{a}'\mathbf{Y}$$

are formed in the left and right places, respectively. The row operations are applied to the bottom left-hand matrix, $\mathbf{Y}'\boldsymbol{a}$, so as to subtract from it multiples of the previous rows and reduce all entries to zero. These row operations are equivalent to a left multiplication by a matrix of the form

$$\begin{bmatrix} L & 0 \\ M & I \end{bmatrix}. \tag{3.6}$$

Multiplication of partitioned matrices gives

$$\begin{bmatrix} L & 0 \\ M & I \end{bmatrix}\begin{bmatrix} a'a & a'\mathrm{Y} \\ \mathrm{Y}'a & \mathrm{Y}'\mathrm{Y} \end{bmatrix} = \begin{bmatrix} L(a'a) & L(a'\mathrm{Y}) \\ M(a'a) + \mathrm{Y}'a & M(a'\mathrm{Y}) + \mathrm{Y}'\mathrm{Y} \end{bmatrix}.$$

Identifying the left-hand parts of this last matrix with the results of the row operations we see that $L(a'a) = I$ leading to $L = (a'a)^{-1}$, as previously noted, and $M(a'a) + \mathrm{Y}'a = 0$ so that $M = -(\mathrm{Y}'a)(a'a)^{-1}$. The bottom right-hand entry is then

$$M(a'\mathrm{Y}) + \mathrm{Y}'\mathrm{Y} = \mathrm{Y}'\mathrm{Y} - (\mathrm{Y}'a)(a'a)^{-1}(a'\mathrm{Y}).$$

We conclude that

$$\begin{bmatrix} a'a & a'\mathrm{Y} \\ \mathrm{Y}'a & \mathrm{Y}'\mathrm{Y} \end{bmatrix} \rightarrow \boxed{\text{row operations}} \rightarrow \begin{bmatrix} I & (a'a)^{-1}a'\mathrm{Y} \\ 0 & \mathrm{Y}'\mathrm{Y} - \mathrm{Y}'a(a'a)^{-1}a'\mathrm{Y} \end{bmatrix}. \tag{3.7}$$

The result (3.7) shows that these row operations have produced the regression parameters in the top right-hand position of the matrix, and the residual sum of squares for fitting the respective model as the bottom right-hand entry since, from (1.4) and (1.8), we obtain

$$S_{\min} = \mathrm{Y}'\mathrm{Y} - \mathrm{Y}'a(a'a)^{-1}a'\mathrm{Y}.$$

Goodnight (1979), in an excellent paper on this and other related methods of carrying out regression analysis, defines the operation on row J as ADJUST(J).

ADJUST (J)

(1) Set the pivot equal to a_{JJ}.
(2) Divide row J by the pivot.
(3) For each other row produce zero in the Jth column by subtracting a multiple of row J from it.

Now it is interesting to note that if we augment the matrix a by adjoining not just a single column vector but an $n \times l$ matrix Y, then the result in line (3.7) still holds. This extension is useful when we consider fitting more than one model. In the right-hand portion of the adjusted matrix, the top block gives the regression parameters, and the diagonal elements of the bottom block give the residual sum of squares for the various models.

As an illustration of this principle, suppose we have data in the form

$$\begin{bmatrix} 1 & x_{11} & x_{21} & x_{31} & x_{41} \\ 1 & x_{12} & x_{22} & x_{32} & x_{42} \\ \vdots & \vdots & \vdots & \vdots & \vdots \end{bmatrix},$$

so that the matrix (3.5) is

$$\begin{bmatrix} n & \sum x_{1i} & \sum x_{2i} & \sum x_{3i} & \sum x_{4i} \\ \sum x_{1i} & \sum x_{1i}^2 & \sum x_{1i}x_{2i} & \sum x_{1i}x_{3i} & \sum x_{1i}x_{4i} \\ \vdots & \vdots & \vdots & \vdots & \vdots \end{bmatrix}.$$

Then ADJUST(1) produces

$$\left[\begin{array}{c|cccc} 1 & \bar{x}_{1.} & \bar{x}_{2.} & \bar{x}_{3.} & \bar{x}_{4.} \\ \hline 0 & \mathrm{CS}(x_1, x_1) & \mathrm{CS}(x_1, x_2) & \mathrm{CS}(x_1, x_3) & \mathrm{CS}(x_1, x_4) \\ \vdots & \vdots & \vdots & \vdots & \vdots \end{array}\right]. \quad (3.8)$$

Thus the right-hand partition gives the parameters for fitting the 'regressions'

$$X_1 = \alpha_1 + \varepsilon,$$
$$X_2 = \alpha_2 + \varepsilon,$$
$$X_3 = \alpha_3 + \varepsilon,$$
$$X_4 = \alpha_4 + \varepsilon,$$

where the ε's are independent normal error, and the diagonal elements of the bottom right-hand partition are the sums of squared deviations for fitting these 'regressions', which in this case are simply the corrected sums of squares. In fact, all that is happening here is that the mean of the variables is being fitted. The appropriate analysis of variance is therefore as follows:

ANOVA

Source	CSS	d.f.
Due to mean	$n\bar{x}^2$	1
Deviations	$\mathrm{CS}(x, x)$	$(n-1)$
Total	$\sum x^2$	n

Usually, we take it for granted that the mean is being fitted, and start from the corrected sum of squares as the 'total'.

If now an ADJUST(2) operation is performed we obtain a result with the following structure

$$\left[\begin{array}{cc:ccc} 1 & 0 & a_2 & a_3 & a_4 \\ 0 & 1 & b_2 & b_3 & b_4 \\ \hdashline 0 & 0 & s_2 & \cdot & \cdot \\ 0 & 0 & \cdot & s_3 & \cdot \\ 0 & 0 & \cdot & \cdot & s_4 \end{array}\right], \tag{3.9}$$

where

$$a_2 = \bar{x}_2 - \frac{\mathrm{CS}(x_1, x_2)}{\mathrm{CS}(x_1, x_1)}\bar{x}_{1}., \qquad b_2 = \frac{\mathrm{CS}(x_1, x_2)}{\mathrm{CS}(x_1, x_1)},$$

$$s_2 = \mathrm{CS}(x_2, x_2) - \frac{\mathrm{CS}(x_1, x_2)^2}{\mathrm{CS}(x_1, x_1)}, \quad \text{etc.}$$

The elements of the top right-hand partition are the parameters for fitting the regressions,

$$\left.\begin{aligned} X_2 &= \alpha_2 + \beta_2 X_1 + \varepsilon, \\ X_3 &= \alpha_3 + \beta_3 X_1 + \varepsilon, \\ X_4 &= \alpha_4 + \beta_4 X_1 + \varepsilon. \end{aligned}\right\} \tag{3.10}$$

The diagonal elements in the bottom right-hand corner are again the sum of squared deviations for fitting these regressions.

In this way, successive ADJUST operations will bring more explanatory variables into the regressions remaining and one such regression is eliminated at each step. See Table 3.3 for a numerical illustration.

At this point you may wonder what the use of these regression equations is, since if, say there is only one response variable, only one regression is of interest in the end. The other 'regressions', such as two of the models in (3.10), represent regressions of one *explanatory* variable on *other explanatory variables.* In answer to this, two points arise.

Firstly, if X_4 is the response variable then the above procedure will

Table 3.3 *Application of ADJUST to multiple regression.*

(a) Data, for example,

x_1	x_2	y
160	91	831.1
271	79	836.1
321	73	902.1
368	67	928.4
415	48	880.7
432	34	889.3
484	21	930.2

x_1 is trawlers operating from UK
x_2 is drifters operating from UK
y is fish landed in thousands of tons
(data are for years 1963–69)

(b) The matrix $\boldsymbol{a}$ is a column of 1's, followed by columns of x_1, x_2 and y. We then have

$$\boldsymbol{a}'\boldsymbol{a} = \begin{matrix} 7 & 2\,451 & 413 & 619.79 \\ 2\,451 & 930\,611 & 128\,830 & 2\,190\,669.3 \\ 413 & 128\,830 & 28\,241 & 361\,782.1 \\ 619.79 & 2\,190\,669.3 & 361\,782.1 & 5\,497\,260 \end{matrix}$$

(c) After ADJUST (1) we have

$$\begin{matrix} 1 & 350.14 & 59 & 885.41 \\ 0 & 72\,410.86 & -15\,779 & 20\,518.89 \\ 0 & -15\,779 & 3\,894 & -3\,894 \\ 0 & 20\,518.89 & -3\,894 & 9\,551.21 \end{matrix}$$

(d) After ADJUST (2) we have

$$\begin{matrix} 1 & 0 & 135.29 & 786.1952 \\ 0 & 1 & -0.2179 & 0.2834 \\ 0 & 0 & 435.6090 & 577.2563 \\ 0 & 0 & 577.2563 & 3\,736.8226 \end{matrix}$$

(e) After ADJUST (3) we have

$$\begin{matrix} 1 & 0 & 0 & 606.9 \\ 0 & 1 & 0 & 0.5721 \\ 0 & 0 & 1 & 1.3251 \\ 0 & 0 & 0 & 2971.86 \end{matrix}$$

(f) The interpretation is straightforward. Thus, in part (d), the regression of x_2 on x_1 is

$$x_2 = 135.29 - 0.2179x_1$$

with a residual sum of squares of 435.6 and the regression of y on x_1 is

$$y = 786.20 + 0.2834x_1$$

with a residual sum of squares of 3736.8.

generate the sums of squares for the regression models of

$$X_4 \text{ on } X_1,$$
$$X_4 \text{ on } X_1 \text{ and } X_2,$$
$$X_4 \text{ on } X_1, X_2 \text{ and } X_3,$$

as the ADJUST operations are done on X_1, X_2 and X_3. Thus, while this procedure gives a reasonably efficient method of calculating the final regression (of X_4 on X_1, X_2 and X_3), it generates other sums of squares *en route*, which are useful in interpretation.

Secondly, suppose, for example, that after doing ADJUST(1) and (2), the sum of squared deviations for the regression of X_3 on X_1 and X_2 was very small. This would indicate that X_3 is very nearly a linear combination of X_1 and X_2, so that the final matrix we are trying to invert to obtain the regression of X_4 on X_1, X_2 and X_3 is nearly singular. This is called the problem of multicollinearity and we take it up more fully in Chapter 4. Here we note that unless our procedure for calculation gives a check on the possible presence of multicollinearity, serious errors can result. The computer will give us results which are likely to be highly sensitive to very small changes in the data, such as to the last digit (see Chapter 4).

In the above discussion we referred to the sum of squared deviations being 'very small'. How small should this be? One answer to this is to calculate the multiple correlation coefficient, which is

$$R^2 = 1 - (\text{sum of squared deviations})/(\text{total corrected sum of squares}).$$

For example, if we apply ADJUST(3) to (3.9) the sum of squared deviations for the regression of X_3 on X_1 and X_2 is then given by the appropriate element of the bottom right-hand partition of the adjusted version of (3.9), and the total corrected sum of squares is given by the corresponding element of (3.8). If this multiple correlation coefficient is greater than, say 0.9999, then all three of the variables X_1, X_2 and X_3 should not be entered into the regression (see Chapter 4).

Many elaborate methods rather similar to the above have been devised for carrying out regression analysis calculations, and an excellent description of them is given in the paper by Goodnight (1979).

Exercise 3.5

1. Write a computer program to carry out the ADJUST(J) operation and check it by using the data of Example 1.2.

3.6 Orthogonality. The QR factorization of a matrix

It is the purpose of this section to explain the concept of orthogonality, which has very many applications in regression. We also look at a method of factorizing a matrix in a somewhat similar way to that of Section 3.3, but such that one of the factors has this important orthogonality property. Many numerical algorithms in use today for matrix calculations employ this particular factorization.

A convenient example for illustrating the theory included in this section is the matrix

$$\boldsymbol{A}=\begin{bmatrix}-564 & -916 & -887\\ 95 & 430 & -865\\ 252 & 888 & -109\end{bmatrix}.$$

It is easily verified that $\boldsymbol{A}$ is the matrix product $\boldsymbol{QR}$, where

$$\boldsymbol{Q}=\begin{bmatrix}-0.9024 & 0.424 & 0.0768\\ 0.152 & 0.48 & -0.864\\ 0.4032 & 0.768 & 0.4976\end{bmatrix}$$

and

$$\boldsymbol{R}=\begin{bmatrix}625 & 1250 & 625\\ 0 & 500 & -875\\ 0 & 0 & 625\end{bmatrix}.$$

We recognize the second factor $\boldsymbol{R}$ as an upper triangular matrix. Concerning $\boldsymbol{Q}$, direct calculation shows that

$$\boldsymbol{Q}'\boldsymbol{Q}=\boldsymbol{I} \tag{3.11}$$

(here $\boldsymbol{Q}'$ is the transpose matrix of $\boldsymbol{Q}$ and $\boldsymbol{I}$ is the 3×3 identity matrix). Such a matrix is known as an orthogonal matrix. The characteristic property of $\boldsymbol{Q}$ is that its columns $\mathbf{q}_1, \mathbf{q}_2, \mathbf{q}_3$ form an *orthonormal system of vectors*, in the sense that

$$\mathbf{q}_i'\mathbf{q}_j=\begin{cases}1 & \text{if } i=j,\\ 0 & \text{otherwise.}\end{cases} \tag{3.12}$$

Quite generally two n-dimensional column vectors $\mathbf{x}$ and $\mathbf{y}$ are *orthogonal* provided

$$\mathbf{x}'\mathbf{y} = 0; \tag{3.13}$$

in terms of the coordinates $x_1, \ldots, x_n$ of $\mathbf{x}$ and $y_1, \ldots, y_n$ of $\mathbf{y}$ the condition is

$$\sum_{i=1}^{n} x_i y_i = \mathbf{x}'\mathbf{y} = 0. \tag{3.14}$$

So that $\boldsymbol{Q}$ is a matrix with the property that each of its columns has unit length and any two of these columns are orthogonal. We shall see subsequently that an arbitrary $n \times n$ matrix $\boldsymbol{A}$ having linearly independent columns may be factored as

$$\boldsymbol{A} = \boldsymbol{Q}\boldsymbol{R} \tag{3.15}$$

with $\boldsymbol{Q}$ orthogonal and $\boldsymbol{R}$ upper triangular.

The *scalar product* of the two n-dimensional column vectors $\mathbf{x}$ and $\mathbf{y}$ is defined to be the numerical quantity

$$\mathbf{x}'\mathbf{y}.$$

In terms of coordinates this is

$$\mathbf{x}'\mathbf{y} = \sum_{i=1}^{n} x_i y_i, \tag{3.16}$$

and notice that $\mathbf{y}'\mathbf{x}$ also equals the scalar product of $\mathbf{x}$ and $\mathbf{y}$. When $n = 2$ or 3 the scalar product coincides with the usual 'dot product' of vectors in two or three dimensions. By means of our generalization of this dot product we are able to extend many of the notions of ordinary vector geometry beyond the context of two- and three-dimensional space, so that they apply to, and are fruitful for, the study of vectors of arbitrary dimensionality.

The length of the vector $\mathbf{x}$ is

$$\left(\sum_{i=1}^{n} x_i^2\right)^{1/2}; \tag{3.17}$$

this is often called the *norm* of the vector and written $\|\mathbf{x}\|$, in terms of the scalar product

$$\|\mathbf{x}\| = (\mathbf{x}'\mathbf{x})^{1/2}. \tag{3.18}$$

A *unit vector* is a vector of length 1, so that $\mathbf{x}$ is a unit vector provided $\mathbf{x}'\mathbf{x} = 1$.

As explained above vectors $\mathbf{x}$ and $\mathbf{y}$ are called *orthogonal* if their scalar product vanishes, that is, if

$$\mathbf{x}'\mathbf{y} = 0. \tag{3.19}$$

Orthogonality generalizes the notions of 'right-angle' and 'perpendicularity' from elementary geometry. When $\mathbf{x}$ and $\mathbf{y}$ are orthogonal vectors

$$\|\mathbf{x} + \mathbf{y}\|^2 = \|\mathbf{x}\|^2 + \|\mathbf{y}\|^2, \tag{3.20}$$

and this result corresponds directly to the theorem of Pythagoras about right-angled triangles. To prove it simply note that

$$\|\mathbf{x} + \mathbf{y}\|^2 = (\mathbf{x} + \mathbf{y})'(\mathbf{x} + \mathbf{y}) = \mathbf{x}'\mathbf{x} + 2\mathbf{x}'\mathbf{y} + \mathbf{y}'\mathbf{y} = \|\mathbf{x}\|^2 + \|\mathbf{y}\|^2,$$

where $\mathbf{x}'\mathbf{y} = 0$ because of orthogonality. Note also that the same method of proof shows much more, namely, if we have any finite set of vectors $\mathbf{a}, \mathbf{b}, \mathbf{c}, \ldots,$ any two of which are orthogonal, then

$$\|\mathbf{a} + \mathbf{b} + \mathbf{c} + \cdots\|^2 = \|\mathbf{a}\|^2 + \|\mathbf{b}\|^2 + \|\mathbf{c}\|^2 + \cdots \tag{3.21}$$

It commonly occurs that we have a linearly independent collection of vectors and we would like to replace them with an equivalent set of vectors which are standardized with respect to the scalar product. By such a standardized set of vectors we mean an *orthonormal system of vectors* $\mathbf{e}_1, \mathbf{e}_2, \ldots, \mathbf{e}_r$; here each member of the system is a unit vector and any pair are orthogonal, that is

$$\mathbf{e}_i'\mathbf{e}_j = \begin{cases} 1 & \text{if } i = j, \\ 0 & \text{otherwise.} \end{cases} \tag{3.22}$$

The system is to be equivalent to the original collection in the sense that they are both to have the same linear span. There is a useful algorithm, known as 'Gram–Schmidt orthogonalization', for achieving this end. New vectors are created one at a time to replace the old and at each stage the replacement vector is constructed so that it is orthogonal to those previously constructed. Suppose we start with linearly independent vectors $\mathbf{a}_1, \mathbf{a}_2, \ldots, \mathbf{a}_r$. If we have already constructed $\mathbf{e}_1, \mathbf{e}_2, \ldots, \mathbf{e}_p$ we let

$$\mathbf{b}_{p+1} = \mathbf{a}_{p+1} - \lambda_1\mathbf{e}_1 - \lambda_2\mathbf{e}_2 - \cdots - \lambda_p\mathbf{e}_p$$

and try to choose scalars $\lambda_1, \lambda_2, \ldots, \lambda_p$ so that $\mathbf{b}_{p+1}$ is orthogonal to $\mathbf{e}_1, \mathbf{e}_2, \ldots, \mathbf{e}_p$. Now

$$\mathbf{e}_1'\mathbf{b}_{p+1} = (\mathbf{e}_1'\mathbf{a}_{p+1}) - \lambda_1(\mathbf{e}_1'\mathbf{e}_1) - \lambda_2(\mathbf{e}_1'\mathbf{e}_2) - \cdots - \lambda_p(\mathbf{e}_1'\mathbf{e}_p),$$

but $\mathbf{e}_1'\mathbf{e}_1 = 1$ whereas $\mathbf{e}_1'\mathbf{e}_2 = \cdots = \mathbf{e}_1'\mathbf{e}_p = 0$, so that

$$\mathbf{e}_1'\mathbf{b}_{p+1} = (\mathbf{e}_1'\mathbf{a}_{p+1}) - \lambda_1.$$

Thus, the choice $\lambda_1 = \mathbf{e}_1'\mathbf{a}_{p+1}$ ensures that $\mathbf{b}_{p+1}$ is orthogonal to $\mathbf{e}_1$. Likewise, by taking $\lambda_2 = \mathbf{e}_2'\mathbf{a}_{p+1}, \ldots, \lambda_p = \mathbf{e}_p'\mathbf{a}_{p+1}$ we guarantee that $\mathbf{b}_{p+1}$ is orthogonal to each of $\mathbf{e}_2, \ldots, \mathbf{e}_p$. Although the resulting vector $\mathbf{b}_{p+1}$ may not be a unit vector, the linear independence of our original collection, $\mathbf{a}_1, \mathbf{a}_2, \ldots, \mathbf{a}_r$, means that $\mathbf{b}_{p+1}$ so constructed is not zero, and we complete this stage of the construction by defining

$$\mathbf{e}_{p+1} = \frac{1}{\|\mathbf{b}_{p+1}\|}\mathbf{b}_{p+1}.$$

The whole process is started by taking $\mathbf{e}_1 = (1/\|\mathbf{a}_1\|)\mathbf{a}_1$. A numerical example is illustrated in Table 3.4, where we apply the procedure to the three vectors which are the columns of the matrix $\boldsymbol{A}$ mentioned at the beginning of this section.

Let us now return to the QR factorization of the matrix $\boldsymbol{A}$. Starting with the columns $\mathbf{a}_1, \mathbf{a}_2, \mathbf{a}_3$ of $\boldsymbol{A}$, Gram–Schmidt orthogonalization has produced an orthonormal system of column vectors $\mathbf{e}_1, \mathbf{e}_2, \mathbf{e}_3$. If we form the matrix $\boldsymbol{Q}$ by taking these three new vectors for its columns then $\boldsymbol{Q}$ is an orthogonal matrix, that is, $\boldsymbol{Q}$ satisfies $\boldsymbol{Q}'\boldsymbol{Q} = \boldsymbol{I}$. Clearly, we can apply this construction with an arbitrary $n \times n$ matrix, having linearly independent columns, in place of $\boldsymbol{A}$ and the outcome will be an $n \times n$ orthogonal matrix. The original columns of $\boldsymbol{A}$ can be expressed in terms of the columns of $\boldsymbol{Q}$ by means of the following equations

$$\begin{aligned}
\mathbf{a}_1 &= \|\mathbf{a}_1\|\mathbf{e}_1 \\
&= 625\mathbf{e}_1, \\
\mathbf{a}_2 &= (\mathbf{e}_1'\mathbf{a}_2)\mathbf{e}_1 + \mathbf{b}_2 = (\mathbf{e}_1'\mathbf{a}_2)\mathbf{e}_1 + \|\mathbf{b}_2\|\mathbf{e}_2 \\
&= 1250\mathbf{e}_1 + 500\mathbf{e}_2, \\
\mathbf{a}_3 &= (\mathbf{e}_1'\mathbf{a}_3)\mathbf{e}_1 + (\mathbf{e}_2'\mathbf{a}_3)\mathbf{e}_2 + \mathbf{b}_3 = (\mathbf{e}_1'\mathbf{a}_3)\mathbf{e}_1 + (\mathbf{e}_2'\mathbf{a}_3)\mathbf{e}_2 + \|\mathbf{b}_3\|\mathbf{e}_3 \\
&= 625\mathbf{e}_1 - 875\mathbf{e}_2 + 625\mathbf{e}_3.
\end{aligned}$$

In matrix terms the foregoing is equivalent to

$$\boldsymbol{A} = \boldsymbol{Q}\boldsymbol{R},$$

where $\boldsymbol{R}$ is the upper triangular matrix built from the coefficients of

Table 3.4 *Gram–Schmidt orthogonalization.*

Initial set of vectors

$$\mathbf{a}_1 = \begin{bmatrix} -564 \\ 95 \\ 252 \end{bmatrix}, \quad \mathbf{a}_2 = \begin{bmatrix} -916 \\ 430 \\ 888 \end{bmatrix}, \quad \mathbf{a}_3 = \begin{bmatrix} -887 \\ -865 \\ -109 \end{bmatrix}.$$

Stage 1

$$\|\mathbf{a}_1\| = \sqrt{((-564)^2 + (95)^2 + (252)^2)} = 625,$$

$$\mathbf{e}_1 = \frac{1}{\|\mathbf{a}_1\|}\mathbf{a}_1 = \begin{bmatrix} -0.9024 \\ 0.152 \\ 0.4032 \end{bmatrix}.$$

Stage 2

$$\mathbf{e}_1'\mathbf{a}_2 = (-0.9024)(-916) + (0.152)(430) + (0.4032)(888) = 1250,$$

$$\mathbf{b}_2 = \mathbf{a}_2 - (\mathbf{e}_1'\mathbf{a}_2)\mathbf{e}_1 = \begin{bmatrix} -916 \\ 430 \\ 888 \end{bmatrix} - (1250)\begin{bmatrix} -0.9024 \\ 0.152 \\ 0.4032 \end{bmatrix} = \begin{bmatrix} 212 \\ 240 \\ 384 \end{bmatrix},$$

$$\|\mathbf{b}_2\| = \sqrt{((212)^2 + (240)^2 + (384)^2)} = 500,$$

$$\mathbf{e}_2 = \frac{1}{\|\mathbf{b}_2\|}\mathbf{b}_2 = \begin{bmatrix} 0.424 \\ 0.48 \\ 0.768 \end{bmatrix}.$$

Stage 3

$$\mathbf{e}_1'\mathbf{a}_3 = (-0.9024)(-887) + (0.152)(-865) + (0.4032)(-109) = 625,$$

$$\mathbf{e}_2'\mathbf{a}_3 = (0.424)(-887) + (0.48)(-865) + (0.768)(-109) = -875,$$

$$\mathbf{b}_3 = \mathbf{a}_3 - (\mathbf{e}_1'\mathbf{a}_3)\mathbf{e}_1 - (\mathbf{e}_2'\mathbf{a}_3)\mathbf{e}_2 = \begin{bmatrix} -887 \\ -865 \\ -109 \end{bmatrix} - (625)\begin{bmatrix} -0.9024 \\ 0.152 \\ 0.4032 \end{bmatrix} - (-875)\begin{bmatrix} 0.424 \\ 0.48 \\ 0.768 \end{bmatrix} = \begin{bmatrix} 48 \\ -540 \\ 311 \end{bmatrix},$$

(Contd.)

Table 3.4 (*Contd.*)

$$\|\mathbf{b}_3\| = \sqrt{((48)^2 + (-540)^2 + (311)^2)} = 625,$$

$$\mathbf{e}_3 = \frac{1}{\|\mathbf{b}_3\|}\mathbf{b}_3$$

$$= \begin{bmatrix} 0.0768 \\ -0.864 \\ 0.4976 \end{bmatrix}.$$

Final set of vectors

$$\mathbf{e}_1 = \begin{bmatrix} -0.9024 \\ 0.152 \\ 0.4032 \end{bmatrix}, \quad \mathbf{e}_2 = \begin{bmatrix} 0.424 \\ 0.48 \\ 0.768 \end{bmatrix}, \quad \mathbf{e}_3 = \begin{bmatrix} 0.0768 \\ -0.864 \\ 0.4976 \end{bmatrix}.$$

$\mathbf{e}_1, \mathbf{e}_2$ and $\mathbf{e}_3$ which appear in the above equations, that is

$$\boldsymbol{R} = \begin{bmatrix} 625 & 1250 & 625 \\ 0 & 500 & -875 \\ 0 & 0 & 625 \end{bmatrix}.$$

Again, this argument is perfectly general and we see that an arbitrary non-singular $n \times n$ matrix may be expressed in the form

$$\boldsymbol{QR}, \tag{3.23}$$

where $\boldsymbol{Q}$ is $n \times n$ orthogonal and $\boldsymbol{R}$ is $n \times n$ upper triangular.

3.7 Orthogonal matrices. The uniqueness of the QR factorization

Before we come to their application we need to pursue a little further the ideas introduced in the previous section.

We have seen that an $n \times n$ matrix $\boldsymbol{Q}$ is orthogonal when

$$\boldsymbol{Q}'\boldsymbol{Q} = \boldsymbol{I};$$

equivalently, when the columns of $\boldsymbol{Q}$ form an orthonormal system of vectors. The condition $\boldsymbol{Q}'\boldsymbol{Q} = \boldsymbol{I}$ implies, in particular, that $\boldsymbol{Q}\mathbf{x} = 0$ only if the vector $\mathbf{x} = 0$. Hence the columns of the orthogonal matrix $\boldsymbol{Q}$ are linearly independent, the rank of $\boldsymbol{Q}$ is n and $\boldsymbol{Q}$ is non-singular. It now follows from the relation $\boldsymbol{Q}'\boldsymbol{Q} = \boldsymbol{I}$ that $\boldsymbol{Q}^{-1} = \boldsymbol{Q}'$. Further

$$\boldsymbol{Q}\boldsymbol{Q}' = \boldsymbol{I};$$

that is, the rows of $\boldsymbol{Q}$ also form an orthogonal system of vectors.

The preceding paragraph implies that, in the case of an orthogonal matrix, the inverse matrix (which equals the transpose) is also orthogonal.

If $\boldsymbol{Q}_1$ and $\boldsymbol{Q}_2$ are orthogonal $n \times n$ matrices then their product $\boldsymbol{Q}_1\boldsymbol{Q}_2$ is orthogonal, because

$$\begin{aligned}(\boldsymbol{Q}_1\boldsymbol{Q}_2)'(\boldsymbol{Q}_1\boldsymbol{Q}_2) &= \boldsymbol{Q}'_2\boldsymbol{Q}'_1\boldsymbol{Q}_1\boldsymbol{Q}_2 \\ &= \boldsymbol{Q}'_2\boldsymbol{Q}_2 \quad (\text{because } \boldsymbol{Q}'_1\boldsymbol{Q}_1 = \boldsymbol{I}) \\ &= \boldsymbol{I}.\end{aligned}$$

A simple calculation easily verifies that the 2×2 matrix

$$\begin{bmatrix} \cos\theta & \sin\theta \\ -\sin\theta & \cos\theta \end{bmatrix}$$

is orthogonal. It is the matrix representation of the linear transformation which rotates two-dimensional space, clockwise, through an angle θ. In n-dimensional space a matrix of the form

$$\begin{bmatrix} 1 & 0 & \cdots & 0 & \cdots & 0 & \cdots & 0 \\ 0 & 1 & \cdots & 0 & \cdots & 0 & \cdots & 0 \\ \vdots & \vdots & \ddots & \vdots & & \vdots & & \vdots \\ 0 & 0 & \cdots & \cos\theta & \cdots & \sin\theta & \cdots & 0 \\ \vdots & \vdots & & \vdots & \ddots & \vdots & & \vdots \\ 0 & 0 & \cdots & -\sin\theta & \cdots & \cos\theta & \cdots & 0 \\ \vdots & \vdots & & \vdots & & \vdots & \ddots & \vdots \\ 0 & 0 & \cdots & 0 & \cdots & 0 & \cdots & 1 \end{bmatrix} \tag{3.24}$$

is orthogonal, and is called a plane rotation (in the (i,j)-plane). (The new matrix is obtained by adapting the $n \times n$ identity matrix, the four entries in the (i,i), (i,j), (j,i) and (j,j) positions are replaced by the corresponding four entries of the 2×2 rotation matrix.) Orthogonal matrices of this type form a very useful supply of basic construction material when it is necessary to build an orthogonal matrix for a specific purpose. We shall see some applications later in the chapter.

Let us look again at the QR factorization of a matrix. If we start with an $n \times m$ matrix $\boldsymbol{a}$, where $n \geq m$ and the columns of $\boldsymbol{a}$ are linearly independent (so that the rank of $\boldsymbol{a}$ is m), then the argument of Section 3.6 (originally applied only to square matrices) shows that $\boldsymbol{a}$ may be factored

$$\boldsymbol{a} = \boldsymbol{QR}, \tag{3.25}$$

where $\boldsymbol{Q}$ is $n \times m$ with the property that each of its columns has unit length and any two of these columns are orthogonal; and $\boldsymbol{R}$ is an upper triangular $m \times m$ matrix having positive diagonal entries. The new feature that we must be aware of when $n > m$ is that, although $\boldsymbol{Q}$ will satisfy $\boldsymbol{Q}'\boldsymbol{Q} =$ the $m \times m$ identity matrix, $\boldsymbol{Q}\boldsymbol{Q}'$ will not equal the $n \times n$ identity matrix.

Given the matrix $\boldsymbol{a}$, the factors $\boldsymbol{Q}$ and $\boldsymbol{R}$ in (3.25) are uniquely determined. For suppose there is a similar factorization

$$\boldsymbol{a} = \boldsymbol{P}\boldsymbol{S},$$

with $\boldsymbol{P}$ and $\boldsymbol{S}$ subject to the same conditions as $\boldsymbol{Q}$ and $\boldsymbol{R}$. Let $\mathbf{a}_1, \mathbf{a}_2, \ldots, \mathbf{a}_m$ denote the columns of $\boldsymbol{a}$, $\mathbf{q}_1, \mathbf{q}_2, \ldots, \mathbf{q}_m$ the columns of $\boldsymbol{Q}$ and $\mathbf{p}_1, \mathbf{p}_2, \ldots, \mathbf{p}_m$ the columns of $\boldsymbol{P}$. Also let the matrix $\boldsymbol{R}$ have entries r_{ij}, so that $r_{ij} = 0$ when $i > j$ (because $\boldsymbol{R}$ is upper triangular) and $r_{ii} > 0$; the entries s_{ij} of the matrix $\boldsymbol{S}$ satisfy the same conditions. Comparing the first columns in the matrix equations $\boldsymbol{a} = \boldsymbol{Q}\boldsymbol{R}$ and $\boldsymbol{a} = \boldsymbol{P}\boldsymbol{S}$ we find that

$$\mathbf{a}_1 = r_{11}\mathbf{q}_1 \quad \text{and} \quad \mathbf{a}_1 = s_{11}\mathbf{p}_1.$$

Since $\|\mathbf{q}_1\| = 1 = \|\mathbf{p}_1\|$ and $r_{11} > 0$, $s_{11} > 0$ we have

$$r_{11} = \|\mathbf{a}_1\| = s_{11}.$$

But then $\mathbf{q}_1 = \mathbf{p}_1$. Looking now at the second columns

$$\mathbf{a}_2 = r_{12}\mathbf{q}_1 + r_{22}\mathbf{q}_2 \quad \text{and} \quad \mathbf{a}_2 = s_{12}\mathbf{p}_1 + s_{22}\mathbf{p}_2$$

so

$$r_{12} = \mathbf{a}_2'\mathbf{q}_1 = \mathbf{a}_2'\mathbf{p}_1 = s_{12},$$

and also

$$r_{22}\mathbf{q}_2 = \mathbf{a}_2 - r_{12}\mathbf{q}_1 = \mathbf{a}_2 - s_{12}\mathbf{p}_1 = s_{22}\mathbf{p}_2,$$

hence $r_{22} = s_{22}$ (because both are positive and $\|\mathbf{q}_2\| = 1 = \|\mathbf{p}_2\|$) and $\mathbf{q}_2 = \mathbf{p}_2$. Continuing in this fashion we easily deduce that $\boldsymbol{Q} = \boldsymbol{P}$ and $\boldsymbol{R} = \boldsymbol{S}$.

3.8 Updating regression

We now use the algebra investigated in the last two sections to develop a procedure for adding rows to a regression model. There will be occasions when it may be undesirable or impossible to wait for all the data before obtaining some regression results. In such circumstances, it is clearly valuable to have a technique which permits us to update the current results as further data becomes available.

Returning to the basic regression model of equation (1.1) suppose the $n \times m$ matrix $\boldsymbol{a}$ is the initial design matrix (which carries the known values of the explanatory variables); here $n \geq m$ and we would usually expect n to be very much larger than m. We assume that the model is of full rank, so that the rank of $\boldsymbol{a}$ equals m.

As we have seen the matrix $\boldsymbol{a}$ may be factored

$$\boldsymbol{a} = \boldsymbol{QR},$$

where $\boldsymbol{Q}$ is $n \times m$ with orthonormal columns and $\boldsymbol{R}$ is upper triangular with positive diagonal entries. If this decomposition is known then to solve the normal equations

$$\boldsymbol{a}'\boldsymbol{a}\boldsymbol{\theta} = \boldsymbol{a}'\mathbf{Y}$$

it suffices to solve first

$$\boldsymbol{R}'\boldsymbol{\phi} = \boldsymbol{a}'\mathbf{Y} \tag{3.26}$$

for $\boldsymbol{\phi}$ and then

$$\boldsymbol{R}\boldsymbol{\theta} = \boldsymbol{\phi} \tag{3.27}$$

for $\boldsymbol{\theta}$. To see this, note that

$$\boldsymbol{a}'\boldsymbol{a} = (\boldsymbol{QR})'(\boldsymbol{QR}) = \boldsymbol{R}'\boldsymbol{Q}'\boldsymbol{QR} = \boldsymbol{R}'\boldsymbol{R}$$

because $\boldsymbol{Q}'\boldsymbol{Q}$ is the $m \times m$ identity matrix, so that if $\boldsymbol{\theta}$ and $\boldsymbol{\phi}$ satisfy equations (3.26)–(3.27) then

$$\boldsymbol{a}'\boldsymbol{a}\boldsymbol{\theta} = \boldsymbol{R}'\boldsymbol{R}\boldsymbol{\theta} = \boldsymbol{R}'\boldsymbol{\phi} = \boldsymbol{a}'\mathbf{Y}.$$

Both systems of equations (3.26) and (3.27) are triangular, so their solution is easily effected; the first by forward substitution because the coefficient matrix $\boldsymbol{R}'$ is lower triangular, the second by backward substitution, the coefficient matrix $\boldsymbol{R}$ being upper triangular. We may also express sums of squares in terms of $\boldsymbol{\phi}$. From equation (1.11)

$$S_{\text{par}} = \boldsymbol{\theta}'\boldsymbol{a}'\mathbf{Y} = \boldsymbol{\theta}'\boldsymbol{R}'\boldsymbol{\phi} = \boldsymbol{\phi}'\boldsymbol{\phi}$$

so that

$$S_{\text{min}} = \mathbf{Y}'\mathbf{Y} - \boldsymbol{\phi}'\boldsymbol{\phi}.$$

Now suppose data from a further observation becomes available.

This is equivalent to being furnished with an extra row $\mathbf{b}$ for the matrix $\boldsymbol{a}$ and an extra entry z for the column vector $\mathbf{Y}$ (which represents values of the response variable). Consider the partitioned $(m+1)\times(m+1)$ matrix

$$\begin{bmatrix} \boldsymbol{R} & \boldsymbol{\phi} \\ \mathbf{b} & z \end{bmatrix}.$$

Suppose we premultiply it by the plane rotation

$$\boldsymbol{P}_1 = \begin{bmatrix} \cos\alpha & 0 & \cdots & \sin\alpha \\ 0 & 1 & & 0 \\ \vdots & \vdots & & \vdots \\ -\sin\alpha & 0 & \cdots & \cos\alpha \end{bmatrix}.$$

With $\boldsymbol{R} = [r_{ij}]$ and $\mathbf{b} = (b_1 b_2 \cdots b_m)$, the $(m+1, 1)$ entry in the matrix product is

$$-r_{11}\sin\alpha + b_1\cos\alpha.$$

If now α is chosen so that

$$\sin\alpha = \frac{b_1}{\sqrt{(r_{11}^2 + b_1^2)}} \quad \text{and} \quad \cos\alpha = \frac{r_{11}}{\sqrt{(r_{11}^2 + b_1^2)}},$$

this quantity is zero and the matrix product

$$\boldsymbol{P}_1 \begin{bmatrix} \boldsymbol{R} & \boldsymbol{\phi} \\ \mathbf{b} & z \end{bmatrix}$$

has a first column with top entry $r_{11}\cos\alpha + b_1\sin\alpha = \sqrt{(r_{11}^2 + b_1^2)}$ and all following entries zero (premultiplication by $\boldsymbol{P}_1$ only affects the entries in the first and last positions). This done we now premultiply by the plane rotation

$$\boldsymbol{P}_2 = \begin{bmatrix} 1 & 0 & \cdots & 0 \\ 0 & \cos\beta & \cdots & \sin\beta \\ \vdots & \vdots & & \vdots \\ 0 & -\sin\beta & \cdots & \cos\beta \end{bmatrix}$$

and choose β so that the second column of $\boldsymbol{P}_2\boldsymbol{P}_1 \begin{bmatrix} \boldsymbol{R} & \boldsymbol{\phi} \\ \mathbf{b} & z \end{bmatrix}$ has zero final entry. Note that $\boldsymbol{P}_2$ does not disturb the first column. Continuing in this way we may construct a sequence of plane rotations

$\boldsymbol{P}_1, \boldsymbol{P}_2, \ldots, \boldsymbol{P}_m$ so that

$$\boldsymbol{P}_m \cdots \boldsymbol{P}_2 \boldsymbol{P}_1 \begin{bmatrix} \boldsymbol{R} & \boldsymbol{\phi} \\ \mathbf{b} & z \end{bmatrix} = \begin{bmatrix} \boldsymbol{S} & \boldsymbol{\psi} \\ 0 & w \end{bmatrix},$$

where $\boldsymbol{S}$ is upper triangular with positive diagonal entries. Then

$$\begin{bmatrix} \boldsymbol{R} & \boldsymbol{\phi} \\ \mathbf{b} & z \end{bmatrix} = \boldsymbol{P}_1^{-1} \boldsymbol{P}_2^{-1} \cdots \boldsymbol{P}_m^{-1} \begin{bmatrix} \boldsymbol{S} & \boldsymbol{\psi} \\ 0 & w \end{bmatrix}.$$

But

$$\begin{bmatrix} \boldsymbol{Q} & 0 \\ 0 & 1 \end{bmatrix} \begin{bmatrix} \boldsymbol{R} & \boldsymbol{\phi} \\ \mathbf{b} & z \end{bmatrix} = \begin{bmatrix} \boldsymbol{QR} & \boldsymbol{Q\phi} \\ \mathbf{b} & z \end{bmatrix} = \begin{bmatrix} \boldsymbol{a} & \boldsymbol{Q\phi} \\ \mathbf{b} & z \end{bmatrix}$$

and so

$$\begin{bmatrix} \boldsymbol{a} & \boldsymbol{Q\phi} \\ \mathbf{b} & z \end{bmatrix} = \begin{bmatrix} \boldsymbol{Q} & 0 \\ 0 & 1 \end{bmatrix} \boldsymbol{P}_1^{-1} \boldsymbol{P}_2^{-1} \cdots \boldsymbol{P}_m^{-1} \begin{bmatrix} \boldsymbol{S} & \boldsymbol{\psi} \\ 0 & w \end{bmatrix}.$$

The matrix $\begin{bmatrix} \boldsymbol{Q} & 0 \\ 0 & 1 \end{bmatrix} \boldsymbol{P}_1^{-1} \boldsymbol{P}_2^{-1} \cdots \boldsymbol{P}_m^{-1}$ is an $(n+1) \times (m+1)$ matrix having orthonormal columns. Suppose we write it as $[\boldsymbol{P} \quad \mathbf{p}]$ where $\boldsymbol{P}$ is $(n+1) \times m$ with orthonormal columns and $\mathbf{p}$ is an $(n+1)$-dimensional unit vector which is orthogonal to the columns of $\boldsymbol{P}$. The matrix equation

$$\begin{bmatrix} \boldsymbol{a} & \boldsymbol{Q\phi} \\ \mathbf{b} & z \end{bmatrix} = [\boldsymbol{P} \quad \mathbf{p}] \begin{bmatrix} \boldsymbol{S} & \boldsymbol{\psi} \\ 0 & w \end{bmatrix}$$

implies

$$\begin{bmatrix} \boldsymbol{a} \\ \mathbf{b} \end{bmatrix} = \boldsymbol{PS},$$

and by uniqueness this is the QR factorization of the matrix $\begin{bmatrix} \boldsymbol{a} \\ \mathbf{b} \end{bmatrix}$. Also

$$\begin{aligned} \boldsymbol{S}'\boldsymbol{\psi} &= \boldsymbol{S}'\boldsymbol{P}'\boldsymbol{P}\boldsymbol{\psi} && \text{(because } \boldsymbol{P}'\boldsymbol{P} = \text{the } m \times m \text{ identity matrix)} \\ &= \boldsymbol{S}'\boldsymbol{P}'(\boldsymbol{P}\boldsymbol{\psi} + w\mathbf{p}) && \text{(because } \mathbf{p} \text{ is orthogonal to the columns of } \boldsymbol{P}) \\ &= \begin{bmatrix} \boldsymbol{a} \\ \mathbf{b} \end{bmatrix}' \begin{bmatrix} \boldsymbol{Q\phi} \\ z \end{bmatrix} \end{aligned}$$

$$
\begin{aligned}
&= \boldsymbol{a}'\boldsymbol{Q}\boldsymbol{\phi} + \mathbf{b}'z \\
&= \boldsymbol{R}'\boldsymbol{Q}'\boldsymbol{Q}\boldsymbol{\phi} + \mathbf{b}'z \\
&= \boldsymbol{R}'\boldsymbol{\phi} + \mathbf{b}'z \\
&= \boldsymbol{a}'\mathbf{Y} + \mathbf{b}'z \\
&= \begin{bmatrix} \boldsymbol{a} \\ \mathbf{b} \end{bmatrix}' \begin{bmatrix} \mathbf{y} \\ z \end{bmatrix}.
\end{aligned}
$$

Thus the matrix $\boldsymbol{S}$ and the vector $\boldsymbol{\psi}$ take over the roles, in the updated situation, originally played by $\boldsymbol{R}$ and $\boldsymbol{\phi}$. The updated estimates of the regression coefficients are determined from

$$\boldsymbol{S}\boldsymbol{\theta} = \boldsymbol{\psi}$$

and the revised sums of squares from

$$S_{\text{par}} = \boldsymbol{\psi}'\boldsymbol{\psi}.$$

In practice, to apply this procedure once the initial $\boldsymbol{R}$ and $\boldsymbol{\phi}$ are determined it is only necessary to calculate the effect of the m plane rotations $\boldsymbol{P}_1, \boldsymbol{P}_2, \ldots, \boldsymbol{P}_m$ on the matrix $\begin{bmatrix} \boldsymbol{R} & \boldsymbol{\phi} \\ \mathbf{b} & z \end{bmatrix}$. Further information may be found in Chambers (1971), where the method is developed in detail. In particular, the author shows how to incorporate the calculation of residuals into this scheme, he assesses the numerical properties of the algorithm and supplies a FORTRAN program to implement it.

As indicated above, the vector $\boldsymbol{\phi}$ is easily calculated when the matrix $\boldsymbol{R}$ is known. However, the theory outlined in Sections 3.6 and 3.7 does not adapt well to the numerical determination of the QR factorization of a matrix, and the matrix $\boldsymbol{R}$ is best found by alternative means. It is otherwise known as the Choleski square root of the matrix $\boldsymbol{a}'\boldsymbol{a}$ and a discussion of numerical procedures for its calculation may be found in Stewart (1973) or Wilkinson and Reinsch (1971).

3.9 The eigenvalue decomposition of a symmetric matrix

Throughout this section we shall suppose $\boldsymbol{A}$ to be an $n \times n$ symmetric matrix. In applications of the following theory to regression it frequently happens that $\boldsymbol{A}$ equals or is some function of $\boldsymbol{a}'\boldsymbol{a}$, where $\boldsymbol{a}$ is the matrix introduced in Section 1.2. More generally, in statistics the need to determine the eigenvalue decomposition of a symmetric matrix arises in canonical correlation, principal components, discrimination and other multivariate methods.

Let us start, by way of example, with a simple candidate for $\boldsymbol{A}$, namely suppose $\boldsymbol{A}$ is the 3×3 diagonal matrix

$$\begin{bmatrix} 5 & 0 & 0 \\ 0 & 3 & 0 \\ 0 & 0 & -1 \end{bmatrix}.$$

The action of this particular $\boldsymbol{A}$ on vectors is very straightforward.

$$\boldsymbol{A}\mathbf{x} = \begin{bmatrix} 5 & 0 & 0 \\ 0 & 3 & 0 \\ 0 & 0 & -1 \end{bmatrix} \begin{bmatrix} x_1 \\ x_2 \\ x_3 \end{bmatrix} = \begin{bmatrix} 5x_1 \\ 3x_2 \\ -x_3 \end{bmatrix}$$

and corresponds to a scaling in each of the directions of the three standard coordinate axes. The scaling factor associated with any one of these directions is the corresponding diagonal entry of $\boldsymbol{A}$. The three individual effects of which the action of $\boldsymbol{A}$ is composed may be isolated, via unit vectors in the three directions as

$$\text{(i) } \boldsymbol{A} \begin{bmatrix} 1 \\ 0 \\ 0 \end{bmatrix} = 5 \begin{bmatrix} 1 \\ 0 \\ 0 \end{bmatrix}, \qquad \text{(ii) } \boldsymbol{A} \begin{bmatrix} 0 \\ 1 \\ 0 \end{bmatrix} = 3 \begin{bmatrix} 0 \\ 1 \\ 0 \end{bmatrix}$$

$$\text{(iii) } \boldsymbol{A} \begin{bmatrix} 0 \\ 0 \\ 1 \end{bmatrix} = (-1) \begin{bmatrix} 0 \\ 0 \\ 1 \end{bmatrix}.$$

It is interesting to determine the inverse of $\boldsymbol{A}$ from this point of view. The inverse is the matrix $\boldsymbol{B}$ which undoes the action of $\boldsymbol{A}$, and is clearly given by

$$\boldsymbol{B}\mathbf{x} = \begin{bmatrix} \frac{1}{5}x_1 \\ \frac{1}{3}x_2 \\ -x_3 \end{bmatrix},$$

that is $\boldsymbol{B} = \boldsymbol{A}^{-1}$ is, of course, the diagonal matrix

$$\begin{bmatrix} \frac{1}{5} & 0 & 0 \\ 0 & \frac{1}{3} & 0 \\ 0 & 0 & -1 \end{bmatrix}.$$

The key fact about symmetric matrices is that the essential features isolated in our analysis of this simple example are always present. Any symmetric matrix behaves exactly like a diagonal matrix, once we have found the appropriate coordinate system.

To illustrate that this is so consider the 2×2 symmetric, but non-diagonal, matrix

$$A = \begin{bmatrix} 0.61 & -0.48 \\ -0.48 & 0.89 \end{bmatrix}$$

Let

$$\mathbf{q}_1 = \begin{bmatrix} 0.8 \\ 0.6 \end{bmatrix} \quad \text{and} \quad \mathbf{q}_2 = \begin{bmatrix} -0.6 \\ 0.8 \end{bmatrix}.$$

It is easy to check that $\mathbf{q}_1$ and $\mathbf{q}_2$ are unit vectors in two-dimensional space and that they are orthogonal (Fig. 3.1).

If we compute $A\mathbf{q}_1$ we find that

$$A\mathbf{q}_1 = 0.25\mathbf{q}_1,$$

so that A scales all vectors in the direction of $\mathbf{q}_1$ by the factor 0.25. Likewise

$$A\mathbf{q}_2 = 1.25\mathbf{q}_2,$$

and A scales all vectors in the direction of $\mathbf{q}_2$ by the factor 1.25.

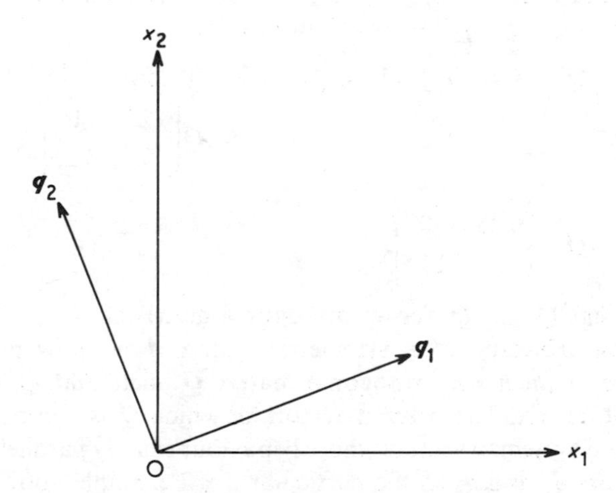

Fig. 3.1 *The orthogonal vectors* $\mathbf{q}_1$ and $\mathbf{q}_2$

Consider the vector

$$\mathbf{x} = \begin{bmatrix} -1 \\ -1 \end{bmatrix}.$$

Expressed in terms of $\mathbf{q}_1$ and $\mathbf{q}_2$ we find

$$\mathbf{x} = -1.4\mathbf{q}_1 - 0.2\mathbf{q}_2$$

(notice that the coefficient of $\mathbf{q}_1$ is $\mathbf{q}_1'\mathbf{x}_1$ and the coefficient of $\mathbf{q}_2$ is $\mathbf{q}_2'\mathbf{x}_2$). Then

$$\begin{aligned} A\mathbf{x} &= -1.4A\mathbf{q}_1 - 0.2A\mathbf{q}_2 \\ &= -1.4(0.25\mathbf{q}_1) - 0.2(1.25\mathbf{q}_2) \\ &= -0.35\mathbf{q}_1 - 0.25\mathbf{q}_2 \\ &= \begin{bmatrix} -0.13 \\ -0.41 \end{bmatrix}. \end{aligned}$$

Thus, once we have expressed our vector $\mathbf{x}$ as a linear combination of $\mathbf{q}_1$ and $\mathbf{q}_2$, $A\mathbf{x}$ is easily determined because of the basic way in which $\mathbf{q}_1$ and $\mathbf{q}_2$ are transformed by A. This is not simply a matter of ease of computation, for notice that this picture of the matrix A gives real insight into the nature of its action on a vector, effectively scaling by 0.25 the component in the $\mathbf{q}_1$ direction and scaling by 1.25 the component in the orthogonal direction $\mathbf{q}_2$.

Notice further that if Q is the 2×2 orthogonal matrix which has the vectors $\mathbf{q}_1$ and $\mathbf{q}_2$ for its columns, then

$$\begin{aligned} AQ = A[\mathbf{q}_1, \mathbf{q}_2] = [A\mathbf{q}_1 \quad A\mathbf{q}_2] &= [0.25\mathbf{q}_1 \quad 1.25\mathbf{q}_2] \\ &= Q\begin{bmatrix} 0.25 & 0 \\ 0 & 1.25 \end{bmatrix} \end{aligned}$$

so that

$$Q'AQ = \begin{bmatrix} 0.25 & 0 \\ 0 & 1.25 \end{bmatrix}.$$

(Recall that $Q^{-1} = Q'$ for an orthogonal matrix.)

For an arbitrary $n \times n$ symmetric matrix A we now prove the existence of an $n \times n$ orthogonal matrix Q such that $Q'AQ$ is a diagonal matrix. The column vectors of which Q is composed will then give a system of vectors whose behaviour exactly parallels that of the vectors $\mathbf{q}_1$ and $\mathbf{q}_2$ in the particular 2×2 example above.

When $A = [a_{ij}]$ is any $n \times n$ symmetric matrix, set

$$\delta(A) = \text{the sum of the squares of the entries above the main diagonal} \\ = \sum_{i<j} a_{ij}^2.$$

The following lemma is a key ingredient in the argument we give to show that it is possible to diagonalize any symmetric matrix. After reading the statement of the lemma some readers may wish to skip its proof (which employs the plane rotation matrices introduced in Section 3.7) and pass directly to the main part of our argument.

Lemma

If the symmetric matrix A is not diagonal there is an orthogonal matrix P such that

$$\delta(P'AP) < \delta(A).$$

Proof:

Let $\mathbf{A} = [a_{ij}]$ and suppose for the moment the element $a_{12} \neq 0$. Let P be the plane rotation

$$P = \begin{bmatrix} \cos\theta & \sin\theta & 0 & \cdots & 0 \\ -\sin\theta & \cos\theta & 0 & \cdots & 0 \\ & & & & \\ 0 & \cdots & 0 & 1 & \cdots & 0 \\ \vdots & & \vdots & \vdots & & \vdots \\ 0 & \cdots & 0 & 0 & \cdots & 1 \end{bmatrix}$$

If $B = P'AP$ then B is identical to A except for elements appearing in the first two rows and columns. More precisely, when $B = [b_{ij}]$,

$$\begin{aligned} b_{11} &= a_{11}\cos^2\theta - 2a_{12}\cos\theta\sin\theta + a_{22}\sin^2\theta, \\ b_{12} = b_{21} &= (a_{11} - a_{22})\cos\theta\sin\theta + a_{12}(\cos^2\theta - \sin^2\theta), \\ b_{22} &= a_{11}\sin^2\theta + 2a_{12}\cos\theta\sin\theta + a_{22}\cos^2\theta, \end{aligned}$$

$$\left.\begin{aligned} b_{1j} = b_{j1} &= a_{1j}\cos\theta - a_{2j}\sin\theta, \\ b_{2j} = b_{j2} &= a_{1j}\sin\theta + a_{2j}\cos\theta, \end{aligned}\right\} \quad \text{when } j > 2,$$

whilst $b_{ij} = a_{ij}$ for $i, j > 2$. So

$$\delta(B) - \delta(A) = b_{12}^2 + \sum_{j>2}(b_{1j}^2 + b_{2j}^2) - a_{12}^2 - \sum_{j>2}(a_{1j}^2 + a_{2j}^2).$$

But note that

$$\begin{aligned} b_{1j}^2 + b_{2j}^2 &= (a_{1j}\cos\theta - a_{2j}\sin\theta)^2 + (a_{1j}\sin\theta + a_{2j}\cos\theta)^2 \\ &= a_{1j}^2(\cos^2\theta + \sin^2\theta) + a_{2j}^2(\sin^2\theta + \cos^2\theta) = a_{1j}^2 + a_{2j}^2, \end{aligned}$$

and hence

$$\delta(\boldsymbol{B}) - \delta(\boldsymbol{A}) = b_{12}^2 - a_{12}^2.$$

If now we choose θ so that $b_{12} = 0$, which is certainly possible because it amounts to finding θ with

$$\tfrac{1}{2}(a_{11} - a_{22})\sin 2\theta + a_{12}\cos 2\theta = 0$$

and this can always be done, then

$$\delta(\boldsymbol{P}'\boldsymbol{A}\boldsymbol{P}) - \delta(\boldsymbol{A}) = \delta(\boldsymbol{B}) - \delta(\boldsymbol{A}) = -a_{12}^2 < 0.$$

The argument proceeds in a similar way if $a_{12} = 0$, because there is some non-zero off-diagonal entry a_{pq}, where $p < q$, and we may then take $\boldsymbol{P}$ to be a plane rotation in the (p, q)-plane.

Equipped with this lemma we now turn to the problem of diagonalizing a given $n \times n$ symmetric matrix $\boldsymbol{A}$. There is a choice of orthogonal matrix $\boldsymbol{Q}$ so that $\delta(\boldsymbol{Q}'\boldsymbol{A}\boldsymbol{Q})$ assumes its minimal value, that is

$$\delta(\boldsymbol{Q}'\boldsymbol{A}\boldsymbol{Q}) \leq \delta(\boldsymbol{X}'\boldsymbol{A}\boldsymbol{X}) \tag{3.28}$$

for all other orthogonal matrices $\boldsymbol{X}$. The technical justification for this statement is as follows:

Let $N = n^2$. When $\boldsymbol{X} = [x_{ij}]$ is an $n \times n$ orthogonal matrix view $\boldsymbol{X}$ as the point in N-dimensional space having coordinates

$$(x_{11}, x_{12}, \ldots, x_{1n}, x_{21}, x_{22}, \ldots, x_{2n}, \ldots, x_{n1}, x_{n2}, \ldots, x_{nn}).$$

The set S of points in N-dimensional space corresponding to the collection of all orthogonal matrices is closed and bounded, because the condition $\boldsymbol{X}'\boldsymbol{X} = \boldsymbol{I}$ on the matrix $\boldsymbol{X}$ translates to the conditions $\sum_{i=1}^{n} x_{ij}^2 = 1$ for $1 \leq j \leq n$ and $\sum_{i=1}^{n} x_{ij}x_{ik} = 0$ for $1 \leq j, k \leq n$, where $j \neq k$, on the coordinates $x_{11}, x_{12}, \ldots, x_{nn}$. For $\boldsymbol{X}$ in S let

$$f(\boldsymbol{X}) = \delta(\boldsymbol{X}'\boldsymbol{A}\boldsymbol{X});$$

f is a continuous real-valued function defined on the closed and bounded subset S of N-dimensional space. In these circumstances the existence theorem for maxima and minima asserts there is an orthogonal matrix $\boldsymbol{Q}$ so that

$$f(\boldsymbol{Q}) = \delta(\boldsymbol{Q}'\boldsymbol{A}\boldsymbol{Q}) \leq f(\boldsymbol{X}) = \delta(\boldsymbol{X}'\boldsymbol{A}\boldsymbol{X})$$

for all other orthogonal matrices $\boldsymbol{X}$. (See, for example, Clark 1982.)

Then $\boldsymbol{Q}'\boldsymbol{A}\boldsymbol{Q}$ is a diagonal matrix for otherwise, according to the lemma, there is an orthogonal matrix $\boldsymbol{P}$ so that

$$\delta(\boldsymbol{P}'\boldsymbol{Q}'\boldsymbol{A}\boldsymbol{Q}\boldsymbol{P}) < \delta(\boldsymbol{Q}'\boldsymbol{A}\boldsymbol{Q})$$

and this contradicts (3.28) when $\boldsymbol{X} = \boldsymbol{Q}\boldsymbol{P}$.

In principle, then, if we are given an $n \times n$ symmetric matrix $\boldsymbol{A}$ there always exists an $n \times n$ orthogonal matrix $\boldsymbol{Q}$ such that $\boldsymbol{Q}'\boldsymbol{A}\boldsymbol{Q}$ is a diagonal matrix, say

$$\boldsymbol{Q}'\boldsymbol{A}\boldsymbol{Q} = \begin{bmatrix} \lambda_1 & & & 0 \\ & \lambda_2 & & \\ & & \ddots & \\ 0 & & & \lambda_n \end{bmatrix}. \tag{3.29}$$

Let $\mathbf{q}_1, \mathbf{q}_2, \ldots, \mathbf{q}_n$ denote the columns of the matrix $\boldsymbol{Q}$. Of course, the orthogonality of $\boldsymbol{Q}$ implies that this system of vectors is orthonormal. Because $\boldsymbol{Q}^{-1} = \boldsymbol{Q}'$ the matrix relation (3.29) is equivalent to

$$\boldsymbol{A}\boldsymbol{Q} = \boldsymbol{Q}\begin{bmatrix} \lambda_1 & & & 0 \\ & \lambda_2 & & \\ & & \ddots & \\ 0 & & & \lambda_n \end{bmatrix}$$

from which we see that each diagonal entry λ_i and corresponding column $\mathbf{q}_i$ of $\boldsymbol{Q}$ are related by

$$\boldsymbol{A}\mathbf{q}_i = \lambda_i \mathbf{q}_i. \tag{3.30}$$

The numbers $\lambda_1, \ldots, \lambda_n$ are known as the *eigenvalues* of the matrix $\boldsymbol{A}$, and the vectors $\mathbf{q}_1, \ldots, \mathbf{q}_n$ as an associated *orthonormal system of eigenvectors*.

The n vectors $\mathbf{q}_1, \mathbf{q}_2, \ldots, \mathbf{q}_n$ are linearly independent, as is any orthonormal system of vectors, so they span the space of n-dimensional column vectors. Consequently, an arbitrary n-dimensional column vector $\mathbf{x}$ may be written as a linear combination of the orthonormal system, say

$$\mathbf{x} = \alpha_1\mathbf{q}_1 + \alpha_2\mathbf{q}_2 + \cdots + \alpha_n\mathbf{q}_n.$$

An important consequence of the orthonormality of the vectors $\mathbf{q}_1, \mathbf{q}_2, \ldots, \mathbf{q}_n$ is that the scalar coefficients $\alpha_1, \alpha_2, \ldots, \alpha_n$ may be very easily determined. For

$$\begin{aligned}\mathbf{q}_1'\mathbf{x} &= \mathbf{q}_1'(\alpha_1\mathbf{q}_1 + \alpha_2\mathbf{q}_2 + \cdots + \alpha_n\mathbf{q}_n) \\ &= \alpha_1\mathbf{q}_1'\mathbf{q}_1 + \alpha_2\mathbf{q}_1'\mathbf{q}_2 + \cdots + \alpha_n\mathbf{q}_1'\mathbf{q}_n,\end{aligned}$$

but orthonormality implies $\mathbf{q}_1'\mathbf{q}_1 = 1$ whereas $\mathbf{q}_1'\mathbf{q}_2 = \cdots = \mathbf{q}_1'\mathbf{q}_n = 0$ and we find that $\alpha_1 = \mathbf{q}_1'\mathbf{x}$. In the same way $\alpha_2 = \mathbf{q}_2'\mathbf{x}, \ldots, \alpha_n = \mathbf{q}_n'x$. Hence the representation of $\mathbf{x}$ as a linear combination of the orthonormal system of vectors $\mathbf{q}_1, \mathbf{q}_2, \ldots, \mathbf{q}_n$ is

$$\mathbf{x} = (\mathbf{q}_1'\mathbf{x})\mathbf{q}_1 + (\mathbf{q}_2'\mathbf{x})\mathbf{q}_2 + \cdots + (\mathbf{q}_n'\mathbf{x})\mathbf{q}_n. \tag{3.31}$$

Applying A to this expression for $\mathbf{x}$ in terms of $\mathbf{q}_1, \mathbf{q}_2, \ldots, \mathbf{q}_n$ gives

$$A\mathbf{x} = (\mathbf{q}_1'\mathbf{x})A\mathbf{q}_1 + (\mathbf{q}_2'\mathbf{x})A\mathbf{q}_2 + \cdots + (\mathbf{q}_n'\mathbf{x})A\mathbf{q}_n$$

and because of the relations (3.30)

$$A\mathbf{x} = \lambda_1(\mathbf{q}_1'\mathbf{x})\mathbf{q}_1 + \lambda_2(\mathbf{q}_2'\mathbf{x})\mathbf{q}_2 + \cdots + \lambda_n(\mathbf{q}_n'\mathbf{x})\mathbf{q}_n. \tag{3.32}$$

In matrix terms equation (3.32) says

$$A = \lambda_1\mathbf{q}_1\mathbf{q}_1' + \lambda_2\mathbf{q}_2\mathbf{q}_2' + \cdots + \lambda_n\mathbf{q}_n\mathbf{q}_n'. \tag{3.33}$$

The two equations (3.29) and (3.33) (which are equivalent presentations of the same information) constitute the eigenvalue decomposition of the $n \times n$ symmetric matrix A. The numbers $\lambda_1, \lambda_2, \ldots, \lambda_n$ represent the strength of the effects of the matrix and the vectors $\mathbf{q}_1, \mathbf{q}_2, \ldots, \mathbf{q}_n$ (equivalently, the matrix Q) describe the directions in which these effects are felt. A non-diagonal symmetric matrix has this information about the strength and the orientation of these effects merged. The two kinds of data are confused and disguised; to understand the action of the matrix it is necessary to disentangle them.

As an illustration of the usefulness of this decomposition note that it follows from either (3.29) or (3.33) that the matrix A is non-singular provided none of the eigenvalues is zero, and that

$$A^{-1} = Q \begin{bmatrix} \lambda_1^{-1} & & & 0 \\ & \lambda_2^{-1} & & \\ & & \ddots & \\ 0 & & & \lambda_n^{-1} \end{bmatrix} Q' \tag{3.34}$$

or

$$A^{-1} = \lambda_1^{-1}\mathbf{q}_1\mathbf{q}_1' + \lambda_2^{-1}\mathbf{q}_2\mathbf{q}_2' + \cdots + \lambda_n^{-1}\mathbf{q}_n\mathbf{q}_n'.$$

For example, if

$$A = \begin{bmatrix} 0.61 & -0.48 \\ -0.48 & 0.89 \end{bmatrix}, \tag{3.35}$$

an example considered earlier in this section, then

$$A^{-1} = (0.25)^{-1}\begin{bmatrix} 0.8 \\ 0.6 \end{bmatrix}[0.8 \quad 0.6] + (1.25)^{-1}\begin{bmatrix} -0.6 \\ 0.8 \end{bmatrix}[-0.6 \quad 0.8]$$

$$= \begin{bmatrix} 2.848 & 1.536 \\ 1.536 & 1.952 \end{bmatrix}.$$

It is usual to order the eigenvalues $\lambda_1, \lambda_2, \ldots, \lambda_n$ as an increasing or decreasing sequence. Although not evident from the proof we have given, it is easily verified that this sequence of numbers appearing in the decompositions (3.29) and (3.33) is uniquely determined by A. The corresponding statement about the associated orthonormal system of eigenvectors is a little more complicated. Clearly, any $\mathbf{q}_i$ may be replaced by its negative without affecting (3.29) or (3.33). Beyond this should it happen that a number of eigenvalues coincide, say $\lambda_i = \lambda_{i+1} = \cdots = \lambda_j$ (where $i < j$) then the vectors $\mathbf{q}_i, \mathbf{q}_{i+1}, \ldots, \mathbf{q}_j$ may be replaced by any orthonormal system of vectors $\mathbf{e}_i, \mathbf{e}_{i+1}, \ldots, \mathbf{e}_j$ in the subspace spanned by $\mathbf{q}_i, \mathbf{q}_{i+1}, \ldots, \mathbf{q}_j$.

Eigenvalues and eigenvectors play a crucial role in the structural analysis of matrices, and they frequently surface as the key explanatory variables in applications of linear algebra. There is an extensive theory, an introduction to which may be found in Strang (1980). Here we have emphasized those aspects which are important for regression and other statistical methods.

Our derivation of the eigenvalue decomposition for a symmetric matrix provides very little indication of how to calculate the eigenvalues and eigenvectors in a particular instance. In numerical terms it is not a simple matter to determine these quantities for a given matrix, though much recent research has been devoted to the problem and very reliable computer algorithms are now available.

For good algorithms in ALGOL see Wilkinson and Reinsch (1971), for FORTRAN see Sparks and Todd (1973), and for APL see Ruhe (1980). In addition to these, algorithms are available in the NAG and EISPACK software libraries.

There is a variety of terminology associated with the concepts we have been looking at (which reflects their widespread usefulness). Commonly 'latent root' and 'latent vector' are used in place of eigenvalue and eigenvector. Other terms used for eigenvalue are 'characteristic value', 'proper value' and 'secular value'.

3.10 The singular value decomposition (SVD) of a matrix

To illustrate the importance of the eigenvalue decomposition of a symmetric matrix described in the last section we use it to develop the singular value decomposition (SVD) of an arbitrary $n \times m$ matrix $\boldsymbol{a}$.

If we put $\boldsymbol{A} = \boldsymbol{a}'\boldsymbol{a}$ then $\boldsymbol{A}$ is a symmetric $m \times m$ matrix. Suppose the eigenvalues of $\boldsymbol{A}$ are $\lambda_1, \ldots, \lambda_m$ and that $\mathbf{q}_1, \ldots, \mathbf{q}_m$ form an associated orthonormal system of eigenvectors. The first point to note in this situation is that each eigenvalue is non-negative. This is so because (3.30) tells us

$$\boldsymbol{A}\mathbf{q}_i = \lambda_i \mathbf{q}_i$$

and therefore

$$\begin{aligned}
\lambda_i &= \mathbf{q}_i'(\boldsymbol{A}\mathbf{q}_i) \qquad \text{(because } \mathbf{q}_i'\mathbf{q} = 1\text{)} \\
&= \mathbf{q}_i'(\boldsymbol{a}'\boldsymbol{a}\mathbf{q}_i) \\
&= (\boldsymbol{a}\mathbf{q}_i)'(\boldsymbol{a}\mathbf{q}_i) \\
&= \|\boldsymbol{a}\mathbf{q}_i\|^2 \geq 0.
\end{aligned}$$

It is usual in this context to suppose that the eigenvalues of $\boldsymbol{A}$ are arranged in decreasing order of magnitude, so that $\lambda_1 \geq \lambda_2 \geq \cdots \geq \lambda_m \geq 0$. A certain number of these eigenvalues may be zero; in fact, the number of non-zero eigenvalues equals the rank of the matrix $\boldsymbol{A}$, which is also the rank of the original matrix $\boldsymbol{a}$. Suppose $\lambda_1, \ldots, \lambda_k$ are non-zero but $\lambda_{k+1} = \lambda_{k+2} = \cdots = \lambda_m = 0$. Put $s_1 = \sqrt{\lambda_1}$,

$s_2 = \sqrt{\lambda_2}, \ldots, s_k = \sqrt{\lambda_k}$ and let $\boldsymbol{\Sigma}$ be the $k \times k$ diagonal matrix

$$\boldsymbol{\Sigma} = \begin{bmatrix} s_1 & & & 0 \\ & s_2 & & \\ & & \ddots & \\ 0 & & & s_k \end{bmatrix}$$

If, for $i = 1$ to k, we let $\mathbf{p}_i = s_i^{-1}\boldsymbol{a}\mathbf{q}_i$, we have

$$\begin{aligned} \mathbf{p}_i\mathbf{p}_j &= (s_i^{-1}\boldsymbol{a}\mathbf{q}_i)'(s_j^{-1}\boldsymbol{a}\mathbf{q}_j) \\ &= (s_i^{-1}s_j^{-1})\mathbf{q}_i'\boldsymbol{a}'\boldsymbol{a}\mathbf{q}_j \\ &= (s_i^{-1}s_j^{-1})\mathbf{q}_i'\boldsymbol{A}\mathbf{q}_j \\ &= (s_i^{-1}s_j^{-1})(\lambda_j\mathbf{q}_i'\mathbf{q}_j) \\ &= (s_i^{-1}s_j^{-1}\lambda_j)\mathbf{q}_i'\mathbf{q}_j \end{aligned}$$

and we see that $\mathbf{p}_i'\mathbf{p}_j = 1$ if $i = j$ but is zero otherwise. Let $\boldsymbol{P}$ be the $n \times k$ matrix whose columns are the vectors $\mathbf{p}_1, \ldots, \mathbf{p}_k$ and let $\boldsymbol{Q}$ be the $m \times k$ matrix whose columns are the vectors $\mathbf{q}_1, \ldots, \mathbf{q}_k$. We have

$$\begin{aligned} \boldsymbol{a}[\mathbf{q}_1, \ldots, \mathbf{q}_m] &= [\boldsymbol{a}\mathbf{a}_1, \ldots, \boldsymbol{a}\mathbf{q}_k\, \boldsymbol{a}\mathbf{q}_{k+1}, \ldots, \boldsymbol{a}\mathbf{q}_m] \\ &= [s_1\mathbf{p}_1, \ldots, s_k\mathbf{p}_k 0, \ldots, 0] \\ &= \boldsymbol{P}[\boldsymbol{\Sigma}|0]. \end{aligned}$$

But we know the matrix $[\mathbf{q}_1, \ldots, \mathbf{q}_m]$ is orthogonal, so $[\mathbf{q}_1, \ldots, \mathbf{q}_m]^{-1} = [\mathbf{q}_1, \ldots, \mathbf{q}_m]'$, and

$$\begin{aligned} \boldsymbol{a} = P[\Sigma|0] \begin{bmatrix} \mathbf{q}_1' \\ \vdots \\ \mathbf{q}_k' \\ \mathbf{q}_{k+1}' \\ \vdots \\ \mathbf{q}_m' \end{bmatrix} &= \boldsymbol{P}[\boldsymbol{\Sigma}|0] \begin{bmatrix} \boldsymbol{Q}' \\ \mathbf{q}_{k+1}' \\ \vdots \\ \mathbf{q}_m' \end{bmatrix} \\ &= \boldsymbol{P\Sigma Q}'. \end{aligned}$$

To summarize:

3.10.1 Singular value decomposition (SVD)

An $n \times m$ matrix $\boldsymbol{a}$ has the factorization

$$\boldsymbol{a} = \boldsymbol{P\Sigma Q}'. \tag{3.36}$$

Here $\boldsymbol{P}$ is $n \times k$, $\boldsymbol{Q}$ is $n \times k$ and both matrices have the property that their columns are unit vectors which are pairwise orthogonal. Further $\boldsymbol{\Sigma}$ is an $k \times k$ diagonal matrix, its diagonal entries are strictly positive and are referred to as the *singular values* of $\boldsymbol{A}$. The integer k is the rank of $\boldsymbol{a}$ and, of course, satisfies $k \leq \min(m, n)$.

There exist very effective numerical procedures for obtaining the SVD of a matrix; consult Chambers (1977) for details.

3.10.2 Application to regression

If we are fitting the model

$$E(\mathbf{Y}) = \boldsymbol{a}\boldsymbol{\theta} \tag{3.37}$$

and we use the SVD

$$\boldsymbol{a} = \boldsymbol{P}\boldsymbol{\Sigma}\boldsymbol{Q}',$$

then

$$E(\mathbf{Y}) = \boldsymbol{P}\boldsymbol{\Sigma}\boldsymbol{Q}'\boldsymbol{\theta}.$$

Denoting the k-dimensional vector $\boldsymbol{\Sigma}\boldsymbol{Q}'\boldsymbol{\theta}$ by $\boldsymbol{\phi}$ the original model becomes

$$E(\mathbf{Y}) = \boldsymbol{P}\boldsymbol{\phi}. \tag{3.38}$$

The least squares solution for $\boldsymbol{\phi}$ is obtained by the usual matrix equation

$$\hat{\boldsymbol{\phi}} = (\boldsymbol{P}'\boldsymbol{P})^{-1}\boldsymbol{P}'\mathbf{Y},$$

but now $\boldsymbol{P}'\boldsymbol{P} = \boldsymbol{I}$ because the columns of the matrix $\boldsymbol{P}$ are unit vectors which are pairwise orthogonal, and so

$$\hat{\boldsymbol{\phi}} = \boldsymbol{P}'\mathbf{Y}.$$

The relation $\boldsymbol{\phi} = \boldsymbol{\Sigma}\boldsymbol{Q}'\boldsymbol{\theta}$ between $\boldsymbol{\theta}$ and $\boldsymbol{\phi}$ also holds for the least squares estimates for $\boldsymbol{\theta}$ and $\boldsymbol{\phi}$, so that

$$\hat{\boldsymbol{\phi}} = \boldsymbol{\Sigma}\boldsymbol{Q}'\hat{\boldsymbol{\theta}}. \tag{3.39}$$

The introduction of the intermediate vector of parameters $\boldsymbol{\phi}$ leads to various insights in regression analysis: see, for example, Mandel (1982). One application is to the problem of 'multicollinearity', a small (though non-zero) singular value for the matrix $\boldsymbol{a}$ indicates the presence of a near-linear relationship in the data set. We take up this important topic in the next chapter.

Further reading

Those wishing to read further about numerical methods in least squares should consult Chambers (1977). For further reading on algebraic matters related to least squares, see Seber (1977) or Rao (1965).

For a more detailed account of the algebra itself, see Strang (1980).

References

Chambers, J. M. (1971) Regression updating. *J. Amer. Statist. Assoc.*, **66**, 744–748.

Chambers, J. M. (1977) *Computational Methods for Data Analysis*. Wiley, New York.

Clark, C. W. (1982) *Elementary Mathematical Analysis*, 2nd edn. Wadsworth Publishers of Canada Ltd, Belmont, CA.

Goodnight, J. H. (1979) A tutorial on the SWEEP operator. *Amer. Statist.*, **33**, 149–158.

Mandel, J. (1982) Use of the singular value decomposition in regression analysis. *Amer. Statist.*, **36**, 15–24.

Rao, C. R. (1965) *Linear Statistical Inference and Its Applications*. Wiley, New York.

Ruhe, A. (1980) Eigenvalues and eigenvectors by Rayleigh quotient iteration. *APL Quote Quad*, **10** (3), 29–30. [Corrigenda. *APL Quote Quad*, **10** (4), 18.]

Seber, G. A. F. (1977) *Linear Regression Analysis*. Wiley, New York.

Sparks, D. N. and Todd, A. D. (1973) Algorithm AS60. Latent roots and vectors of a symmetric matrix. *Appl. Statist.*, **22**, 260–265.

Stewart, G. W. (1973) *Introduction to Matrix Computations*. Academic Press, London.

Strang, G. (1980) *Linear Algebra with Its Applications*, 2nd edn. Academic Press, London.

Wilkinson, J. H. and Reinsch, C. (1971) *Linear Algebra*. Vol. II of *Handbook for Automatic Computation* (eds F. L. Bauer, A. S. Householder, F. W. J. Olver, H. Rutishauser, K. Samelson, E. Stiegel). Springer-Verlag, New York.

Further reading

Clarke, M. R. B. (1981) Algorithm AS163. A Gauss algorithm for moving from one linear model to another without going back to the data. *Appl. Statist.*, **30**, 198–203.

Clarke, M. R. B. (1982) The Gauss–Jordan SWEEP operator with detection of collinearity. *Appl. Statist.*, **31**, 166–168.

CHAPTER 4

Multicollinearity

4.1 Introduction

This chapter deals with the problem of multicollinearity, also commonly termed collinearity and ill-conditioning. The problem arises when there exist *near*-linear dependencies among the vectors of explanatory variables, so that if the Gaussian elimination of Section 3.2 were used to solve the least squares equations then at least one pivot would be near-zero. The effect of multicollinearity is to inflate the variance of the least squares estimator and possibly any predictions made, and also to restrict the generality and applicability of the estimated model. Therefore, when multicollinearities occur they should be investigated thoroughly and, if they prove harmful, an effort should be made to deal with them appropriately (see below). Throughout this chapter we shall use the representation of the linear model given in (1.13):

$$E(\mathbf{Y}) = \alpha \mathbf{1} + X\boldsymbol{\beta},$$

where the columns of X ($n \times p$) consist of centred explanatory variables, i.e. $\mathbf{1}'X = 0$. We also assume for convenience that the explanatory variables have been standardized, i.e. $\mathbf{x}_i'\mathbf{x}_i = 1$, where $\mathbf{x}_i$ is the ith column of X. Although we shall be investigating *near*-linear dependencies in the explanatory variables, we shall be assuming that there are no exact linear dependencies; this is equivalent to assuming that X has full rank p, and thus that unique solutions of the least squares equations exist. In the presence of exact linear dependencies the generalized inverse methods described in Chapter 5 are appropriate.

Three possible sources of multicollinearity are given in Mason *et al.* (1975):

(1) Due to physical constraints on the model or in the population. There may be some reason why there is a constraint on the model.

For example, the contents of certain constituents in a chemical process may sum to a constant or near constant.

(2) Due to sampling techniques. Here the experimenter, possibly unwittingly, may have sampled a subspace of the p-dimensional space of the explanatory variables. For example, in plant operations this may be due to the system being necessarily run at near-optimal conditions.

(3) Due to an over-defined model. For example, there may be as many, or more explanatory variables than observations or else the model may be simply over-parametrized. There may also be an unnecessarily complicated model including, for example, many quadratic and cross-product terms.

4.2 Effects of multicollinearities

We consider now the relationships between multicollinearities and the behaviour of the least squares estimator of $\boldsymbol{\beta}$. We begin by establishing that if multicollinearities exist then at least one of the eigenvalues of $X'X$ will be small, where a more precise indication of what is meant by 'small' will be given later. Suppose the set of vectors $\mathbf{w}_1, \mathbf{w}_2, \ldots, \mathbf{w}_m$ are linearly dependent, then there exists a set of scalars $b_1, b_2, \ldots, b_m$, not all zero, such that

$$\sum_{i=1}^{m} b_i \mathbf{w}_i = \mathbf{0}.$$

In the same spirit we may say that the p vectors $\mathbf{x}_1, \mathbf{x}_2, \ldots, \mathbf{x}_p$, which constitute the columns of X, exhibit *near*-linear dependency if there exists a set of scalars $c_1, c_2, \ldots, c_p$ such that

$$\sum_{i=1}^{p} c_i \mathbf{x}_i = \boldsymbol{\delta}, \tag{4.1}$$

where $\boldsymbol{\delta}$ is 'small'. To avoid trivial solutions, and for consistency, we constrain $\sum c_i^2 = 1$. Note that $\boldsymbol{\delta}$ is a vector and so there is no unique way of defining its 'smallness'. However, the norm $\|\boldsymbol{\delta}\|$ defined in (3.18) is a measure of size that is both natural and convenient in the present circumstances, and so we say that $\boldsymbol{\delta}$ is small if

$$\|\boldsymbol{\delta}\| = (\boldsymbol{\delta}'\boldsymbol{\delta})^{1/2} < \varepsilon, \tag{4.2}$$

for some small value of ε. Combining (4.1) and (4.2) we get

$$\varepsilon > \|\boldsymbol{\delta}\| = \|\sum c_i \mathbf{x}_i\| = (\mathbf{c}'X'X\mathbf{c})^{1/2},$$

where $\mathbf{c}' = (c_1, c_2, \ldots, c_p)$, that is

$$\mathbf{c}'X'X\mathbf{c} = \lambda < \varepsilon^2. \tag{4.3}$$

Let $\lambda_1, \lambda_2, \ldots, \lambda_p$ be the eigenvalues of $X'X$ with corresponding orthonormal eigenvectors, $\mathbf{v}_1, \mathbf{v}_2, \ldots, \mathbf{v}_p$ and let V be the $p \times p$ matrix whose ith column is $\mathbf{v}_i$. Then since $X'X$ is a non-singular real symmetric matrix (see Section 3.9), we have

$$V'X'XV = \Lambda,$$

where Λ is a diagonal matrix with the eigenvalues of $X'X$ on the diagonal. We can then put $\mathbf{c} = V\boldsymbol{\gamma}$ for some suitable $\boldsymbol{\gamma}$ so that

$$\begin{aligned} \mathbf{c}'\mathbf{X}'X\mathbf{c} &= \boldsymbol{\gamma}'V'X'XV\boldsymbol{\gamma} \\ &= \boldsymbol{\gamma}'\Lambda\boldsymbol{\gamma} \\ &= \sum_{i=1}^{p} \gamma_i^2 \lambda_i \\ &= \lambda. \end{aligned}$$

Now

$$\sum_{i=1}^{p} \gamma_i^2 \lambda_i \geq \underset{1 \leq i \leq p}{\text{Min}}\,(\lambda_i) \sum_{i=1}^{p} \gamma_i^2 = \underset{1 \leq i \leq p}{\text{Min}}\,(\lambda_i)$$

(since $\sum \gamma_i^2 = \boldsymbol{\gamma}'\boldsymbol{\gamma} = \mathbf{c}'V'V\mathbf{c} = \mathbf{c}'\mathbf{c} = 1$).
Hence

$$\text{Min}\,(\lambda_i) \leq \lambda.$$

Equality will only hold when all $\lambda_i = 1$ and so the matrix X is orthogonal or $\mathbf{c}$ is the eigenvector corresponding to $\text{Min}\,(\lambda_i)$. So, in general, the smallest eigenvalue of $X'X$ will be less than the size of the linear combination $X\mathbf{c}$.

We now investigate the effects of multicollinearities on the least squares estimator, $\hat{\boldsymbol{\beta}} = (X'X)^{-1}X'\mathbf{y}$, of $\boldsymbol{\beta}$. From the eigenvalue decomposition of $X'X$ we have that

$$(X'X)^{-1} = V\Lambda^{-1}V' = \sum \lambda_i^{-1}\mathbf{v}_i\mathbf{v}_i'$$

and so we can write

$$\hat{\boldsymbol{\beta}} = \sum \lambda_i^{-1}\,\mathrm{d}_i\mathbf{v}_i, \tag{4.4}$$

where $\mathrm{d}_i = \mathbf{v}_i'X'\mathbf{y}$. For convenience, we assume that $\lambda_{\min} = \lambda_p$; then we see from (4.4) that $\hat{\boldsymbol{\beta}}$ will tend to be dominated by $\mathbf{v}_p$. Further, since $\mathbf{v}_p$

is an eigenvector of $X'X$, $X'X\mathbf{v}_p = \lambda_p \mathbf{v}_p$, so that

$$\mathbf{v}_p' X' X \mathbf{v}_p = \lambda_p \mathbf{v}_p \mathbf{v}_p' = \lambda_p$$

and hence

$$\| X\mathbf{v}_p \|^2 = (X\mathbf{v}_p)'(X\mathbf{v}_p) = \lambda_p < \lambda$$

implying that $X\mathbf{v}_p$ is small. Those columns of X which correspond to non-negligible elements of $\mathbf{v}_p$ are therefore the original explanatory variables which are involved in the multicollinearity. If there is more than one multicollinearity, then there would be more than one small eigenvalue of $X'X$ and the variables involved in each would be identified in the same way as above using the appropriate eigenvector.

We now consider the variance–covariance matrix of $\hat{\boldsymbol{\beta}}$. This can be written from Section 1.6 as

$$V(\hat{\boldsymbol{\beta}}) = \sigma^2 (X'X)^{-1}.$$

Again, using the eigenvalue decomposition of $X'X$ we can write

$$\begin{aligned} V(\hat{\boldsymbol{\beta}}) &= \sigma^2 V \Lambda^{-1} V' \\ &= \sigma^2 \sum \lambda_i \mathbf{v}_i \mathbf{v}_i', \end{aligned}$$

and it can be seen from this that those elements of $\hat{\boldsymbol{\beta}}$ which correspond to the non-negligible elements of $\mathbf{v}_p$ will have large variances and corresponding covariances. The mean square error of $\hat{\boldsymbol{\beta}}$ is

$$\begin{aligned} \text{MSE}(\hat{\boldsymbol{\beta}}) &= E\{(\hat{\boldsymbol{\beta}} - \boldsymbol{\beta})'(\hat{\boldsymbol{\beta}} - \boldsymbol{\beta})\} \\ &= \text{tr}\{V(\hat{\boldsymbol{\beta}})\} \\ &= \sigma^2 \, \text{tr}\{V \Lambda^{-1} V'\} \\ &= \sigma^2 \sum \lambda_i^{-1}. \end{aligned}$$

Thus, even though the least squares estimator of $\boldsymbol{\beta}$ is the minimum variance unbiased linear estimator, its mean square error will still be large if multicollinearities exist among the explanatory variables. The variance of a predicted point $\mathbf{p}$ is

$$\begin{aligned} V(\mathbf{p}'\hat{\boldsymbol{\beta}}) &= \mathbf{p}' V(\hat{\boldsymbol{\beta}}) \mathbf{p} \\ &= \sigma^2 \mathbf{p}' V \Lambda^{-1} V' \mathbf{p} \\ &= \sigma^2 \sum \tilde{\rho}_i^2 \lambda_i^{-1}, \end{aligned}$$

where $\mathbf{p}'\mathbf{V} = \tilde{\mathbf{p}}' = (\tilde{\mathrm{p}}_1, \tilde{\mathrm{p}}_2, \ldots, \tilde{\mathrm{p}}_p)$. If the prediction point $\mathbf{p}$ has a large coefficient, $\tilde{\mathrm{p}}_p$, corresponding to the eigenvector, $\mathbf{v}_p$, its variance will be large. In conclusion, we see that relatively imprecise estimation will be obtained in the direction of eigenvectors that correspond to multicollinearities and relatively precise estimation is possible in the directions of the remaining eigenvectors. For further details of estimability, see Silvey (1969).

4.3 Multicollinearity measures

If there are only two variables involved in a multicollinear relationship then they will have a high correlation, thus examination of the correlation matrix will reveal simple linear near-dependencies. However, this approach has two drawbacks. Firstly, the question arises as to what constitutes a high correlation; secondly and, more importantly, when three or more variables are involved in the relationships the correlations between the relevant pairs of variables need not be large.

Let us now consider the $\boldsymbol{X}$ matrix partitioned with the ith variable in the first column and the remainder $\boldsymbol{X}_i^*$ in the following columns. Then we have

$$(\boldsymbol{X}'\boldsymbol{X}) = \begin{bmatrix} \mathbf{x}_i'\mathbf{x}_i & \mathbf{x}_i'\boldsymbol{X}_i^* \\ \boldsymbol{X}_i^{*\prime}\mathbf{x}_i & \boldsymbol{X}_i^{*\prime}\boldsymbol{X}_i^* \end{bmatrix},$$

and the inverse will be of the form

$$(\boldsymbol{X}'\boldsymbol{X})^{-1} = \begin{bmatrix} c_{ii} & \boldsymbol{D}_1' \\ \boldsymbol{D}_1 & \boldsymbol{D}_2 \end{bmatrix}$$

for some scalar c_{ii} and matrices $\boldsymbol{D}_1$ and $\boldsymbol{D}_2$ with $\boldsymbol{D}_2$ symmetric. Note that since $(\boldsymbol{X}'\boldsymbol{X})$ is symmetric, so must be its inverse. It can be shown that

$$c_{ii} = (\mathbf{x}_i'\mathbf{x}_i - \mathbf{x}_i'\boldsymbol{X}_i^*(\boldsymbol{X}_i^{*\prime}\boldsymbol{X}_i^*)^{-1}\boldsymbol{X}_i^{*\prime}\mathbf{x}_i)^{-1}.$$

Recall that $\boldsymbol{X}$ has been centred so that the first term in the brackets is the corrected sum of squares of the ith variable and the second term is the regression sum of squares for regressing $\mathbf{x}_i$ on all other variables. Since the columns of $\boldsymbol{X}$ have unit length we have

$$c_{ii} = (1 - R_i^2)^{-1}, \tag{4.5}$$

where R_i^2 is the multiple correlation coefficient of $\mathbf{x}_i$ regressed on all the other explanatory variables.

If the ith variable is involved in a multicollinearity clearly R_i^2 will be close to unity. The multiple correlation coefficients can therefore be used to indicate the variables involved in multicollinearities. However, the number of multicollinearities is not suggested and there is still the issue of when is R_i^2 close to unity.

The values c_{ii} are called variance inflation factors (VIFs), see Farrar and Glauber (1967), and warrant a special mention. The variance of the ith parameter estimate is

$$V(\hat{\beta}_i) = \sigma^2 c_{ii}. \tag{4.6}$$

If the explanatory variables were mutually orthogonal then we would have

$$\boldsymbol{X}'\boldsymbol{X} = (\boldsymbol{X}'\boldsymbol{X})^{-1} = \boldsymbol{I}_p,$$

where $\boldsymbol{I}_p$ is the $p \times p$ identity matrix.

Such a matrix is perfectly conditioned and has no multicollinearities. This represents the 'ideal' situation in regression, and we then have all of the R_i^2 are zero and all of the VIFs unity. As $\boldsymbol{X}'\boldsymbol{X}$ becomes more and more ill-conditioned so some of the R_i^2 will become closer to unity and so from (4.5) the corresponding VIFs will increase in size. Also (4.6) shows that the VIFs measure the increase in the variance of the ith parameter estimate due entirely to multicollinearities in the explanatory variables.

Attention is usually focused on the largest VIF and a rule-of-thumb is that the largest VIF should not be larger than ten. We give some examples of VIFs later for Data Set 1.1.

The determinant of a square matrix is equal to the product of its eigenvalues and this will be zero if the matrix is singular. Since we are considering $\boldsymbol{X}$ to be standardized, so that $\sum \lambda_i = p$, the determinant can be used to measure the degree of ill-conditioning in the data. Note that if $\boldsymbol{X}$ were not standardized then when $\boldsymbol{X}'\boldsymbol{X}$ was orthogonal, the ideal situation, the determinant would be

$$|\boldsymbol{X}'\boldsymbol{X}| = |\alpha \boldsymbol{I}_p| = \alpha^p$$

for some constant α. Clearly, this can be made arbitrarily small by rescaling the columns of $\boldsymbol{X}$ so as to decrease the size of α. For standardized $\boldsymbol{X}$ the determinant will be unity if the columns of $\boldsymbol{X}$ are orthogonal and between zero and unity otherwise. Once again the problem of what constitutes small appears when interpreting the size of the determinant. Farrar and Glauber (1967) proposed a test of multicollinearity based on the determinant of $X'X$. However, as Kumar (1975) points out, since the test is derived under the assumption that the rows of X, before standardization, are independently multivariate normally distributed, its use is strictly limited. Obviously, the determinant can only indicate the presence or lack of multicollinearities and not, when appropriate, the number of dependencies. An indication of the number of relationships can be obtained by examining the partial products of the eigenvalues

$$\prod_{i=1}^{j} \lambda_i, \qquad j = 1, 2, \ldots, p,$$

where the λ_i are assumed ordered into decreasing magnitude. The number of small partial products will indicate the number of multicollinearities. This technique does not identify the variables involved in the relationships.

However, it was shown in Section 4.2 that the large elements of the eigenvectors corresponding to small eigenvalues do identify these variables. Thus, examination of the eigenvectors and eigenvalues of the correlation matrix yield most of the required information when investigating multicollinearities; see also Gunst and Mason (1977a). A more sophisticated approach is possible which can yield further information and this uses the *condition number* and associated *condition index* of the matrix $X'X$. We describe this now, referring the reader to Stewart (1973) for fuller details and proofs of the following results.

Much of the material that follows is highly technical and may be omitted if desired. The reader is referred to the summary at the end of this section for the pertinent details.

Definition 4.1

A vector norm on $\mathbb{R}^n$ is a real valued function v: $\mathbb{R}^n \to \mathbb{R}$ that satisfies the following conditions:

(1) $\mathbf{x} \neq \mathbf{0} \Rightarrow v(\mathbf{x}) > 0$;

(2) $v(\alpha \mathbf{x}) = |\alpha| v(\mathbf{x}), \qquad \alpha \in \mathbb{R}$;

(3) $v(\mathbf{x} + \mathbf{y}) \leq v(\mathbf{x}) + v(\mathbf{y})$.

EXAMPLE 4.1

$$\|\mathbf{x}\|_1 = \sum_{i=1}^{n} |x_i|,$$

$\|\mathbf{x}\|_2 = \sqrt{\left(\sum_{i=1}^{n} x_i^2\right)} = \sqrt{(\mathbf{x}'\mathbf{x})}$ (Euclidean Norm defined in Section 3.6).

Definition 4.2

A matrix norm is a function ν: $\mathbb{R}^{m \times n} \to \mathbb{R}$ that satisfies the conditions

(1) $A \neq 0 \Rightarrow \nu(A) > 0, \qquad A \in \mathbb{R}^{m \times n};$
(2) $\nu(\alpha A) = |\alpha| \nu(A), \qquad A \in \mathbb{R}^{m \times n}, \quad \alpha \in \mathbb{R};$
(3) $\nu(A + B) \leq \nu(A) + \nu(B), \qquad A, B \in \mathbb{R}^{m \times n}.$

Such a norm is called consistent if, for conformable A and B, it satisfies the extra condition,

(4) $\nu(A \cdot B) \leq \nu(A) \cdot \nu(B) \qquad A, B \in \mathbb{R}^{m \times n}.$

It can be shown that given a vector norm $\|\cdot\|$ it is possible to define a matrix norm consistent with $\|\cdot\|$.

EXAMPLE 4.2

Suppose A is an $m \times n$ matrix with $A = (a_{ij})$, $i = 1, 2, \ldots, m$ and $j = 1, 2, \ldots, n$. Then

$$\|A\|_1 = \max\left(\sum_{i=1}^{n} a_{ij}: j = 1, 2, \ldots, n\right),$$

$$\|A\|_2 = \sqrt{(\text{largest eigenvalue of } A'A)},$$

and so

$$\|A'A\|_2 = \text{largest eigenvalue of } A'A.$$

Definition 4.3

The condition number of an $n \times n$ non-singular matrix A is

$$\kappa(A) = \|A\| \, \|A^{-1}\|.$$

EXAMPLE 4.3

Let the eigenvalues of $X'X$ be ordered with $\lambda_1 \geq \lambda_2 \geq \cdots \geq \lambda_p$, then using $\|\cdot\|$

$$\kappa(X'X) = \lambda_1/\lambda_p,$$

the ratio of the largest to smallest eigenvalue.

Theorem 4.1

Let A be approximated by $A + \Delta A$ for some suitably small matrix Δ. Then the relative error in $(A + \Delta A)^{-1}$, is

$$\frac{\|A^{-1} - (A + \Delta A)^{-1}\|}{\|A^{-1}\|} \leq \frac{\kappa(A)\dfrac{\|\Delta A\|}{\|A\|}}{1 - \kappa(A)\dfrac{\|\Delta A\|}{\|A\|}}.$$

For ΔA sufficiently small the right-hand side of the inequality can be approximated by $\kappa(A)\|\Delta A\|/\|A\|$. Thus Theorem 4.1 states that the relative error in $(A+\Delta A)^{-1}$ is not greater than $\kappa(A)$ times the relative error in $A+\Delta A$. The proof of this theorem is given in Stewart (1973, Chapter 4).

EXAMPLE 4.4
Suppose $X'X$ is calculated to t significant digits in the sense that the relative error in $X'X$ is 10^{-t}. Then if $\kappa(X'X)=10^k$ the relative error in $(X'X)$ may be as large as $10^{-t}\cdot 10^k = 10^{k-t}$.

Theorem 4.2
Let A and ΔA be as for Theorem 4.1. Then the solution to the system of linear equations

$$A\boldsymbol{\beta} = \mathbf{y} \tag{4.7}$$

satisfies

$$\frac{\|\Delta\boldsymbol{\beta}\|}{\|\boldsymbol{\beta}+\Delta\boldsymbol{\beta}\|} \leq \kappa(A)\frac{\|\Delta A\|}{\|A\|},$$

where $\Delta\boldsymbol{\beta}$ is the error in the computed solution to (4.7) due to using the approximation $(A+\Delta A)$ to A. Further, if $\mathbf{y}$ is approximated by $\mathbf{y}+\Delta\mathbf{y}$ in (4.7) then

$$\frac{\|\Delta\boldsymbol{\beta}\|}{\|\boldsymbol{\beta}\|} \leq \kappa(A)\frac{\|\Delta\mathbf{y}\|}{\|\mathbf{y}\|}.$$

This theorem shows how the condition number of the $X'X$ matrix in least squares regression furnishes an upper bound to the relative error in the solution vector $\boldsymbol{\beta}$ induced by errors in the design matrix X and the response variable $\mathbf{y}$.

Definition 4.4
The condition index of the correlation matrix $X'X$ is the set of p-values

$$\kappa_i(X'X) = \frac{\max\limits_{1\leq j\leq p}\{\lambda_j\}}{\lambda_i},$$

where λ_i is an eigenvalue of $X'X$.

The condition index helps in identifying the number of multicollinearities in the data. The number of large values of the κ_i indicates the number of multicollinearities. This can be justified by using the following theorem (see Lawson and Hanson, 1974, Chap. 5).

Theorem 4.3
Let $\boldsymbol{A}$ be an $m \times m$ matrix and suppose $\boldsymbol{A}_{(k)}$ is created by deleting the kth column and row from $\boldsymbol{A}$. Then the ordered eigenvalues $\lambda_{i(k)}$ of $\boldsymbol{A}_{(k)}$ interlace with the ordered eigenvalues λ_i of $\boldsymbol{A}$. That is

$$\lambda_1 \geq \lambda_{1(k)} \geq \lambda_2 \geq \lambda_{2(k)} \geq \cdots \geq \lambda_{m-1} \geq \lambda_{m-1(k)} \geq \lambda_m.$$

Corollary
The condition number of $\boldsymbol{A}_{(k)}$ lies in the interval

$$\kappa(\boldsymbol{A}) = \kappa_m(\boldsymbol{A}) \geq \kappa(\boldsymbol{A}_{(k)}) \geq \lambda_2/\lambda_{m-1}.$$

Further, by repeated application of this result, it can be shown that the ith largest condition index of $\boldsymbol{A}_{(\kappa)}, \kappa_{m-i}(\boldsymbol{A}_{(\kappa)})$ is not less than λ_i/λ_{m-i}.

If we put $\boldsymbol{A} = \boldsymbol{X}'\boldsymbol{X}$, a correlation matrix, then the sum of its eigenvalues is equal to its trace, that is m. Consequently, the largest eigenvalue of a submatrix cannot be less than unity. In practice, it tends to decrease in size slowly from λ_1 down to unity as more and more rows and columns are removed. So an approximate lower bound for the condition number of any submatrix of $\boldsymbol{X}'\boldsymbol{X}$ with i rows and columns removed is $\lambda_1/\lambda_{p-i} = \kappa(\boldsymbol{X}'\boldsymbol{X})$, the ith largest condition index of $\boldsymbol{X}'\boldsymbol{X}$.

A decision of when the condition number of a matrix is large will depend upon the accuracy of the data and the accuracy to which the arithmetic calculations are made. As an example, if the data were measured to three significant figures then Theorems 4.1 and 4.2 would suggest that a condition number of the order $10^3 = 1000$ is very serious.

Summary

The condition number (with respect to the Euclidean norm) of the correlation matrix $\boldsymbol{X}'\boldsymbol{X}$ is defined to be

$$\kappa(\boldsymbol{X}'\boldsymbol{X}) = \lambda_1/\lambda_p,$$

where λ_i are the ordered eigenvalues of this matrix. The order of magnitude of the condition number indicates an upper bound to the number of decimal digits of accuracy lost in the production of

$(\boldsymbol{X}'\boldsymbol{X})^{-1}$ and $\hat{\boldsymbol{\beta}}$ induced by rounding errors in the design matrix $\boldsymbol{X}$ and the response variable $\mathbf{y}$. These errors may be due to the finite representation of numbers in computers or to rounding of the data at source, etc. As an example a condition number of $2145 \simeq 2 \times 10^3$ provides an upper bound of three decimal digits loss of accuracy due to ill-conditioning and so if the data were measured to five significant digits the estimates of $\boldsymbol{\beta}$ would be unreliable beyond the second digit. Condition numbers as small as 1000 can be considered serious.

The condition index of $\boldsymbol{X}'\boldsymbol{X}$ is the set of p-values

$$\kappa_i(\boldsymbol{X}'\boldsymbol{X}) = \frac{\max\limits_{1 \leq j \leq p} (\lambda_j)}{\lambda_i}.$$

The jth largest value of κ_i is an approximate upper bound to the condition number for the correlation matrix formed by deleting j columns of $\boldsymbol{X}$. Thus there are as many multicollinearities in $\boldsymbol{X}$ as there are large values of κ_i.

Data Set 1.1

In order to illustrate multicollinearities we will consider an extension to the model in Example 1.1. The model to be fitted is to include quadratic and cross-product terms in the explanatory variables. There can be high correlations between the original variables and the

Table 4.1 *Some measures of multicollinearity for Data Set 1.1.*

Variable	*VIF*	*mult. corr. coeff.*	*Corr. with resp.*
x1	2 856 748.97	0.999 408	0.945 038
x2	10 956.14	0.990 446	0.370 035
x3	2 017 162.54	0.999 296	−0.913 978
x1**2	2 501 944.63	0.999 368	0.944 925
x2**2	65.73	0.876 659	0.314 531
x3**2	12 667.10	0.991 115	−0.852 947
x1.x2	9 802.90	0.989 900	0.465 560
x1.x3	1 428 091.89	0.999 163	−0.919 328
x2.x3	240.36	0.935 499	−0.543 990

Condition index

Dimension	1	2	3	4	5	6	7	8	9
Index	1	2	33	47	273	1979	51 015	164 808	50 202 670

squared and cross-product variables which are potential sources of multicollinearity. Transformations exist which reduce these correlations and so improve the conditioning of the correlation matrix. More details can be found in Chapter 12. Table 4.1 lists the variance inflation factors (VIFs), the multiple correlation coefficients, the correlations of the explanatory variables with the response and the condition index. The quadratic terms are denoted x**2, etc., and the cross-product terms x1.x2, etc. The size of the VIFs in particular suggests severe problems. The condition number, 50 202 670, is of the order 10^7 and so indicates that unless the data are measured to more than about seven significant figures, rounding errors in the data could alter the regression estimates in the first significant digit. The condition index suggests that there are four or five multicollinearities. This is a common problem with quadratic models.

4.4 Solutions to multicollinearity

It is sometimes possible to remove multicollinearities by appropriate transformations of the explanatory variables. This is most likely to be the case in multiple quadratic regression where cross-product and squared variables are used, for example, in Data Set 1.1. Since multiple quadratic regression is considered in general in Chapter 12, we defer until then discussion of this approach. Otherwise, the most satisfactory solution to multicollinearity is to take extra observations in the direction or directions of the multicollinearities. Silvey (1969) considered ways of adding a single point to the design matrix in certain optimal ways; unfortunately, this is frequently impractical or impossible.

An alternative approach is to delete appropriate variables from the model. Since multicollinearities are caused by near-linear dependencies amongst some of the variables, deletion of one variable from each of the sets of columns involved in each multicollinearity will overcome the problem. However, if we delete the ith column of X from the model this is effectively assuming that the unknown ith parameter of $\boldsymbol{\beta}$, β_i is zero. If $\beta_i \neq 0$ then the least squares estimator of $\boldsymbol{\beta}$ thus produced will be biased with the amount of bias depending on the size of β_i. It should be noted that in this case the estimators of $\beta_j, j \neq i$, will also be biased unless the ith column of X is orthogonal to the remaining columns; see Draper and Smith (1966) and Chapter 10.

Although the least squares estimator of $\boldsymbol{\beta}$ is the minimum variance unbiased estimator (assuming the correct model) improvements in the sense of mean square error of $\boldsymbol{\beta}$ can be made by using biased estimators. Variable rejection, as just mentioned, is one biased method. Others, and ways of using them as variable rejection techniques, will be discussed in the remaining sections of this chapter; see also Hocking (1976).

4.5 Principal component regression

The idea of principal components has already been discussed in Chapter 2. We give a brief recap here for the present context. As in Section 4.2 we have the eigenvalue decomposition $\boldsymbol{X}'\boldsymbol{X} = \boldsymbol{V}\boldsymbol{\Lambda}\boldsymbol{V}'$; we then write the linear model (1.3)

$$\begin{aligned} E(\mathbf{Y}) &= \alpha\mathbf{1} + \boldsymbol{X}\boldsymbol{\beta} \\ &= \alpha\mathbf{1} + \boldsymbol{X}\boldsymbol{V}\boldsymbol{V}'\boldsymbol{\beta} \qquad \text{(because } \boldsymbol{V}\boldsymbol{V}' = \boldsymbol{I}_p) \\ &= \alpha\mathbf{1} + (\boldsymbol{X}\boldsymbol{V})(\boldsymbol{V}'\boldsymbol{\beta}) \\ &= \alpha\mathbf{1} + \boldsymbol{Z}\boldsymbol{\gamma}, \end{aligned} \tag{4.8}$$

with $\boldsymbol{Z} = \boldsymbol{X}\boldsymbol{V}$ the matrix with principal components stored down the columns, and $\boldsymbol{\gamma} = \boldsymbol{V}'\boldsymbol{\beta}$ the vector of unknown parameters. The least squares estimate of $\boldsymbol{\gamma}$ is

$$\begin{aligned} \hat{\boldsymbol{\gamma}} &= (\boldsymbol{Z}'\boldsymbol{Z})^{-1}\boldsymbol{Z}'\mathbf{Y} \\ &= \boldsymbol{\Lambda}^{-1}\boldsymbol{Z}'\mathbf{Y} \end{aligned}$$

and the corresponding estimator of $\boldsymbol{\beta}$ is

$$\hat{\boldsymbol{\beta}} = \boldsymbol{V}\hat{\boldsymbol{\gamma}}.$$

Principal component regression estimators of $\boldsymbol{\gamma}$ are formed by deleting $(p - r)$ columns of $\boldsymbol{Z}$ to give $\boldsymbol{Z}_{(r)}$, say, and then using least squares to give an estimator $\tilde{\boldsymbol{\gamma}}$ for $\boldsymbol{\gamma}$ of

$$\tilde{\boldsymbol{\gamma}} = (\boldsymbol{Z}'_{(r)}\boldsymbol{Z}_{(r)})^{-1}\boldsymbol{Z}'_{(r)}Y.$$

The principal component estimator of $\boldsymbol{\beta}$ is then

$$\begin{aligned} \tilde{\boldsymbol{\beta}} &= \boldsymbol{V}_{(r)}\boldsymbol{\gamma} \\ &= \boldsymbol{V}_{(r)}(\boldsymbol{Z}'_{(r)}\boldsymbol{Z}_{(r)})^{-1}\boldsymbol{Z}'_{(r)}\mathbf{Y}, \end{aligned} \tag{4.9}$$

where $\boldsymbol{V}_{(r)}$ is formed by deleting the appropriate columns of $\boldsymbol{V}$ so that $\boldsymbol{Z}_{(r)} = \boldsymbol{X}\boldsymbol{V}_{(r)}$. Asssume also that $\boldsymbol{Z}_{(r)}$ was formed by deleting the last

$(p-r)$ columns of $\boldsymbol{Z}$. Then

$$V(\tilde{\boldsymbol{\beta}}) = \sigma^2 \operatorname{tr}\{(\boldsymbol{Z}'_{(r)}\boldsymbol{Z}_{(r)})^{-1}\}$$
$$= \sigma^2 \sum_{i=1}^{r} \frac{1}{\lambda_i},$$

and so clearly the greatest reduction in the variance of the principal component estimator over least squares is achieved by deleting the $(p-r)$ principal components corresponding to the smallest eigenvalues of $\boldsymbol{X}'\boldsymbol{X}$.

Principal component regression can be viewed as the use of a restricted least squares estimator (see Johnson *et al.*, 1973). Equation (4.9) is the restricted least squares estimator for the model (4.8) subject to the $(p-r)$ restrictions

$$\boldsymbol{V}_{(-r)}\boldsymbol{\beta} = \mathbf{0},$$

where $\boldsymbol{V}_{(-r)} = (\mathbf{v}_{r+1}, \mathbf{v}_{r+2}, \ldots, \mathbf{v}_p)$. Fomby *et al.* (1978) in a generalization of the result of Greenberg (1975) show that the variance for the principal component estimator obtained by deleting the $(p-r)$ components associated with the smallest eigenvalues is at least as small as the variance for any other restricted least squares estimator with an equal or fewer number of restrictions. Thus, the principal component estimator provides a benchmark for the potential gain in variance reduction obtainable by linear restrictions on the coefficients of the parameters.

Garnham (1979) shows that the squared bias due to using the principal component estimator (4.9) is

$$\sum_{i=r+1}^{p} \gamma_i^2$$

and so the mean square error is

$$\text{MSE}(\tilde{\boldsymbol{\beta}}) = \sigma^2 \sum_{i=1}^{r} \frac{1}{\lambda_i} + \sum_{i=r+1}^{p} \gamma_i^2.$$

Consideration of variance alone will mean that components are deleted regardless of whether or not they are important (Massy, 1965). In such cases, the mean square error for $\hat{\boldsymbol{\beta}}$ can be large and predictions of the response poor. So there is a need to consider the correlation between a principal component and the response (see, for example, Hill *et al.* 1977; Lott, 1973).

One criticism of principal component regression is that deletion of some components will not, in general, cause the deletion of any of the original variables and so the final equation is as complex as the one formed using ordinary least squares. However, principal component analysis can be used as a criterion for variable selection in such a way as to remove multicollinearities from the explanatory variables. Three methods taken from Jolliffe (1972) will now be discussed.

All three of these methods begin with a principal component analysis on all p variables in which the eigenvalues and vectors of $\boldsymbol{X}'\boldsymbol{X}$ are obtained. For the first two methods those eigenvectors corresponding to eigenvalues less than some constant λ_0 (typically $0.6 \leq \lambda_0 \leq 0.7$) are selected. One of the original variables is then associated with each of these eigenvectors in turn, starting with the one corresponding to the smallest eigenvalue, by examining the coefficients in the eigenvectors of greatest magnitude. If there are $(p - r)$ eigenvalues less than λ_0 then the $(p - r)$ variables thus identified are deleted. The first method, J1, stops at this stage, while the second method, J2, repeats this process with the r remaining variables until a principal component analysis is performed in which all of the eigenvalues are greater than λ_0. The third method, J3, is a backwards version of the above algorithm. In it the $(p - r)$ eigenvectors corresponding to the eigenvalues greater than λ_0 are selected and variables are successively associated with the principal components as before, but beginning with the component corresponding to the largest eigenvalue. In this way $(p - r)$ variables are retained. Jolliffe (1972) and (1973) finds J1 and J3 to be satisfactory methods, within the context of principal component analysis although they occasionally retain too few variables. However, use of the condition index above, with an appropriate cut-off value, to indicate the number of variables for deletion may overcome this problem when considering variable deletion to overcome multicollinearities.

The eigenvalues and eigenvectors of the correlation matrix associated with Data Set 1.1 are given in Table 4.2. Suppose that after examination of the condition index given in Table 4.1 we wish to decide which four variables we should delete in order to improve the conditioning of the correlation matrix. By examining the eigenvectors in Table 4.2 corresponding to the four smallest eigenvalues, but bearing in mind that the linear terms should not be deleted unless the appropriate cross-product and squared terms have already been removed (see Chapter 12), we see that four possible candidates for deletion are x1**2, x1.x2, x1.x3 and x3**2. The condition number of

Table 4.2 *Eigenvectors of the correlation matrix for Data Set 1.1.*

Latent value	*Variable*								
	x1	*x2*	*x3*	*x1**2*	*x2**2*	*x3**2*	*x1.x2*	*x1.x3*	*x2.x3*
5.705 211	0.4059	0.1526	0.4126	0.4038	0.1443	−0.3997	0.2000	−0.4129	−0.2960
2.979 178	0.0788	−0.5384	−0.0748	0.0792	−0.5369	−0.0621	−0.5065	−0.0763	−0.3709
0.170 889	−0.4586	0.0162	−0.2268	−0.5142	0.0073	−0.6554	−0.0876	−0.1769	0.0702
0.120 805	−0.2050	0.0282	0.1416	−0.2114	0.3116	0.1127	0.1472	0.1481	−0.8597
0.020 889	−0.1533	0.4057	−0.1258	−0.1335	−0.7088	0.2479	0.3914	−0.2095	−0.1315
0.002 882	0.2074	0.1023	0.3436	0.1171	−0.3020	−0.5554	0.2314	0.5985	−0.0578
0.000 112	0.3667	0.4395	−0.4590	−0.3221	0.0088	0.1113	−0.4326	0.3937	−0.0578
0.000 035	0.2267	−0.5635	−0.4239	−0.3199	0.0096	0.1022	0.5293	0.2268	0.0829
0.000 000	−0.5695	−0.0022	−0.4779	0.5328	0.0008	0.0371	0.0011	0.4024	0.0007

Percentage of variation accounted for by principal components.

PC *number*	*Variation* (%)	*Cumulative variation* (%)
1	63.39	63.39
2	33.10	96.49
3	1.90	98.39
4	1.34	99.73
5	0.23	99.97
6	0.03	100.00
7	0.00	100.00
8	0.00	100.00
9	0.00	100.00

Table 4.3 *Principal component estimates for Data Set 1.1.*

Variable	*Least squares*	*Number of eigenvectors deleted* 1	2	3	4	5
x1	5.324	−0.309	0.165	0.050	0.068	0.075
x2	19.244	18.931	2.162	0.196	0.328	0.082
x3	13 766.321	1719.609	−537.905	−170.647	−91.613	−77.951
x1**2	−1.93e−3	2.61e−4	−1.61e−5	2.59e−5	3.03e−5	3.26e−5
x2**2	−0.030	−0.026	−0.016	−0.018	−0.031	−0.016
x3**2	−11 581.683	−2875.933	2186.189	1358.287	170.358	−80.023
x1.x2	0.014	−0.014	−2.6e−3	−7.22e−5	1.62e−4	−2.49e−5
x1.x3	10.577	−1.069	0.063	−0.232	−0.103	−0.081
x2.x3	21.035	−19.673	16.080	19.829	18.751	19.908
Constant	−3.62e−3	3.811	−133.815	−55.318	−92.836	−102.563

the reduced model is then 218. Alternatively, we may choose the five variables to retain by looking at the eigenvectors corresponding to the five largest eigenvalues. If we also give a priority to retaining the linear terms then we may decide to retain x3, x2, x1, x3**2, and x2**2. That is to remove x1**2, x1.x2, x1.x3 and x2.x3. The condition number for this reduced model is 1720. Both of these models have vastly improved the conditioning of the resulting problem, and indeed the first method gave a slightly smaller condition number than the fifth condition index for the full model (the value was 273) suggested was possible.

The methods used here were not the same as those given in Jolliffe (1972) since he suggested deleting as many variables as there are eigenvalues less than about 0.7 (with possible recursion in the process). This would have resulted in only two variables being left in the model which seems a bit drastic, initially at least, since the aim was only to reduce the ill-conditioning amongst the explanatory variables.

Principal component estimates for deletion of the eigenvectors corresponding to the smallest up to the five smallest eigenvalues are given in Table 4.3. The values of some of the parameters change drastically as more bias is introduced.

4.6 Latent root regression

Latent root regression is similar to principal component regression with the difference that it operates on the correlation matrix of the response together with the explanatory variables, rather than just on the explanatory variables. The method was originally proposed by Webster *et al.* (1974) and independently by Hawkins (1973) and the reader is referred to these papers for a more detailed algebraic treatment.

Suppose we let A_* be the correlation matrix of the response and explanatory variables with eigenvalues η_j and eigenvectors $\mathbf{t}_j$ for $j = 0, 1, \ldots, p$. Write also

$$\mathbf{t}'_j = (t_{0j}, t_{1j}, \ldots, t_{pj}),$$

with t_{0j} the coefficient of the jth eigenvector of A corresponding to the response and t_{ij} the coefficient corresponding to the ith explanatory variable ($i = 1, 2, \ldots, p$). Then if none of the η_j are small the data set is not ill-conditioned and so one can use ordinary least squares. If this is not the case then one can consider creating latent root regression estimates from linear combinations of the vectors $\boldsymbol{t}_i$ with the element t_{0i} removed, denoted by $\boldsymbol{t}_i^0$ say. Suppose that η_j is small then there are four possibilities.

(1) If $\eta_j = 0$ and $t_{0j} = 0$ then $\boldsymbol{A}_*$ has a non-predictive singularity. That is, a singularity that involves only the explanatory variables and not the response. Thus the jth component must be removed before the least squares estimates can be computed.
(2) If $\eta_j = 0$ and $t_{0j} \neq 0$ then $\boldsymbol{A}_*$ has a predictive singularity and so a perfect predictor exists and can be calculated using $\boldsymbol{t}_j$.
(3) If η_j is small but non-zero and t_{0j} is small then $\boldsymbol{A}_*$ has a non-predictive near-singularity and so $\boldsymbol{t}_j^0$ can be removed from the regression, hopefully without adversely affecting the goodness of fit by much.
(4) If η_j is small and non-zero and t_{0j} is not small then $\boldsymbol{A}_*$ has a predictive near-singularity and so $\boldsymbol{t}_j^0$ can be removed from the fit.

This procedure still has a number of problems. Firstly, a decision must be taken as to what constitutes a small η_j or t_{0j}. However, use of the condition index mentioned earlier can indicate the number of near and exact multicollinearities amongst the explanatory variables and this information may then be compared with the results from the latent root regression analysis. Secondly, it is possible to delete $\mathbf{t}_j$ when it is parallel to $\boldsymbol{\beta}$, although it is unlikely that t_{0j} will be small in this case.

Gunst *et al.* (1976) and Gunst and Mason (1977b) performed a number of simulation experiments to compare least squares with latent root regression. Several orientations of $\boldsymbol{\beta}$ and $\mathbf{v}_p$ were examined. There did not appear to be a great improvement in least squares over latent root regression when $\boldsymbol{\beta}$ and $\mathbf{v}_p$ were nearly parallel. However, when $\boldsymbol{\beta}$ and $\mathbf{v}_p$ were nearly orthogonal the latent root regression estimates were generally superior.

The eigenvalues and vectors η_j and $\mathbf{t}_j$ can also be used to suggest explanatory variables for deletion or retention in a similar fashion to those used in principal component regression. Variables with large coefficients in the $\mathbf{t}_j$ that define predictive near-singularities should be retained whilst those with small coefficients in the $\mathbf{t}_j$ that define non-predictive near-singularities are candidates for deletion. Unfortunately, this process is more complicated than the corresponding one using the principal components.

Table 4.4 *Eigenvectors of augmented correlation matrix for Data Set 1.1.*

Latent value	*Variable*									
	resp.	*x1*	*x2*	*x3*	*x1**2*	*x2**2*	*x3**2*	*x1.x2*	*x1.x3*	*x2.x3*
6.586 901	0.3693	0.3795	0.1451	−0.3831	0.3778	0.1353	−0.3693	0.1885	−0.3837	−0.2676
2.980 172	−0.0162	0.0839	−0.5363	−0.0802	0.0842	−0.5346	−0.0675	−0.5037	−0.0817	−0.3752
0.245 298	−0.6169	−0.2403	0.0621	−0.1364	−0.2735	0.2282	−0.3762	0.0888	−0.1048	−0.5019
0.157 746	0.1776	−0.3000	−0.0111	−0.2349	−0.3419	−0.1767	−0.5697	−0.1644	−0.1985	0.5330
0.021 861	−0.1560	0.2410	−0.3889	0·0910	0.2244	0.6377	−0.2497	−0.3792	0.1723	0.2552
0.005 526	0.5960	−0.3598	−0.1164	0.0295	−0.3338	0.4251	0.1722	−0.2110	−0.0504	−0.3635
0.002 378	0.2624	0.0666	0.0921	0.3916	−0.0144	0.1573	−0.5273	0.1423	0.6259	−0.2234
0.000 108	0.0178	−0.3962	−0.3704	0.5150	0.3498	−0.0057	−0.1299	0.3669	−0.4064	0.0359
0.000 010	−0.0501	−0.1870	0.6133	0.3241	0.3210	−0.0204	−0.0671	−0.5762	−0.1955	−0.0561
0.000 000	0.0027	−0.5633	−0.0269	−0.4901	0.5207	0.0018	0.0394	0.0242	0.4104	0.0024

Condition index

Dimension	1	2	3	4	5	6	7	8	9	10
Index	1	2	27	42	301	1192	2770	61 235	672 575	70 957 805

Data Set 1.1 (continued)

The eigenvectors and eigenvalues of the augmented matrix for Data Set 1.1 are given in Table 4.4. Interpretation of these values in the manner prescribed above is not easy due to a lack of any guidelines on what constitutes a small coefficient of the response in the eigenvectors. However, suppose that, on examination of the condition index for this augmented correlation matrix (see Table 4.4), there appears to be five near-singularities. The coefficient of the response for the sixth and seventh eigenvectors appears quite large in relation to the other coefficients of these eigenvectors. This leaves three non-predictive near-singularities that we can use to suggest three variables for deletion. Once again we wish to retain the linear terms and so on examination of the final three eigenvectors in Table 4.4 variables x1**2, x1.x2 and x1.x3 seem appropriate ones to delete. The condition number of the correlation matrix for this reduced model (not including the response) is 2740, a considerable improvement.

4.7 Ridge regression

Ridge regression, as proposed by Hoerl and Kennard (1970a, b), attempts to overcome the problems of multicollinearities in the data by adding a small positive constant, k, to the diagonal terms of the matrix $\boldsymbol{X}'\boldsymbol{X}$. That is the ridge regression estimator obtained by solving

$$(\boldsymbol{X}'\boldsymbol{X} + k\boldsymbol{I}) = \boldsymbol{X}'\mathbf{y}$$

giving

$$\begin{aligned}\boldsymbol{\beta}^* &= (\boldsymbol{X}'\boldsymbol{X} + k\boldsymbol{I})^{-1}\boldsymbol{X}'\mathbf{y} \\ &= \boldsymbol{W}\boldsymbol{X}'\mathbf{y} \\ &= \boldsymbol{W}\boldsymbol{X}'\boldsymbol{X}\hat{\boldsymbol{\beta}}. \qquad (4.10)\end{aligned}$$

The advantage of using ridge regression can be seen by considering the eigenvalues $\lambda_1, \lambda_2, \ldots, \lambda_p$ and corresponding eigenvectors $\mathbf{v}_1, \mathbf{v}_2, \ldots, \mathbf{v}_p$ of $\boldsymbol{X}'\boldsymbol{X}$. Now

$$\begin{aligned}(\boldsymbol{X}'\boldsymbol{X} + k\boldsymbol{I})v_i &= \boldsymbol{X}'\boldsymbol{X}\mathbf{v}_i + k\mathbf{v}_i \\ &= (\lambda_i + k)\mathbf{v}_i,\end{aligned}$$

and hence,

$$\boldsymbol{W}\mathbf{v}_i = (\lambda_i + k)^{-1}\mathbf{v}_i,$$

and so

$$(\boldsymbol{X}'\boldsymbol{X} + k\boldsymbol{I})^{-1} = \sum_{i=1}^{p} (\lambda_i + k)^{-1}\mathbf{v}_i\mathbf{v}_i',$$

showing that the eigenvalues of $\boldsymbol{W}$ are $(\lambda_i + k)^{-1}, i = 1, 2, \ldots, p$. If $\boldsymbol{X}'\boldsymbol{X}$ is nearly singular with λ_p small then the smallest eigenvalue of $(\boldsymbol{X}'\boldsymbol{X} + k\boldsymbol{I})$ will be $\lambda_p + k$ and this later matrix will not be so close to singularity. Clearly, ridge regression discounts chiefly the contribution to the estimator of the eigenvectors most associated with multicollinearities.

We now find the mean square error of $\boldsymbol{\beta}^*$ in terms of the eigenvalues, λ_i and k. We have

$$\text{MSE}(\boldsymbol{\beta}^*) = \text{tr}\{\boldsymbol{V}(\boldsymbol{\beta}^*)\} + \{E(\boldsymbol{\beta}^*) - \beta\}'\{E(\boldsymbol{\beta}^*) - \boldsymbol{\beta}\}.$$

Now, since $\boldsymbol{W}$ and $\boldsymbol{X}'\boldsymbol{X}$ have the same eigenvectors, we can write $\boldsymbol{X}'\boldsymbol{X} = \boldsymbol{V}\boldsymbol{\Lambda}\boldsymbol{V}'$ and $\boldsymbol{W} = \boldsymbol{V}(\boldsymbol{\Lambda} + k\boldsymbol{I})^{-1}\boldsymbol{V}'$, where $\boldsymbol{V}$ and $\boldsymbol{\Lambda}$ are the matrices of eigenvectors and eigenvalues introduced in Section 4.2. Hence

$$\begin{aligned} \boldsymbol{V}(\boldsymbol{\beta}^*) &= \sigma^2 \boldsymbol{W}\boldsymbol{X}'\boldsymbol{X}\boldsymbol{W} \\ &= \sigma^2 \boldsymbol{V}(\boldsymbol{\Lambda} + k\boldsymbol{I}_p)^{-1}\boldsymbol{V}'\boldsymbol{V}\boldsymbol{\Lambda}\boldsymbol{V}'\boldsymbol{V}(\boldsymbol{\Lambda} + k\boldsymbol{I}_p)^{-1}\boldsymbol{V}' \\ &= \sigma^2 \boldsymbol{V}\boldsymbol{\Lambda}_*\boldsymbol{V}', \end{aligned}$$

where $\boldsymbol{\Lambda}_*$ is a diagonal matrix with $(\boldsymbol{\Lambda}_*)_{ii} = \lambda_i(\lambda_i + k)^{-2}$, and so $\text{tr}\,\boldsymbol{V}(\boldsymbol{\beta}^*) = \sum \lambda_i(\lambda_i + k)^{-2}$. Using (4.10)

$$E(\beta^*) = \boldsymbol{W}\boldsymbol{X}'\boldsymbol{X}E(\hat{\boldsymbol{\beta}}) = \boldsymbol{W}\boldsymbol{X}'\boldsymbol{X}\boldsymbol{\beta}$$

and

$$\begin{aligned} \{E(\boldsymbol{\beta}^*) - \boldsymbol{\beta}\}'\{E(\boldsymbol{\beta}^*) - \boldsymbol{\beta}\} &= \boldsymbol{\beta}'\{\boldsymbol{I}_p - (\boldsymbol{X}'\boldsymbol{X})\boldsymbol{W}\}\{\boldsymbol{I}_p - \boldsymbol{W}(\boldsymbol{X}'\boldsymbol{X})\}\boldsymbol{\beta} \\ &= \boldsymbol{\beta}'\boldsymbol{V}\boldsymbol{\Lambda}_+\boldsymbol{V}'\boldsymbol{\beta}, \end{aligned}$$

where $\boldsymbol{\Lambda}_+$ is a diagonal matrix with $(\boldsymbol{\Lambda}_+)_{ii} = k^2(\lambda_i + k)^{-2}$. Setting $\boldsymbol{\gamma}' = (\gamma_1, \gamma_2, \ldots, \gamma_p) = \boldsymbol{\beta}'\boldsymbol{V}$, we therefore have

$$\begin{aligned} \text{MSE}(\boldsymbol{\beta}^*) &= \sum_{i=1}^{p} \lambda_i^2(\lambda_i + k)^{-2} + \sum_{i=1}^{p} \gamma_i k^2(\lambda_i + k)^{-2} \\ &= \sum (\lambda_i^2 + \gamma_i k^2)(\lambda_i + k)^{-2}. \end{aligned}$$

It is now fairly easy to show that the variance of $\boldsymbol{\beta}^*$ is a decreasing function of k while the bias is an increasing function of k. Hoerl and Kennard (1970a) show that there exists a k such that the mean square error for the ridge estimator is less than that for the least squares estimator. This is the main justification for the use of ridge regression; see also Marquardt and Snee (1975). It can also be shown that

$$\boldsymbol{\beta}^{*\prime}\boldsymbol{\beta}^* < \hat{\boldsymbol{\beta}}'\hat{\boldsymbol{\beta}} \quad \text{for all } k > 0 \quad \text{and that} \quad \boldsymbol{\beta}^{*\prime}\boldsymbol{\beta}^* \to 0$$

as k increases. That is, the ridge regression estimate shrinks towards the origin as k increases.

A problem not so far discussed is the estimation of the ridge parameter k. One popular, although subjective, method involves the use of the 'ridge trace'. The ridge trace is formed by plotting $\boldsymbol{\beta}^*$ against k for increasing values of k. When $k=0$ we have the least squares estimates. Having formed the ridge trace it can then be examined and a value of k chosen at the point where the estimates have 'settled down'. The variance inflation factors mentioned earlier can also be used by choosing k such that the largest VIF is greater than one and less than ten. Many other methods for choosing k have been proposed; see, for example, Dempster *et al.* (1977).

Plotting of the ridge trace can be made easier by noting that

$$\begin{aligned}\boldsymbol{\beta}^* &= (\boldsymbol{X}'\boldsymbol{X}+k\boldsymbol{I})^{-1}\boldsymbol{X}'\mathbf{y}\\ &= \sum_{i=1}^{p}\frac{1}{\lambda_i+k}\mathbf{v}_i\mathbf{v}_i'\boldsymbol{X}'\mathbf{y}\\ &= \sum_{i=1}^{p}\frac{1}{\lambda_i+k}d_i\mathbf{v}_i,\end{aligned}$$

where $d_i=\mathbf{v}_i'\boldsymbol{X}'\mathbf{y}$. Thus, once the eigenvectors and values of $\boldsymbol{X}'\boldsymbol{X}$ have been obtained, ridge regression estimates (and consequently ordinary least squares estimates) can easily be created for any value of k. Hawkins (1975) gives an expression for the ridge estimator in terms of the eigenvectors and values of the augmented correlation matrix $\boldsymbol{A}$ used in latent root regression (see Section 4.6).

The ridge trace can also be used to suggest variables for deletion. Any variables whose parameter estimates are unstable with either sign changes or that quickly decrease to zero are candidates for deletion. Variables that hold their predicting power, or become important as a little bias is added to the model, should be retained.

Data Set 1.1 (continued)

Some ridge regression estimates for Data Set 1.1 are produced in Table 4.5. Even though the values of k used are very small many of the parameter estimates change drastically in size and some even change sign. In particular, the parameter estimates for x1.x2, x1.x3 and x2.x3 change sign while the parameter estimates of x1 and x1**2 both change by more than a factor of ten. Indeed, the only parameters

Table 4.5 *Ridge regression estimates for Data Set 1.1.*

	Ridge parameter (k)			
Variable	*0*	*4e–7*	*6e–7*	*8e–7*
x1	5.324	0.942	0.596	0.402
x2	19.244	18.802	18.685	18.577
x3	13 766.321	4360.475	3601.493	3169.678
x1**2	–1.93e–3	–2.26e–4	–9.15e–5	–1.68e–5
x2**2	–0.030	–0.027	–0.027	–0.027
x3**2	–11 581.683	–4747.376	–4180.703	–3850.676
x1.x2	0.014	–0.014	–0.014	–0.014
x1.x3	10.577	–3.161	–2.565	–2.228
x2.x3	21.035	–19.553	–19.261	–19.009
Constant –	–3.62e–3	–798.641	–574.752	–449.153

whose estimates remain fairly stable are x2 and x2**2. Not surprisingly the constant also fluctuates vastly. The signs and order of magnitude of these estimates are similar to those for principal component regression after deletion of one eigenvector (Table 4.3).

4.8 A suggested procedure

There is no easy solution to the problem of multicollinearities. The best solution is to take extra observations at points in the directions of the multicollinearities. Unfortunately, this is frequently impractical or impossible. A suggested procedure for such cases is given below. This procedure requires that the investigator has access to the variance inflation factors or preferably the condition index for the regression model being fitted. The procedure is as follows:

(1) Examine the condition number of the explanatory variable correlation matrix. If this number is larger than about 100 conclude that there are multicollinearities. Alternatively, examine the variance inflation factors and if any are larger than about ten conclude multicollinearities exist.

(2) If possible use the condition index to decide on the number of multicollinearities, say q. This can be deduced from the condition index by noting the number of values of this index larger than about 100.

(3) Delete one variable from each of the multicollinearities. This can be done by using the eigenvectors of the correlation matrix as outlined in Section 4.5. Alternatively use the ridge trace (Section 4.7) and try deleting variables that have parameter estimates that quickly decrease to zero as the bias induced by the ridge parameter k is increased. If the investigator only has access to the variance inflation factors then he or she should delete one of the variables with a large value for this statistic and then repeat steps (2) and (3).
(4) Compare the newly estimated model with the original full model to see how the parameter estimates have changed and to see if the fit is significantly worse. If the fit is worse a decision will need to be made as to whether the reduction in goodness of fit, or alternatively, the harm caused by multicollinearity will have to be tolerated.

The above is simply the outline of a possible method for dealing with multicollinearities. Even this procedure is fairly vague and the investigator will have to make decisions such as which variables to delete from the model. Frequently, deletion of one possible set of variables will be preferable, in some sense, over another possible set for *a priori* reasons.

Exercise 4

1. Use the techniques given in this chapter to analyse the data sets given in the Appendix, especially Data Set A.1.

References

Dempster, P., Schatsoff, M. and Wermuth, D. (1977) A simulation study of alternatives to ordinary least squares. *J. Amer. Statist. Assoc.*, **72**, 77–91.
Draper, N. R. and Smith, H. (1966) *Applied Regression Analysis.* Wiley, New York.
Farrar, D. E. and Glauber, R. R. (1967) Multicollinearity in regression analysis: the problem revisited. *Rev. Econ. Statist.*, **49**, 92–107.
Fomby, T. B., Hill R. C. and Johnson, S. R. (1978) An optimal property of principal components in the context of restricted least squares. *J. Amer. Statist. Assoc.*, **73**, 191–193.
Garnham, N. F. J. (1979) Some aspects of the use of principal components in multiple regression. Dissertation for M.Sc. in Statistics at University of Kent at Canterbury.

Greenberg, E. (1975) Minimum variance properties of principal components regression. *J. Amer. Statist. Assoc.*, **70**, 194–197.

Gunst, R. F. and Mason, R. L. (1977a) Advantages of examining multicollinearities in regression analysis. *Biometrics*, **33**, 249–260.

Gunst, R. F. and Mason, R. L. (1977b) Biased estimation in regression: an evaluation using mean square error. *J. Amer. Statist. Assoc.*, **72**, 616–628.

Gunst, R. F. Webster, J. T. and Mason, R. L. (1976) A comparison of least squares and latent root regression estimators. *Technometrics*, **18**, 75–83.

Hawkins, D. M. (1973) On the investigation of alternative regressions by principal component analysis. *Appl. Statist.*, **22**, 275–286.

Hawkins, D. M. (1975) Relations between ridge regression and eigenanalysis of the augmented correlation matrix. *Technometrics*, **17**, 477–480.

Hill, R. C., Fomby, T. B. and Johnson, S. R. (1977) Component selection norms for principal components regression. *Commun. Statist.*, A, **6**, 309–317.

Hocking, R. R. (1976) The analysis and selection of variables in linear regression. *Biometrics*, **32**, 1–49.

Hoerl, A. E. and Kennard, R. W. (1970a) Ridge regression: biased estimation for non-orthogonal problems. *Technometrics*, **12**, 55–67.

Hoerl, A. E. and Kennard, R. W. (1970b) Ridge regression: application to non-orthogonal problems. *Technometrics*, **12**, ~~591–612~~. 69-82.

Johnson, S. R., Reimer, S. C. and Rothrock, T. P. (1973) Principal components and the problem of multicollinearity. *Metroeconomica*, **25**, 306–317.

Jolliffe, I. T. (1972) Discarding variables in a principal component analysis. I: Artificial data. *Appl. Statist.*, **21**, 160–173.

Jolliffe, I. T. (1973) Discarding variables in a principal component analysis, II: Real data. *Appl. Statist.*, **22**, 21–31.

Kumar, T. K. (1975) Multicollinearity in regression analysis. *Rev. Econ. Statist.*, **57**, 365–366.

Lawson, C. L. and Hanson, R. J. (1974) *Solving Least Squares Problems*. Prentice-Hall, Englewood Cliffs, NJ.

Lott, W. F. (1973) The optimal set of principal component restrictions on a least squares regression. *Commun. Statist.* A, **2**, 449–464.

Marquardt, D. W. and Snee, R. D. (1975) Ridge regression in practice. *Amer. Statist.*, **29**, 3–19.

Mason, R. L., Gunst, R. F. and Webster, J. T. (1975) Regression analysis and problems of multicollinearity. *Commun. Statist.* A, **4** (3), 277–292.

Massy, W. F. (1965) Principal components regression in exploratory statistical research. *J. Amer. Statist. Assoc.*, **60**, 234–256.

Silvey, S. D. (1969) Multicollinearity and imprecise estimation. *J. Roy. Statist. Soc.* B, **31**, 539–552.

Stewart, G. W. (1973) *Introduction to Matrix Computations*. Academic Press, New York.

Webster, J. T., Gunst, R. F. and Mason, R. L. (1974) Latent root regression analysis. *Technometrics*, **16**, 513–522.

Marquardt 1970 Technom. V. 12
p. 69-82

CHAPTER 5

Generalized inverse regression

5.1 Introduction

We have just been studying the problems which can arise due to multicollinearities in regression. Examples quite frequently occur when there is an exact, rather than an approximate, linear relationship between the explanatory variables, and the problem we then face is that the matrix in the normal equations which we wish to invert is singular. Examples will be given below. In fact, it is possible to go through with much of least squares theory using what we call *generalized inverses*. Multiple regression using generalized inverses is a useful technique for both practical and theoretical purposes, as we shall see below. First, we consider an example.

EXAMPLE 5.1

(Cochran and Cox 1957, modified). In a certain agricultural experiment three different levels of application of potash to a cotton crop were to be compared. The levels were 36, 90 and 144 lb per acre, and these are labelled T_1, T_2 and T_3 below. A randomized blocks design was used because of fertility differences over the field. The field layout and results were as in Table 5.1.

Experiments of the type shown in Example 5.1 can be readily modelled in the form (1.1). We must include in the model parameters for both block and treatment effects which we can denote as

$$(\beta_1, \beta_2, \beta_3) \quad \text{and} \quad (\alpha_1, \alpha_2, \alpha_3),$$

Table 5.1 *Results of fertilizer experiment.*

Block 1	T_2 7.46	T_1 7.62	T_3 7.76
Block 2	T_3 7.73	T_1 8.00	T_2 7.68
Block 3	T_1 7.93	T_3 7.74	T_2 7.21

respectively. We can then state the model in the form

$$E(\boldsymbol{Y}) = \boldsymbol{a\theta},$$

where

$$\boldsymbol{\theta}' = (\mu, \beta_1, \beta_2, \beta_3, \alpha_1, \alpha_2, \alpha_3) \tag{5.1}$$

and where μ is an overall mean. There is redundancy in this parametrization, for example, adding an amount δ to μ and subtracting it from α_1, α_2 and α_3 leaves the expected values unchanged. Now if the treatments are ordered within the blocks we see that

$$\boldsymbol{a} = \begin{bmatrix} 1 & 1 & 0 & 0 & 1 & 0 & 0 \\ 1 & 1 & 0 & 0 & 0 & 1 & 0 \\ 1 & 1 & 0 & 0 & 0 & 0 & 1 \\ 1 & 0 & 1 & 0 & 1 & 0 & 0 \\ 1 & 0 & 1 & 0 & 0 & 1 & 0 \\ 1 & 0 & 1 & 0 & 0 & 0 & 1 \\ 1 & 0 & 0 & 1 & 1 & 0 & 0 \\ 1 & 0 & 0 & 1 & 0 & 1 & 0 \\ 1 & 0 & 0 & 1 & 0 & 0 & 1 \end{bmatrix}. \tag{5.2}$$

For the normal equations we need $\boldsymbol{A} = \boldsymbol{a}'\boldsymbol{a}$, which is as follows:

$$\boldsymbol{A} = \begin{bmatrix} 9 & 3 & 3 & 3 & 3 & 3 & 3 \\ 3 & 3 & 0 & 0 & 1 & 1 & 1 \\ 3 & 0 & 3 & 0 & 1 & 1 & 1 \\ 3 & 0 & 0 & 3 & 1 & 1 & 1 \\ 3 & 1 & 1 & 1 & 3 & 0 & 0 \\ 3 & 1 & 1 & 1 & 0 & 3 & 0 \\ 3 & 1 & 1 & 1 & 0 & 0 & 3 \end{bmatrix}. \tag{5.3}$$

Although this is a 7×7 matrix it only has rank 5, since the sum of columns 2 to 4 or 5 to 7 is equal to column 1 in both cases. The ordinary method of solving the normal equations, as in (1.4) and (1.5), is therefore not open to us.

In this particular case we can get round the problem very easily by using the constraints

$$\sum_1^3 \alpha_i = \sum_1^3 \beta_j = 0$$

and eliminating, say α_3 and β_3.

However, such a procedure has a number of disadvantages. It forces us to model in terms of a structure which is not natural to the problem. Furthermore, the actual constraints to use in a more complicated analysis could be a matter of argument. In any case, it is not necessary to use constraints and we shall proceed without avoiding the problem of the singular matrix. We set out the theory for doing this in the following sections. The basic idea is to replace the inverse matrix by a 'generalized inverse', and this plays a rôle for a singular matrix similar to that of the ordinary inverse in the case of an invertible matrix.

In practice, the singular matrix case arises in several situations:

(1) It arises in the analysis of designed experiments, such as Example 5.1, when we wish to use a natural parametrization, rather than avoid the problem by using constraints.
(2) In some situations the singular matrix case arises naturally, as in the analysis of mixtures, when the sum of all the components is 100%. One illustration of this is in the analysis of experiments on mixtures of propellants for rockets.
(3) Sometimes in psychological tests on subjects, linear combinations of the variables are constructed and added in for the analysis. The linear combinations used often have some special interpretation.
(4) The singular matrix case can arise simply due to extreme multicollinearity.

The theory of generalized inverses is a very large topic, and entire volumes have been written on the subject. In this chapter our aim is to cover the following points. We shall define generalized inverses and lay some background theory, after which we shall go through the theory of least squares using generalized inverses. Then we shall study the construction and practical use of generalized inverses. Finally, we shall give a basis for theoretical developments of the analysis of variance using generalized inverses.

5.2 Generalized inverses

A generalized inverse of an $n \times n$ matrix $\boldsymbol{A}$ is an $n \times n$ matrix $\boldsymbol{Z}$ such that

$$\boldsymbol{A}\boldsymbol{Z}\boldsymbol{A} = \boldsymbol{A}. \tag{5.4}$$

EXAMPLE 5.2
For the matrix (5.3) a generalized inverse is

$$Z = \begin{bmatrix} \frac{1}{9} & 0 & 0 & 0 & 0 & 0 & 0 \\ 0 & \frac{2}{9} & -\frac{1}{9} & -\frac{1}{9} & 0 & 0 & 0 \\ 0 & -\frac{1}{9} & \frac{2}{9} & -\frac{1}{9} & 0 & 0 & 0 \\ 0 & -\frac{1}{9} & -\frac{1}{9} & \frac{2}{9} & 0 & 0 & 0 \\ 0 & 0 & 0 & 0 & \frac{2}{9} & -\frac{1}{9} & -\frac{1}{9} \\ 0 & 0 & 0 & 0 & -\frac{1}{9} & \frac{2}{9} & -\frac{1}{9} \\ 0 & 0 & 0 & 0 & -\frac{1}{9} & -\frac{1}{9} & \frac{2}{9} \end{bmatrix}.$$

It is readily checked that (5.4) holds for this example.

If A is invertible with inverse A^{-1}, then $Z = A^{-1}$ so that in this case a generalized inverse is necessarily the ordinary inverse. When A is singular, that is, not invertible, the ordinary inverse does not exist but a generalized inverse will exist and can often be used in place of the inverse. Unlike the ordinary inverse, a generalized inverse is not unique.

EXAMPLE 5.3
If

$$A = \begin{bmatrix} 3 & 2 \\ 6 & 4 \end{bmatrix}$$

a simple calculation verifies that

$$Z = \begin{bmatrix} 1 & 2 \\ -1 & -3 \end{bmatrix} \quad \text{and} \quad Z_1 = \begin{bmatrix} 3 & 1 \\ 5 & -6 \end{bmatrix}$$

are both generalized inverses of A.

Suppose we wish to solve a system of linear equations

$$A\mathbf{x} = \mathbf{b}. \tag{5.5}$$

When the matrix A is invertible there is a solution $\mathbf{x}$ of the system whatever the vector $\mathbf{b}$; that solution is given by

$$\mathbf{x} = A^{-1}\mathbf{b},$$

and is unique. In the contrary case, when the matrix A is singular it can happen that for some $\mathbf{b}$ there is no solution $\mathbf{x}$. Should there be a solution the system is said to be *consistent*. The condition for a consistent system is

$$\boldsymbol{A}\boldsymbol{Z}\mathbf{b} = \mathbf{b}. \tag{5.6}$$

This is because if the system is consistent we know (in principle) that some $\mathbf{x}_0$ satisfies $\boldsymbol{A}\mathbf{x}_0 = \mathbf{b}$, so that

$$\boldsymbol{A}\boldsymbol{Z}\mathbf{b} = \boldsymbol{A}\boldsymbol{Z}\boldsymbol{A}\mathbf{x}_0 = \boldsymbol{A}\mathbf{x}_0 = \mathbf{b}.$$

But then, clearly, $\mathbf{x} = \boldsymbol{Z}\mathbf{b}$ is a solution of the system of linear equations. This solution is not unique and we show below that the general solution is

$$\mathbf{x} = \boldsymbol{Z}\mathbf{b} + (\boldsymbol{I} - \boldsymbol{Z}\boldsymbol{A})\mathbf{v}, \tag{5.7}$$

where $\mathbf{v}$ is an arbitrary vector.

EXAMPLE 5.4

Consider the system of equations

$$\begin{aligned} x_1 - 2x_2 + 3x_3 &= 8, \\ 2x_1 + 4x_2 - 2x_3 &= 4, \\ 3x_1 + 2x_2 + x_3 &= 12. \end{aligned}$$

Here we need a generalized inverse of

$$\boldsymbol{A} = \begin{bmatrix} 1 & -2 & 3 \\ 2 & 4 & -2 \\ 3 & 2 & 1 \end{bmatrix}.$$

A generalized inverse is

$$\boldsymbol{Z} = \frac{1}{72}\begin{bmatrix} 7 & 4 & 11 \\ -4 & 8 & 4 \\ 11 & -4 & 7 \end{bmatrix}.$$

(See the following section for details of how $\boldsymbol{Z}$ is constructed.) It can be readily checked that condition (5.6) holds, and a particular solution is

$$\mathbf{x} = \boldsymbol{Z}\mathbf{b} = \frac{1}{72}\begin{bmatrix} 7 & 4 & 11 \\ -4 & 8 & 4 \\ 11 & -4 & 7 \end{bmatrix}\begin{bmatrix} 8 \\ 4 \\ 12 \end{bmatrix} = \begin{bmatrix} \frac{17}{6} \\ \frac{2}{3} \\ \frac{13}{6} \end{bmatrix}.$$

In order to arrive at the general solution we need $\boldsymbol{I} - \boldsymbol{Z}\boldsymbol{A}$ and computation gives

$$\boldsymbol{I} - \boldsymbol{Z}\boldsymbol{A} = \begin{bmatrix} \frac{1}{3} & -\frac{1}{3} & -\frac{1}{3} \\ -\frac{1}{3} & \frac{1}{3} & \frac{1}{3} \\ -\frac{1}{3} & \frac{1}{3} & \frac{1}{3} \end{bmatrix}.$$

Applying (5.7) we see that the general solution of the given system of linear equations is

$$x_1 = \tfrac{17}{6} + \tfrac{1}{3}(v_1 - v_2 - v_3) = \tfrac{17}{6} + \delta,$$
$$x_2 = \tfrac{2}{3} - \tfrac{1}{3}(v_1 - v_2 - v_3) = \tfrac{2}{3} - \delta,$$
$$x_3 = \tfrac{13}{6} - \tfrac{1}{3}(v_1 - v_2 - v_3) = \tfrac{13}{6} - \delta,$$

where δ may take any value (and arises as the combination $\delta = \tfrac{1}{3}(v_1 - v_2 - v_3)$ of the coordinates of the arbitrary vector $\mathbf{v}$).

Whenever generalized inverses are applied in multiple regression, this arbitrariness of the vector $\mathbf{v}$ in (5.7) occurs, but it has a natural interpretation. For example, in Example 5.1, it accounts for the fact that in model (5.1) we can add any arbitrary amount δ to μ, and subtract it from α_1, α_2 and α_3, and leave all expected values unchanged.

Before finishing this section we shall show that (5.7) is a formula for the general solution of (5.5).

On the one hand, (5.7) is a solution of (5.5) since, on premultiplying by $\boldsymbol{A}$, we obtain

$$\boldsymbol{A}(\boldsymbol{Z}\mathbf{b} + (\boldsymbol{I} - \boldsymbol{Z}\boldsymbol{A})\mathbf{v}) = \boldsymbol{A}\boldsymbol{Z}\mathbf{b} + \boldsymbol{A}(\boldsymbol{I} - \boldsymbol{Z}\boldsymbol{A})\mathbf{v} = \mathbf{b}$$

because of (5.4) and (5.6).

On the other hand, we must show that if $\mathbf{x}$ is *any* solution of (5.5), it can be expressed in the form (5.7). Let us put $\mathbf{v} = \mathbf{x} - \boldsymbol{Z}\mathbf{b}$. Then

$$\boldsymbol{A}\mathbf{v} = \boldsymbol{A}\mathbf{x} - \boldsymbol{A}\boldsymbol{Z}\mathbf{b} = \mathbf{b} - \mathbf{b} = 0.$$

Hence it follows that

$$\boldsymbol{Z}\boldsymbol{A}\mathbf{v} = 0$$

and so

$$\mathbf{v} = (\boldsymbol{I} - \boldsymbol{Z}\boldsymbol{A})\mathbf{v}.$$

But then

$$\mathbf{x} = \boldsymbol{Z}\mathbf{b} + \mathbf{v} = \boldsymbol{Z}\mathbf{b} + (\boldsymbol{I} - \boldsymbol{Z}\boldsymbol{A})\mathbf{v}$$

and we have expressed $\mathbf{x}$ in the form (5.7).

Exercises 5.2

1. The coefficient matrix for the system of equations

$$x_1 + x_3 = b_1,$$
$$x_1 + x_4 = b_2,$$

$$x_2 + x_3 = b_3,$$
$$x_2 + x_4 = b_4,$$

is

$$A = \begin{bmatrix} 1 & 0 & 1 & 0 \\ 1 & 0 & 0 & 1 \\ 0 & 1 & 1 & 0 \\ 0 & 1 & 0 & 1 \end{bmatrix}.$$

Verify that

$$Z = \frac{1}{8}\begin{bmatrix} 3 & 3 & -1 & -1 \\ -1 & -1 & 3 & 3 \\ 3 & -1 & 3 & -1 \\ -1 & 3 & -1 & 3 \end{bmatrix}$$

is a generalized inverse of $\boldsymbol{A}$. Calculate the matrices $\boldsymbol{AZ}$ and $\boldsymbol{ZA}$. Hence show that the given system of equations is consistent precisely when $b_1 + b_4 = b_2 + b_3$ and that the general solution is

$$x_1 = \tfrac{1}{8}(3b_1 + 3b_2 - b_3 - b_4 + \delta),$$
$$x_2 = \tfrac{1}{8}(-b_1 - b_2 + 3b_3 + 3b_4 + \delta),$$
$$x_3 = \tfrac{1}{8}(3b_1 - b_2 + 3b_3 - b_4 - \delta),$$
$$x_4 = \tfrac{1}{8}(-b_1 + 3b_2 - b_3 + 3b_4 - \delta),$$

where δ may take any value.

2. Verify that

$$Z_1 = \begin{bmatrix} 0 & 1 & 0 & 0 \\ -1 & 1 & 1 & 0 \\ 1 & -1 & 0 & 0 \\ 0 & 0 & 0 & 0 \end{bmatrix}$$

is also a generalized inverse of the matrix $\boldsymbol{A}$ in Exercise 5.2.1. Show that this alternative to $\boldsymbol{Z}$ gives rise to the same consistency condition and the same general solution for the original equation system.

5.3 The Moore–Penrose generalized inverse

We have seen that to a given singular matrix there will correspond more than one generalized inverse. When we come to apply generalized inverses to regression later in the chapter, we shall find that this lack of uniqueness can be adequately accommodated. For the moment, however, we wish to pursue another possibility. If to the defining condition (5.4) of a generalized inverse we add some further, very natural, conditions then the resulting generalized inverse matrix is specified uniquely. We give a construction for this particular matrix and thereby show that every singular matrix has a generalized inverse.

Suppose the matrix $\boldsymbol{A}$ is $n \times n$ and has rank $k \leq n$. We can choose k columns of $\boldsymbol{A}$ which are linearly independent and which generate the remaining columns of $\boldsymbol{A}$ through appropriate linear combinations. Let $\boldsymbol{B}$ be the $n \times k$ matrix formed from this generating set of $\boldsymbol{A}$'s columns. Then

$$\boldsymbol{A} = \boldsymbol{BC},$$

where $\boldsymbol{C}$ is a $k \times n$ matrix and the entries in a particular column of $\boldsymbol{C}$ arise as the scalar multiples needed to express the corresponding column of matrix $\boldsymbol{A}$ as a linear combination of the columns of $\boldsymbol{B}$.

EXAMPLE 5.5

Suppose $\boldsymbol{A}$ is the matrix appearing in Example 5.4, that is,

$$\boldsymbol{A} = \begin{bmatrix} 1 & -2 & 3 \\ 2 & 4 & -2 \\ 3 & 2 & 1 \end{bmatrix}.$$

The first two columns of $\boldsymbol{A}$ are linearly independent, but the third column equals the first minus the second. In this case

$$\boldsymbol{B} = \begin{bmatrix} 1 & -2 \\ 2 & 4 \\ 3 & 2 \end{bmatrix} \quad \text{and} \quad \boldsymbol{C} = \begin{bmatrix} 1 & 0 & 1 \\ 0 & 1 & -1 \end{bmatrix}.$$

The $k \times k$ matrix $\boldsymbol{B'B}$ has the same rank as the matrix $\boldsymbol{B}$, namely k, consequently $\boldsymbol{B'B}$ is invertible. For the same reason the $k \times k$ matrix $\boldsymbol{CC'}$ is invertible. If we set

$$\boldsymbol{Z} = \boldsymbol{C'}(\boldsymbol{CC'})^{-1}(\boldsymbol{B'B})^{-1}\boldsymbol{B'}, \tag{5.8}$$

then

$$\boldsymbol{AZ} = \boldsymbol{BCC'}(\boldsymbol{CC'})^{-1}(\boldsymbol{B'B})^{-1}\boldsymbol{B'} = \boldsymbol{B}(\boldsymbol{B'B})^{-1}\boldsymbol{B'}$$

and

$$\boldsymbol{AZA} = \boldsymbol{B}(\boldsymbol{B'B})^{-1}\boldsymbol{B'BC} = \boldsymbol{BC} = \boldsymbol{A}.$$

That is, the matrix $\boldsymbol{Z}$ defined in (5.8) is a generalized inverse of $\boldsymbol{A}$.

EXAMPLE 5.5 (Continued)
Straightforward matrix calculations show that

$$B'B = \begin{bmatrix} 14 & 12 \\ 12 & 24 \end{bmatrix},$$

$$(B'B)^{-1} = \frac{1}{96}\begin{bmatrix} 12 & -6 \\ -6 & 7 \end{bmatrix},$$

$$(B'B)^{-1}B' = \frac{1}{24}\begin{bmatrix} 6 & 0 & 6 \\ -5 & 4 & -1 \end{bmatrix},$$

$$CC' = \begin{bmatrix} 2 & -1 \\ -1 & 2 \end{bmatrix},$$

$$(CC')^{-1} = \frac{1}{3}\begin{bmatrix} 2 & 1 \\ 1 & 2 \end{bmatrix},$$

$$C'(CC')^{-1} = \frac{1}{3}\begin{bmatrix} 2 & 1 \\ 1 & 2 \\ 1 & -1 \end{bmatrix},$$

$$Z = C'(CC')^{-1}(B'B)^{-1}B' = \frac{1}{72}\begin{bmatrix} 7 & 4 & 11 \\ -4 & 8 & 4 \\ 11 & -4 & 7 \end{bmatrix}.$$

Note that this is the generalized inverse used previously in Example 5.4.

Besides satisfying $AZA = A$ the matrix defined in (5.8) also has a number of additional properties. In particular:

AZ is symmetric: (5.9)

$$AZ = (BC)C'(CC')^{-1}(B'B)^{-1}B' = B(B'B)^{-1}B',$$

ZA is symmetric: (5.10)

$$ZA = C'(CC')^{-1}(B'B)^{-1}B'(BC) = C'(CC')^{-1}C,$$

$ZAZ = Z$: (5.11)

$$\begin{aligned} ZAZ &= C'(CC')^{-1}(B'B)^{-1}B'(BC)C'(CC')^{-1}(B'B)^{-1}B' \\ &= C(CC')^{-1}(B'B)^{-1}B' = Z. \end{aligned}$$

Suppose Z_1 is a second generalized inverse of A which also satisfies these three conditions. Then

$$\begin{aligned} Z &= ZAZ && \text{(because } Z \text{ satisfies condition (5.11))} \\ &= A'Z'Z && \text{(because } Z \text{ satisfies condition (5.10))} \end{aligned}$$

$$
\begin{aligned}
&= A'Z_1'A'Z'Z && \text{(because } Z_1 \text{ is a generalized inverse of } A)\\
&= Z_1AZAZ && \text{(because } Z \text{ and } Z_1 \text{ satisfy condition (5.10))}\\
&= Z_1AZ && \text{(because } Z \text{ is a generalized inverse of } A)\\
&= Z_1AZ_1AZ && \text{(because } Z_1 \text{ is a generalized inverse of } A)\\
&= Z_1Z_1'A'Z'A' && \text{(because } Z \text{ and } Z_1 \text{ satisfy condition (5.9))}\\
&= Z_1Z_1'A' && \text{(because } Z \text{ is a generalized inverse of } A)\\
&= Z_1AZ_1 && \text{(because } Z_1 \text{ satisfies condition (5.9))}\\
&= Z_1 && \text{(because } Z_1 \text{ satisfies condition (5.11)).}
\end{aligned}
$$

Consequently the matrix Z of (5.8) is the unique generalized inverse of A satisfying the further conditions (5.9), (5.10) and (5.11). This particular generalized inverse is known as the *Moore–Penrose generalized inverse*, and we denote it subsequently by $A^{\div}$. Of course, if A is invertible $A^{\div} = A^{-1}$.

Exercise 5.3

1. Show that the Moore–Penrose generalized inverse of the coefficient matrix A for the equation system in Exercise 5.2.1 is the matrix Z given there. Verify by calculating the matrix products that $ZAZ = Z$.

5.4 Some theory involving generalized inverses

In this section we shall establish a number of results which we shall need for our development of the theory of regression. In that theory we will be involved with the generalized inverse Z of a matrix

$$A = a'a,$$

where a is an $n \times m$ matrix of rank k.

Result 5.1
The matrix $P = aZa'$ does not depend on which generalized inverse Z of $A = a'a$ is chosen.

To prove this we employ the following lemma:

Lemma
If X is an $n \times m$ matrix such that $X'X = 0$ then $X = 0$.

Proof

Suppose $\boldsymbol{X}$ has entries x_{ij}. The first diagonal entry of $\boldsymbol{X}'\boldsymbol{X}$ equals

$$x_{11}^2 + x_{21}^2 + \cdots + x_{n1}^2;$$

but when $\boldsymbol{X}'\boldsymbol{X} = 0$ this entry is necessarily zero, so that $x_{11} = x_{21} = \cdots = x_{n_1} = 0$; that is, the first column of $\boldsymbol{X}$ consists entirely of zeros. In the same way, inspection of each of the remaining diagonal entries of $\boldsymbol{X}'\boldsymbol{X}$ shows that no other column of $\boldsymbol{X}$ can contain a non-zero entry. It follows that $\boldsymbol{X} = 0$.

Proof of Result 5.1

If $\boldsymbol{Z}$ is any generalized inverse of $\boldsymbol{a}'\boldsymbol{a}$, then

$$\boldsymbol{a}'\boldsymbol{a}\boldsymbol{Z}\boldsymbol{a}'\boldsymbol{a} = \boldsymbol{a}'\boldsymbol{a}.$$

The first important point to note is that

$$\boldsymbol{a}\boldsymbol{Z}\boldsymbol{a}'\boldsymbol{a} = \boldsymbol{a}. \tag{5.12}$$

To see this let

$$\boldsymbol{X} = \boldsymbol{a}\boldsymbol{Z}\boldsymbol{a}'\boldsymbol{a} - \boldsymbol{a},$$

then

$$\begin{aligned}\boldsymbol{X}'\boldsymbol{X} &= (\boldsymbol{a}\boldsymbol{Z}\boldsymbol{a}'\boldsymbol{a} - \boldsymbol{a})'(\boldsymbol{a}\boldsymbol{Z}\boldsymbol{a}'\boldsymbol{a} - \boldsymbol{a}) \\ &= (\boldsymbol{a}'\boldsymbol{a}\boldsymbol{Z}'\boldsymbol{a}' - \boldsymbol{a}')(\boldsymbol{a}\boldsymbol{Z}\boldsymbol{a}'\boldsymbol{a} - \boldsymbol{a}) \\ &= \boldsymbol{a}'\boldsymbol{a}\boldsymbol{Z}'(\boldsymbol{a}'\boldsymbol{a}\boldsymbol{Z}\boldsymbol{a}'\boldsymbol{a} - \boldsymbol{a}'\boldsymbol{a}) - (\boldsymbol{a}'\boldsymbol{a}\boldsymbol{Z}\boldsymbol{a}'\boldsymbol{a} - \boldsymbol{a}'\boldsymbol{a}) \\ &= 0.\end{aligned}$$

We may now use the lemma to conclude that $\boldsymbol{X} = 0$ but this establishes (5.12).

Suppose $\boldsymbol{P} = \boldsymbol{a}\boldsymbol{Z}\boldsymbol{a}'$. If $\boldsymbol{Z}_1$ is a second generalized inverse of $\boldsymbol{A} = \boldsymbol{a}'\boldsymbol{a}$, we show that $\boldsymbol{P} = \boldsymbol{a}\boldsymbol{Z}_1\boldsymbol{a}$ also. To do this we again employ the lemma, this time let

$$\boldsymbol{X} = (\boldsymbol{P} - \boldsymbol{a}\boldsymbol{Z}_1\boldsymbol{a}')'.$$

We now have

$$\begin{aligned}\boldsymbol{X}'\boldsymbol{X} &= \boldsymbol{a}(\boldsymbol{Z} - \boldsymbol{Z}_1)\boldsymbol{a}'\boldsymbol{a}(\boldsymbol{Z} - \boldsymbol{Z}_1)'\boldsymbol{a}' \\ &= (\boldsymbol{a}\boldsymbol{Z}\boldsymbol{a}'\boldsymbol{a} - \boldsymbol{a}\boldsymbol{Z}_1\boldsymbol{a}'\boldsymbol{a})(\boldsymbol{Z} - \boldsymbol{Z}_1)'\boldsymbol{a}'.\end{aligned}$$

But $\boldsymbol{a}\boldsymbol{Z}\boldsymbol{a}'\boldsymbol{a}$ and $\boldsymbol{a}\boldsymbol{Z}_1\boldsymbol{a}'\boldsymbol{a}$ both equal $\boldsymbol{a}$, so $\boldsymbol{X}'\boldsymbol{X} = 0$. Because of the lemma, we may conclude that $\boldsymbol{X} = 0$ and so

$$\boldsymbol{P} = \boldsymbol{a}\boldsymbol{Z}_1\boldsymbol{a}'.$$

Result 5.2

$$\boldsymbol{P} = \boldsymbol{P}' = \boldsymbol{P}^2.$$

Proof
When $\boldsymbol{Z}$ is a generalized inverse of $\boldsymbol{A} = \boldsymbol{a}'\boldsymbol{a}$ the transpose $\boldsymbol{Z}'$ is also a generalized inverse because $\boldsymbol{A}$ is symmetric. From the previous result is follows that $\boldsymbol{P} = \boldsymbol{a}\boldsymbol{Z}'\boldsymbol{a}'$. But

$$\boldsymbol{P}' = (\boldsymbol{a}\boldsymbol{Z}\boldsymbol{a}')' = \boldsymbol{a}\boldsymbol{Z}'\boldsymbol{a}',$$

so that $\boldsymbol{P} = \boldsymbol{P}'$. In addition

$$\boldsymbol{P}^2 = (\boldsymbol{a}\boldsymbol{Z}\boldsymbol{a}')(\boldsymbol{a}\boldsymbol{Z}\boldsymbol{a}') = (\boldsymbol{a}\boldsymbol{Z}\boldsymbol{a}'\boldsymbol{a})\boldsymbol{Z}\boldsymbol{a}' = \boldsymbol{a}\boldsymbol{Z}\boldsymbol{a}' = \boldsymbol{P}.$$

The matrix $\boldsymbol{P}$ is known as the 'range projection' of our original matrix $\boldsymbol{a}$. Key properties which relate $\boldsymbol{P}$ to $\boldsymbol{a}$ are the following:

Result 5.3
A vector $\mathbf{y}$ satisfies $\boldsymbol{P}\mathbf{y} = \mathbf{y}$ if and only if the linear equation system $\boldsymbol{a}\boldsymbol{\theta} = \mathbf{y}$ has a solution for $\boldsymbol{\theta}$.

Result 5.4
A vector $\mathbf{y}$ satisfies $\boldsymbol{P}\mathbf{y} = 0$ if and only if $\boldsymbol{a}'\mathbf{y} = 0$.

Proof of result 5.3
Clearly if $\mathbf{y} = \boldsymbol{a}\boldsymbol{\theta}$ then

$$\boldsymbol{P}\mathbf{y} = \boldsymbol{a}\boldsymbol{Z}\boldsymbol{a}'\mathbf{y} = \boldsymbol{a}\boldsymbol{Z}\boldsymbol{a}'\boldsymbol{a}\boldsymbol{\theta} = \mathbf{a}\boldsymbol{\theta} = \mathbf{y}.$$

But, on the other hand, if $\boldsymbol{P}\mathbf{y} = \mathbf{y}$, put $\boldsymbol{\theta} = \boldsymbol{Z}\boldsymbol{a}'\mathbf{y}$; we have then

$$\boldsymbol{a}\boldsymbol{\theta} = \boldsymbol{a}\boldsymbol{Z}\boldsymbol{a}'\mathbf{y} = \boldsymbol{P}\mathbf{y} = \mathbf{y}.$$

Proof of Result 5.4
Suppose $\boldsymbol{P}\mathbf{y} = 0$, Then $\boldsymbol{a}'\boldsymbol{a}\boldsymbol{Z}\boldsymbol{a}'\mathbf{y} = \boldsymbol{a}'(\boldsymbol{a}\boldsymbol{Z}\boldsymbol{a}')\mathbf{y} = 0$ because $\boldsymbol{P} = \boldsymbol{a}\boldsymbol{Z}\boldsymbol{a}'$. But $\boldsymbol{a}' = \boldsymbol{a}'\boldsymbol{a}\boldsymbol{Z}\boldsymbol{a}'$, as we see by first replacing $\boldsymbol{Z}$ with $\boldsymbol{Z}'$ in equation (5.12), which we may do since $\boldsymbol{Z}'$ is also a generalized inverse of $\boldsymbol{a}'\boldsymbol{a}$ and then transposing. Hence $\boldsymbol{a}'\mathbf{y} = 0$.

Conversely, $\boldsymbol{a}'\mathbf{y} = 0$ clearly implies $\boldsymbol{P}\mathbf{y} = \mathbf{a}\mathbf{Z}\mathbf{a}'\mathbf{y} = 0$.

5.5 Least squares with a generalized inverse

We are concerned with the model

$$E(\mathbf{Y}) = \boldsymbol{a}\boldsymbol{\theta},$$
$$V(\mathbf{Y}) = \sigma^2\boldsymbol{I},$$

where $\mathbf{Y}$ is an n-dimensional vector of random variables representing the response variable, $\boldsymbol{a}$ is an $n \times m$ matrix of known constants and $\boldsymbol{\theta}$ is an m-dimensional vector of explanatory parameters. When we are supplied with a data vector $\mathbf{y}$ corresponding to an observation of the response variable $\mathbf{Y}$, the principle of least squares may be used to fit the model by determining a value of $\boldsymbol{\theta}$ to minimize

$$S(\boldsymbol{\theta}) = (\mathbf{y} - \boldsymbol{a}\boldsymbol{\theta})'(\mathbf{y} - \boldsymbol{a}\boldsymbol{\theta}). \tag{5.13}$$

Now suppose $\hat{\boldsymbol{\theta}}$ satisfies

$$\boldsymbol{a}'\boldsymbol{a}\hat{\boldsymbol{\theta}} = \boldsymbol{a}'\mathbf{y}; \tag{5.14}$$

we note below that such $\hat{\boldsymbol{\theta}}$ certainly exist but that $\hat{\boldsymbol{\theta}}$ need not be uniquely determined. Then

$$\begin{aligned} S(\boldsymbol{\theta}) &= \{(\mathbf{y} - \boldsymbol{a}\hat{\boldsymbol{\theta}}) + \boldsymbol{a}(\hat{\boldsymbol{\theta}} - \boldsymbol{\theta})\}'\{(\mathbf{y} - \boldsymbol{a}\hat{\boldsymbol{\theta}}) + \boldsymbol{a}(\hat{\boldsymbol{\theta}} - \boldsymbol{\theta})\} \\ &= S(\hat{\boldsymbol{\theta}}) + 2(\hat{\boldsymbol{\theta}} - \boldsymbol{\theta})'\boldsymbol{a}'(\mathbf{y} - \boldsymbol{a}\hat{\boldsymbol{\theta}}) + (\hat{\boldsymbol{\theta}} - \boldsymbol{\theta})'\boldsymbol{a}'\boldsymbol{a}(\hat{\boldsymbol{\theta}} - \boldsymbol{\theta}) \\ &= S(\hat{\boldsymbol{\theta}}) + \| \boldsymbol{a}(\hat{\boldsymbol{\theta}} - \boldsymbol{\theta}) \|^2 \\ &\geq S(\hat{\boldsymbol{\theta}}). \end{aligned}$$

Therefore the minimum of S is attained for any solution of (5.14).

A particular solution is

$$\hat{\boldsymbol{\theta}} = \boldsymbol{Z}\boldsymbol{a}'\mathbf{y}, \tag{5.15}$$

where $\boldsymbol{Z}$ is a generalized inverse of $\boldsymbol{A} = \boldsymbol{a}'\boldsymbol{a}$. The general solution is

$$\hat{\boldsymbol{\theta}} = \boldsymbol{Z}\boldsymbol{a}'\mathbf{y} + (\boldsymbol{I} - \boldsymbol{Z}\boldsymbol{A})\mathbf{v}, \tag{5.16}$$

where $\mathbf{v}$ is an arbitrary vector.

The minimized sum of squares, which we usually denote $S_{\min}$, is obtained by inserting (5.15) into (5.13)

$$\begin{aligned} S_{\min} &= S(\hat{\boldsymbol{\theta}}) \\ &= \mathbf{y}'\mathbf{y} - 2\mathbf{y}'\boldsymbol{a}\boldsymbol{Z}\boldsymbol{a}'\mathbf{y} + \mathbf{y}'\boldsymbol{a}\boldsymbol{Z}'\boldsymbol{a}'\boldsymbol{a}\boldsymbol{Z}\boldsymbol{a}'\mathbf{y} \\ &= \mathbf{y}'\mathbf{y} - 2\mathbf{y}'\boldsymbol{P}\mathbf{y} + \mathbf{y}'\boldsymbol{P}\mathbf{y}, \end{aligned}$$

where $\boldsymbol{P} = \boldsymbol{a}\boldsymbol{Z}\boldsymbol{a}'$ is the matrix we met in Section 5.4; that is,

$$S_{\min} = \mathbf{y}'\mathbf{y} - \mathbf{y}'\boldsymbol{P}\mathbf{y}. \tag{5.17}$$

Of course, the sum of squares $S_{\min}$ does not depend on choosing any particular solution for $\hat{\boldsymbol{\theta}}$ since, as we have seen, it represents the minimum value for $S(\boldsymbol{\theta})$. Notice that this is confirmed by Result 5.1

and formula (5.17). Thus, although we cannot calculate a unique solution for $\hat{\boldsymbol{\theta}}$, we can calculate the sum of squares.

Since $\hat{\boldsymbol{\theta}}$ is, in general, not unique, there is little interest in calculating any of the solutions. However, certain linear combinations of the parameters within $\boldsymbol{\theta}$ are often meaningful. For example, in Example 5.1 the three treatments are equally spaced levels of application of potash so that analysis by orthogonal polynomials is relevant. The linear combinations $\alpha_1 - \alpha_3$ and $\alpha_1 - 2\alpha_2 + \alpha_3$ represent the linear and quadratic effects. For a linear function of the parameters

$$f = \mathbf{c}'\boldsymbol{\theta}$$

to be meaningful it should take the same value for all solutions $\hat{\boldsymbol{\theta}}$ in (5.16). Such a linear combination of the model parameters is known as an *identifiable contrast*. The condition on the vector $\mathbf{c}$ which ensures this is

$$\mathbf{c}'(\boldsymbol{Za}'\mathbf{y} + (\boldsymbol{I} - \boldsymbol{ZA})\mathbf{v}) = \mathbf{c}'(\boldsymbol{Za}'\mathbf{y})$$

for all $\mathbf{v}$, that is

$$\mathbf{c}'(\boldsymbol{I} - \boldsymbol{ZA}) = 0. \tag{5.18}$$

The value $\hat{f}$ of f when θ minimizes S is

$$\hat{f} = \mathbf{c}'\hat{\boldsymbol{\theta}} = \mathbf{c}'\boldsymbol{Za}'\mathbf{y}. \tag{5.19}$$

Of course, the data vector $\mathbf{y}$ represents one particular value of the stochastic response variable $\mathbf{Y}$. Accordingly, we may view $S_{\min}$ and $\hat{f}$ not simply as numerical quantities dependent on the data but as random variables which represent estimators of those quantities. So, in place of (5.17) and (5.19), we may write

$$S_{\min} = \mathbf{Y}'\mathbf{Y} - \mathbf{Y}'\boldsymbol{P}\mathbf{Y} \tag{5.20}$$

and

$$\hat{f} = \mathbf{c}'\boldsymbol{Za}'\mathbf{Y}. \tag{5.21}$$

We investigate the properties of $S_{\min}$ as a random variable below. Regarding $\hat{f}$ note that

$$\begin{aligned} E(\hat{f}) &= \mathbf{c}'\boldsymbol{Za}'E(\mathbf{Y}) \\ &= \mathbf{c}'\boldsymbol{Za}'\boldsymbol{a}\boldsymbol{\theta} \\ &= \mathbf{c}'\boldsymbol{ZA}\boldsymbol{\theta} \\ &= \mathbf{c}'\boldsymbol{\theta} \quad \text{(because (5.18) implies } \mathbf{c}' = \mathbf{c}'\boldsymbol{ZA}) \\ &= f, \end{aligned} \tag{5.22}$$

that is $\hat{f}$ is unbiased, and

$$\begin{aligned} V(\hat{f}) &= \mathbf{c}'\boldsymbol{Za}' \operatorname{var}(\mathbf{Y})\boldsymbol{aZ}'\mathbf{c} \\ &= \sigma^2\mathbf{c}'\boldsymbol{Za}'\boldsymbol{aZ}'\mathbf{c} \\ &= \sigma^2\mathbf{c}'\boldsymbol{Z}\mathbf{c}, \end{aligned} \tag{5.23}$$

where we have used $\mathbf{c} = \boldsymbol{AZ}'\mathbf{c}$ which follows by transposing (5.18).

The vector of residuals $\mathbf{r}$ is

$$\mathbf{r} = \mathbf{y} - \boldsymbol{a}\hat{\boldsymbol{\theta}} = \mathbf{y} - \boldsymbol{aZa}'\mathbf{y} = (\boldsymbol{I} - \boldsymbol{P})\mathbf{y},$$

which is independent of the particular generalized inverse $\boldsymbol{Z}$ used, and if $\mathbf{R}$ is the corresponding random variable

$$\mathbf{R} = (\boldsymbol{I} - \boldsymbol{P})\mathbf{Y}. \tag{5.24}$$

Notice that we have the usual relation between $S_{\min}$ and $\mathbf{R}$, namely

$$S_{\min} = \mathbf{R}'\mathbf{R}. \tag{5.25}$$

We also have

$$E(\mathbf{R}) = 0, \tag{5.26}$$

$$V(\mathbf{R}) = \sigma^2(\boldsymbol{I} - \boldsymbol{P}), \tag{5.27}$$

$$C(\hat{f}, \mathbf{R}) = 0. \tag{5.28}$$

These results may be proved as follows:

$$E(\mathbf{R}) = (\boldsymbol{I} - \boldsymbol{P})E(\mathbf{Y}) = (\boldsymbol{I} - \boldsymbol{P})\boldsymbol{a}\boldsymbol{\theta} = 0$$

because (5.12) implies $\boldsymbol{a} = \boldsymbol{aZa}'\boldsymbol{a} = \boldsymbol{Pa}$;

$$V(\mathbf{R}) = (\boldsymbol{I} - \boldsymbol{P})V(\mathbf{Y})(\boldsymbol{I} - \boldsymbol{P}) = \sigma^2(\boldsymbol{I} - \boldsymbol{P}),$$

where we have used $(\boldsymbol{I} - \boldsymbol{P}) = (\boldsymbol{I} - \boldsymbol{P})' = (\boldsymbol{I} - \boldsymbol{P})^2$, which is easily derived from Result 5.2;

$$C(\hat{f}, \mathbf{R}) = C(\boldsymbol{c}'\boldsymbol{Za}'\mathbf{Y}, (\boldsymbol{I} - \boldsymbol{P})\mathbf{Y}) = \sigma^2\boldsymbol{c}'\boldsymbol{Za}'(\boldsymbol{I} - \boldsymbol{P}) = 0.$$

Notice, in particular, that equations (5.24) and (5.27) show that standardized residuals can be calculated independently of the actual generalized inverse chosen. The results which we have obtained for the residuals exactly parallel those that apply in the full rank case. They are essential to ensure that the residuals can be used in the usual way to study such matters as deviations from the model assumptions and including outliers.

We now analyse the random variable $S_{\min}$. Suppose $\mathbf{q}_1, \mathbf{q}_2, \ldots, \mathbf{q}_n$ is an orthonormal system of vectors in the space of n-dimensional column vectors, chosen so that the columns of the matrix $\boldsymbol{a}$ are linear combinations of the vectors $\mathbf{q}_1, \ldots, \mathbf{q}_k$ and vice versa. (The Gram–Schmidt algorithm outlined in Section 3.6 shows that such vectors $\mathbf{q}_1, \ldots, \mathbf{q}_n$ can be determined, at least in principle.) Then it follows from Results 5.3 and 5.4 that for $1 \leq i \leq k$

$$\boldsymbol{P}\mathbf{q}_i = \mathbf{q}_i,$$

whereas for $k + 1 \leq i \leq n$

$$\boldsymbol{P}\mathbf{q}_i = 0.$$

Define random variables $\eta_1, \eta_2, \ldots, \eta_n$ by

$$\eta_i = \mathbf{q}_i'\mathbf{Y},$$

so that by (3.31)

$$\mathbf{Y} = \eta_1\mathbf{q}_1 + \eta_2\mathbf{q}_2 + \cdots + \eta_n\mathbf{q}_n. \tag{5.29}$$

Then

$$\begin{aligned} S_{\min} &= \mathbf{Y}'\mathbf{Y} - \mathbf{Y}'\boldsymbol{P}\boldsymbol{Y} \\ &= (\eta_1^2 + \cdots + \eta_n^2) - (\eta_1^2 + \cdots + \eta_k^2) \\ &= \eta_{k+1}^2 + \cdots + \eta_n^2. \end{aligned} \tag{5.30}$$

Suppose now $\boldsymbol{Q}$ is the orthogonal matrix having for its columns the vectors $\mathbf{q}_1, \mathbf{q}_2, \ldots, \mathbf{q}_n$ and $\boldsymbol{\eta}$ is the n-dimensional vector with coordinates the random variables $\eta_1, \eta_2, \ldots, \eta_n$. In terms of $\boldsymbol{Q}$ relation (5.29) becomes

$$\mathbf{Y} = \boldsymbol{Q}\boldsymbol{\eta}$$

and since $\boldsymbol{Q}$ is orthogonal, $\boldsymbol{Q}^{-1} = \boldsymbol{Q}'$, so that

$$\boldsymbol{\eta} = \boldsymbol{Q}'\mathbf{Y}.$$

Hence

$$E(\boldsymbol{\eta}) = \boldsymbol{Q}'E(\mathbf{Y}) = \boldsymbol{Q}'\boldsymbol{a}\boldsymbol{\theta} \tag{5.31}$$

and

$$V(\boldsymbol{\eta}) = \boldsymbol{Q}'V(\mathbf{Y})\boldsymbol{Q} = \boldsymbol{Q}'\sigma^2\boldsymbol{I}\boldsymbol{Q} = \sigma^2\boldsymbol{I}. \tag{5.32}$$

In particular, these formulae imply that $E(\boldsymbol{\eta}_i) = 0$ when $k + 1 \leq i \leq n$, any pair of $\eta_1, \eta_2, \ldots, \eta_n$ are uncorrelated, and each has variance σ^2. From (5.30) we now see that

$$E(S_{\min}) = (n - k)\sigma^2.$$

The sum of squares due to the fitted parameters, which we usually denote S_{par}, is

$$\begin{aligned} S_{\text{par}} &= \mathbf{Y}'\mathbf{Y} - S_{\text{min}} \\ &= \mathbf{Y}'\boldsymbol{P}\mathbf{Y} \\ &= \eta_1^2 + \cdots + \eta_k^2. \end{aligned}$$

Accordingly

$$E(S_{\text{par}}) = E(\eta_1^2) + \cdots + E(\eta_k^2).$$

But for $1 \leq i \leq k$ we know from (5.31) that $E(\eta_i) = \mathbf{q}_i'\boldsymbol{a}\boldsymbol{\theta}$ and from (5.32) that $\operatorname{var}(\eta_i) = \sigma^2$. Hence

$$E(\eta_i^2) = V(\eta_i^2) + E(\eta_i)^2 = \sigma^2 + (\mathbf{q}_i'\boldsymbol{a}\boldsymbol{\theta})^2$$

and

$$E(S_{\text{par}}) = k\sigma^2 + (\mathbf{q}_1'\boldsymbol{a}\boldsymbol{\theta})^2 + \cdots + (\mathbf{q}_k'\boldsymbol{a}\boldsymbol{\theta})^2.$$

Because the columns of $\boldsymbol{a}$ are linear combinations of $\mathbf{q}_1, \ldots, \mathbf{q}_k$ only

$$(\mathbf{q}_1'\boldsymbol{a}\boldsymbol{\theta})^2 + \cdots + (\mathbf{q}_k'\boldsymbol{a}\boldsymbol{\theta})^2 = \boldsymbol{\theta}'\boldsymbol{a}'\boldsymbol{a}\boldsymbol{\theta}$$

by (3.31) and (3.21). So we obtain

$$E(S_{\text{par}}) = k\sigma^2 + \boldsymbol{\theta}'\boldsymbol{a}'\boldsymbol{a}\boldsymbol{\theta}. \tag{5.33}$$

We are thus able to set down an analysis of variance as shown in Table 5.2.

Up to this point there has been no assumption of normality. If the components of $\mathbf{Y}$ are independently and normally distributed, then so are $\eta_1, \ldots, \eta_n$ (since we have already noted that they are uncorrelated). Now

$$\begin{aligned} \hat{f} &= \mathbf{c}'\boldsymbol{Z}\boldsymbol{a}'\mathbf{Y} = \mathbf{c}'\boldsymbol{Z}\boldsymbol{a}'(\eta_1\mathbf{q}_1 + \cdots + \eta_n\mathbf{q}_n) \\ &= \mathbf{c}'\boldsymbol{Z}\boldsymbol{a}'(\eta_1\mathbf{q}_1 + \cdots + \eta_k\mathbf{q}_k) \\ &= (\mathbf{c}'\boldsymbol{Z}\boldsymbol{a}'\mathbf{q}_1)\eta_1 + \cdots + (\mathbf{c}'\boldsymbol{Z}\boldsymbol{a}'\mathbf{q}_k)\eta_k \end{aligned}$$

and according to (5.30)

$$S_{\text{min}} = \eta_{k+1}^2 + \cdots + \eta_n^2.$$

Hence the random variables $\hat{f}$ and S_{min} are independent. Standard distribution results imply that $\hat{f}$ is normally distributed and that $(1/\sigma^2)S_{\text{min}}$ has a χ^2-distribution on $n-k$ degrees of freedom. We may therefore apply the t-distribution for hypothesis tests and confidence intervals to obtain information about the identifiable contrast $f = \mathbf{c}'\boldsymbol{\theta}$.

Table 5.2 *Analysis of variance.*

Meaning	*SS*	*d.f.*	*Mean square*	*Expected mean square*
Due to fitted model	S_{par}	k	S_{par}/k	$\sigma^2 + \frac{1}{k}\boldsymbol{\theta}'\boldsymbol{a}'\boldsymbol{a}\boldsymbol{\theta}$
Deviations from model	S_{min}	$n-k$	$S_{\text{min}}/(n-k)$	σ^2
Total	$Y'Y$	n		

Finally, it is sometimes helpful, when our design matrix has deficient rank, to uncover the reasons for this. For example, in (5.2) we have

$$\text{(col. 1)} = \text{(col. 2)} + \text{(col. 3)} + \text{(col. 4)}$$

and

$$\text{(col. 1)} = \text{(col. 5)} + \text{(col. 6)} + \text{(col. 7)}. \tag{5.34}$$

This means that, for Example 5.1, fitting separate constraints to either all the treatments or all the blocks includes the fitting of an overall mean. If a singular matrix occurs accidentally, it is clearly very helpful to have this pointed out to us. Technically, the design matrix $\boldsymbol{a}$ has a null space and relationships such as (5.34) correspond to a basis for it. Some of the algorithms used for calculating generalized inverses can produce relationships such as (5.34) as a by-product.

We have developed the theory of least squares for models linear in the unknown parameters when the models are not of full rank. At this point statistical methods, for testing hypotheses or for constructing confidence intervals, apply as in the full rank case. We can also test the significance of the fitted model by the usual F-test. Clearly, in most cases, we shall first want to fit a mean, and work with corrected sums of squares, rather than the uncorrected sums of squares presented in Table 5.2. The details for all of this are exactly as in the full rank case and so we omit them; see Wetherill (1981).

To summarize, we have established that when our design matrix is singular we can still calculate fitted values, residuals, sums of squares, and carry out significance tests, even though there is no unique solution for the unknown explanatory parameters. In practice, therefore, a singular matrix is no reason to stop a regression analysis. The sums of squares and the residuals can be calculated and various tests and diagnostics can be run.

5.6 Algorithms for constructing a generalized inverse

There are many ways of constructing a generalized inverse and we do not propose to review them here. It is sufficient for our purposes if some simple and/or efficient algorithms are given for the symmetric case.

A very simple way to construct a generalized inverse, if the rank is known, is as follows. Suppose the matrix has rank k and is partitioned in the form

$$\boldsymbol{A} = \begin{bmatrix} \boldsymbol{A}_{11} & \boldsymbol{A}_{12} \\ \boldsymbol{A}_{21} & \boldsymbol{A}_{22} \end{bmatrix},$$

where $\boldsymbol{A}_{11}$ is $k \times k$ and is non-singular, then a generalized inverse is

$$\boldsymbol{Z} = \begin{bmatrix} \boldsymbol{A}_{11}^{-1} & 0 \\ 0 & 0 \end{bmatrix}.$$

It may be necessary to rearrange the rows and columns of $\boldsymbol{A}$ to achieve a non-singular top left-hand corner; if so, the generalized inverse is then the matrix obtained after performing on $\boldsymbol{Z}$ those interchanges of rows and columns needed to restore $\boldsymbol{A}$.

In the case of the matrices which arise in regression and are of the form

$$\boldsymbol{A} = \boldsymbol{a}'\boldsymbol{a},$$

where $\boldsymbol{a}$ is $n \times m$, another way to obtain a generalized inverse is to use constraints to make the parameters identifiable. This method is particularly appropriate in the analysis of designed experiments. Suppose $\boldsymbol{C}$ is a $p \times m$ matrix with the two properties

(1) the rank of the matrix $\begin{bmatrix} \boldsymbol{a} \\ \boldsymbol{C} \end{bmatrix}$ is m;

(2) if a linear combination of the rows of $\boldsymbol{a}$ and $\boldsymbol{C}$ is zero, then the parts of this linear combination coming from $\boldsymbol{a}$ and $\boldsymbol{C}$ are each separately zero, that is, no non-trivial linear combination of rows from $\boldsymbol{a}$ is a linear combination of rows from $\boldsymbol{C}$.

(If the rank of $\boldsymbol{a}$ is k and $p = m - k$ then condition (1) implies condition

(2).) As an illustration, in Example 5.1, we could let

$$C=\begin{bmatrix} 0 & 1 & 1 & 1 & 0 & 0 & 0 \\ 0 & 0 & 0 & 0 & 1 & 1 & 1 \end{bmatrix}.$$

A generalized inverse for $\boldsymbol{A}$ is then

$$\boldsymbol{Z}=(\boldsymbol{A}+\boldsymbol{C}'\boldsymbol{C})^{-1}, \tag{5.35}$$

and $\boldsymbol{Z}$ also satisfies $\boldsymbol{CZa}'=0$; by means of $\boldsymbol{Z}$ we obtain the unique least squares solution $\hat{\boldsymbol{\theta}}$ which satisfies $\boldsymbol{C}\hat{\boldsymbol{\theta}}=0$.

The methods mentioned above will be satisfactory to use for purposes of exploring the idea of generalized inverses. Clearly, good efficient algorithms are required for use in a general computer program, when the rank of the matrix $\boldsymbol{A}$ may not be known. Healy (1968) describes the two algorithms given above, and he also gives a Choleski square-root method. Dubrulle (1975) gives an algorithm based on the Householder triangularization procedure and Seber (1977) gives a detailed description of several methods, including the Choleski and Householder methods. In the following paragraphs we describe a further algorithm, given by von Hohenbalken and Riddell (1979), which has a number of advantages when used for regression calculations. It may be viewed as a practical way of implementing the theoretical construction of the Moore–Penrose generalized inverse outlined in Section 5.3.

If $\boldsymbol{B}$ is a matrix whose columns form a basis for the column space of $\boldsymbol{A}$, then Boot (1963) and Chipman (1964) gave a formula for the Moore–Penrose generalized inverse of $\boldsymbol{A}$ as

$$\boldsymbol{A}^{+}=\boldsymbol{A}'\boldsymbol{B}(\boldsymbol{B}'\boldsymbol{A}\boldsymbol{A}'\boldsymbol{B})^{-1}\boldsymbol{B}'. \tag{5.36}$$

von Hohenbalken and Riddell (1979) obtain a suitable matrix $\boldsymbol{B}$ as follows.

(1) The columns of $\boldsymbol{A}$ are normalized to unit length.
(2) A non-zero column of $\boldsymbol{A}$ is used as a trial basis, $\boldsymbol{B}_1$.
(3) A matrix $\boldsymbol{\Delta}$ is formed of the difference between $\boldsymbol{A}$ and its projection in $\boldsymbol{B}_1$.

$$\boldsymbol{\Delta}=\boldsymbol{A}-\{\boldsymbol{B}_1(\boldsymbol{B}_1'\boldsymbol{B}_1)^{-1}\boldsymbol{B}_1'\}\boldsymbol{A}.$$

(4) Form the vector $\boldsymbol{S}$, comprising the sum of squares of the column elements of the difference matrix $\boldsymbol{\Delta}$.

(5) The column having the largest non-zero sum of squared differences is entered into the basis, and step (3) is repeated.

This procedure stops when the vector **S** contains elements which are near-zero as tested by the computer.

This algorithm is very quick and an extremely compact APL code for it is given by the authors. It should be noted, however, that steps (4) and (5) effectively determine the rank of $\boldsymbol{A}$. This method of determining rank may not necessarily agree with others. For example, if 'sweep' routines are used in a regression program, then the routines will fail to produce an answer under certain special conditions, related to the way in which the 'sweep' algorithms are written. Apparent differences in rank determination can easily occur in regression problems with high multicollinearity. There is no obvious way in which one method of rank determination can be said to be superior to others, but in a good regression program the near and exact multicollinearities will all be highlighted.

The von Hohenbalken and Riddell algorithm has another advantage in terms of regression. Suppose in Example 5.1 we wish to test whether the treatment effects are significantly different from zero. First, we carry out regression using the full model (5.1), and effectively determine a basis for the matrix $\boldsymbol{A}$ defined in (5.3). Following this we delete from the basis any columns related to the treatment effects – which are the last three. The algorithm can now be restarted with this trial basis. This operation will save many iterations in a large problem. For Example 5.1 it works out as follows:

Basis for full model. Columns: 1, 2, 3, 5, 6.

Columns related to treatment effects: 5, 6, 7.

Trial basis to re-enter algorithm: 1, 2, 3.

In fact, the trial basis is the new basis and no more iterations are performed. This enables the calculation of adjusted sums of squares to be performed relatively quickly.

We conclude this section by briefly indicating a proof of (5.35). The algebra is more difficult than that previously encountered and the reader may wish to pass over it. The matrix

$$\boldsymbol{a}'\boldsymbol{a} + \boldsymbol{C}'\boldsymbol{C} = \begin{bmatrix} \boldsymbol{a} \\ \boldsymbol{C} \end{bmatrix}' \begin{bmatrix} \boldsymbol{a} \\ \boldsymbol{C} \end{bmatrix}$$

is invertible because of condition (1) specifying the rank of the matrix $\begin{bmatrix} \boldsymbol{a} \\ \boldsymbol{C} \end{bmatrix}$ (see Strang (1980), p. 109); let $\boldsymbol{Z} = (\boldsymbol{a}'\boldsymbol{a} + \boldsymbol{C}'\boldsymbol{C})^{-1}$. Condition (2) implies that if $\boldsymbol{X}$ is an $s \times n$ matrix and $\boldsymbol{Y}$ is an $s \times p$ matrix which satisfy

$$\boldsymbol{X}\boldsymbol{a} + \boldsymbol{Y}\boldsymbol{C} = 0,$$

then $\boldsymbol{X}\boldsymbol{a} = 0$ and $\boldsymbol{Y}\boldsymbol{C} = 0$. First, setting $\boldsymbol{X} = (\boldsymbol{a}'\boldsymbol{a} + \boldsymbol{C}'\boldsymbol{C})\boldsymbol{Z}\boldsymbol{a}' - \boldsymbol{a}'$ and $\boldsymbol{Y} = (\boldsymbol{a}'\boldsymbol{a} + \boldsymbol{C}'\boldsymbol{C})\boldsymbol{Z}\boldsymbol{C}' - \boldsymbol{C}'$ we deduce that

$$(\boldsymbol{a}'\boldsymbol{a} + \boldsymbol{C}'\boldsymbol{C})\boldsymbol{Z}\boldsymbol{a}'\boldsymbol{a} = \boldsymbol{a}'\boldsymbol{a}.$$

Transposing this relation and then setting $\boldsymbol{X} = (\boldsymbol{a}'\boldsymbol{a}\boldsymbol{Z}\boldsymbol{a}' - \boldsymbol{a}')$ and $\boldsymbol{Y} = \boldsymbol{a}'\boldsymbol{a}\boldsymbol{Z}\boldsymbol{C}'$ we find that

$$\boldsymbol{a}'\boldsymbol{a}\boldsymbol{Z}\boldsymbol{a}'\boldsymbol{a} = \boldsymbol{a}'\boldsymbol{a}$$

and

$$\boldsymbol{a}'\boldsymbol{a}\boldsymbol{Z}\boldsymbol{C}'\boldsymbol{C} = 0.$$

Finally, we obtain $\boldsymbol{C}\boldsymbol{Z}\boldsymbol{a}' = 0$ by repeated application of the lemma in Section 5.4.

5.7 Testing reduced models

In many applications of least squares we wish to calculate sums of squares for partial models, in order to see whether or not the omitted variables contribute significantly, or else simply to obtain the adjusted sums of squares due to these omitted variables. For example, in the analysis of variance of a two-way layout, the (adjusted) sum of squares for each factor is obtained by omitting it and then by differencing the sums of squares for the full and partial models. In this section we justify the significance tests used in this situation, and we give a method for 'building up' a generalized inverse by successively adding terms to the model.

Suppose that we are fitting the model

$$E(\mathbf{Y}) = \boldsymbol{a}\boldsymbol{\theta} = [\boldsymbol{a}_1 \boldsymbol{a}_2]\begin{bmatrix} \boldsymbol{\phi} \\ \boldsymbol{\psi} \end{bmatrix}, \tag{5.37}$$

$$V(\mathbf{Y}) = \sigma^2 \boldsymbol{I},$$

where the ranks of $\boldsymbol{a}_1$, $\boldsymbol{a}_2$ and $\boldsymbol{a}$ are k_1, k_2 and k so that we must have

$k_1, k_2 \leq k$. With this notation a reduced model is

$$E(\mathbf{Y}) = \boldsymbol{a}_1\boldsymbol{\phi},$$
$$V(\mathbf{Y}) = \sigma^2\boldsymbol{I}.$$

Following the method of Section 5.5, we may now choose the orthonormal basis $\mathbf{q}_1, \ldots, \mathbf{q}_k$ for the column space of $\boldsymbol{a}$ so that $\mathbf{q}_1, \ldots, \mathbf{q}_{k_1}$ span the column space of $\boldsymbol{a}_1$. Let $\boldsymbol{P}_1$ be the range projection for this reduced model, as in Section 5.4, then we find that

$$\begin{aligned} S_{\text{par}} - S_{\text{par}}^{(1)} &= \mathbf{Y}'(\boldsymbol{P} - \boldsymbol{P}_1)\mathbf{Y} \\ &= \eta_{k_1+1}^2 + \cdots + \eta_k^2, \end{aligned} \tag{5.38}$$

where $S_{\text{par}}^{(1)}$ is the sum of squares for fitting the reduced model. This shows that under the normality assumption $\{S_{\text{par}} - S_{\text{par}}^{(1)}\}$ and $S_{\min}$ are statistically independent. Further, we see that

$$\begin{aligned} E\{S_{\text{par}} - S_{\text{par}}^{(1)}\} &= (k - k_1)\sigma^2 + \|(\boldsymbol{I} - \boldsymbol{P}_1)\boldsymbol{a}\boldsymbol{\theta}\|^2 \\ &= (k - k_1)\sigma^2 + \|(\boldsymbol{I} - \boldsymbol{P}_1)\boldsymbol{a}_2\boldsymbol{\psi}\|^2. \end{aligned} \tag{5.39}$$

It follows from this result that $\{S_{\text{par}} - S_{\text{par}}^{(1)}\}/\sigma^2$ has a χ^2-distribution with $(k - k_1)$ degrees of freedom if

$$(\boldsymbol{I} - \boldsymbol{P}_1)\boldsymbol{a}_2\boldsymbol{\psi} = 0,$$

and the meaning of this condition is that $a_2\boldsymbol{\psi} = a_1\tilde{\boldsymbol{\phi}}$ for some $\tilde{\boldsymbol{\phi}}$, but then

$$\boldsymbol{a}\boldsymbol{\theta} = \boldsymbol{a}_1\boldsymbol{\phi} + \boldsymbol{a}_2\boldsymbol{\psi} = \boldsymbol{a}_1(\boldsymbol{\phi} + \tilde{\boldsymbol{\phi}}),$$

so that the reduced model is adequate for fitting the data.

These results justify the usual test procedure of referring the ratio

$$(n - k)\{S_{\text{par}} - S_{\text{par}}^{(1)}\}/(k - k_1)S_{\min}$$

to the F-distribution on $\{(k - k_1), (n - k)\}$ degrees of freedom.

5.7.1 Building generalized inverses in stages

In Section 5.6 we gave a method of calculating a generalized inverse for the solution of a partial model, given that we have a generalized inverse from the full model. Here we give a method of going the other way. That is, we are able to obtain a generalized inverse for the full model given a generalized inverse for the partial model.

Suppose we have a partitioned matrix

$$A = \begin{bmatrix} A_{11} & A_{12} \\ A_{21} & A_{22} \end{bmatrix}, \tag{5.40}$$

and suppose that Z_1 is a generalized inverse of A_{11} and that the following conditions are satisfied

$$A_{11}Z_1A_{12} = A_{12},$$
$$A_{21}Z_1A_{11} = A_{21}.$$

These conditions are automatically satisfied either if A_{11} is invertible, or else if

$$A = a'a \quad \text{and} \quad a = [a_1 a_2]$$

so that

$$A_{11} = a_1'a_1, \qquad A_{12} = A_{21}' = a_1'a_2, \qquad A_{22} = a_2'a_2.$$

It then follows that

$$Z = \begin{bmatrix} I & -Z_1A_{12} \\ O & I \end{bmatrix}\begin{bmatrix} Z_1 & O \\ O & G \end{bmatrix}\begin{bmatrix} I & O \\ -A_{21}Z_1 & I \end{bmatrix}, \tag{5.41}$$

where G is a generalized inverse for $A_{22} - A_{21}Z_1A_{12}$, is a generalized inverse for A. This result can be readily checked by multiplying out AZA.

If the matrix A_1 is, say, $r \times r$ and we are adding, say, two variables then G is a 2×2 matrix and the calculation of the generalized inverse G is relatively quick.

It therefore follows that either by using the methods of this section for adding variables, or by using the method given at the end of Section 5.6 for removing variables, we can calculate an adjusted sum of squares table such as in shown in Table 5.3 where $k_1 \leq k$. This adjusted sum of squares calculation is the basis of the 'stepwise', etc., selection of variable methods which are discussed in Chapter 11.

Table 5.3 *Adjusted sum of squares.*

Source	*d.f.*
Due to $X_1, \ldots, X_p$	k
Due to $X_1, \ldots, X_p$ ignoring X_i, X_j, X_k	k_1
Due to X_i, X_j, X_k, adjusting for $X_1, \ldots, X_p$	$(k - k_1)$

Clearly, unless we have orthogonality, the sums of squares for, say, (x_1, x_2) and (x_3, x_4) are not independent.

5.8* Analysis of variance

5.8.1 One-way layout

We now apply the theory just outlined to the analysis of variance of a one-way layout. We suppose that we have l treatments and that we have run an experiment in which the ith treatment is replicated m_i times. We shall also suppose that the necessary randomizations of the design have been carried out. We therefore have data

$$y_{ij}, \qquad i = 1, 2, \ldots, l; \quad j = 1, 2, \ldots, m_i,$$

and we put

$$n = l + 1, \qquad m = \sum m_i. \tag{5.42}$$

The model we wish to fit is

$$E(Y_{ij}) = \mu + \alpha_i, \tag{5.43}$$

$$V(Y_{ij}) = \sigma^2, \tag{5.44}$$

with the Y_{ij} assumed to be statistically independent.

Model 1

First, we fit only the overall mean so we replace (5.43) by

$$E(Y_{ij}) = \mu. \tag{5.45}$$

Then if we write the $m \times 1$ vector $\mathbf{a}_1$,

$$\mathbf{a}_1' = (1, 1, \ldots, 1)$$

the model is

$$E(\mathbf{Y}) = \boldsymbol{a}_1 \mu.$$

We then have

$$\boldsymbol{a}_1' \boldsymbol{a}_1 = m$$

for which an inverse is $1/m$. The general form of the minimized sum of squares is

$$S_{\min} = \| \mathbf{y} - \boldsymbol{P}\mathbf{y} \|^2, \tag{5.46}$$

where $\boldsymbol{P} = \boldsymbol{a}_1 \boldsymbol{Z}_1 \boldsymbol{a}'_1$ and $\boldsymbol{Z}_1$ is an inverse of $\boldsymbol{a}'_1 \boldsymbol{a}_1$. For this model $\boldsymbol{P}_1 \mathbf{y}$ is an $m \times 1$ vector, with elements $\bar{y}_{..}$,

$$(\boldsymbol{P}_1 \mathbf{y})' = (\bar{y}_{..}, \dots, \bar{y}_{..}),$$
$$\bar{y}_{..} = \sum_i \sum_j y_{ij}/m.$$

This leads to

$$S^{(1)}_{\min} = \sum_i \sum_j (y_{ij} - \bar{y}_{..})^2. \qquad \square$$

This quantity has a $\sigma^2\chi^2$-distribution on $m-1$ degrees of freedom if Model 1 is valid.

Model 2

We now introduce parameters for the treatments and use expression (5.43). In matrix form this can be put as

$$E(\mathbf{Y}) = (\boldsymbol{a}_1 \boldsymbol{a}_2)\begin{pmatrix} \mu \\ \boldsymbol{\alpha} \end{pmatrix},$$

where

$$\boldsymbol{a}_1 = \begin{bmatrix} 1 \\ 1 \\ \vdots \\ 1 \end{bmatrix}, \qquad \boldsymbol{a}_2 = \begin{bmatrix} 1 & 0 & \cdots & 0 \\ 1 & 0 & \cdots & 0 \\ \vdots & \vdots & & \vdots \\ 1 & 0 & \cdots & 0 \\ 0 & 1 & \cdots & 0 \\ \vdots & \vdots & & \vdots \\ 0 & 1 & & 0 \\ \vdots & \vdots & & \vdots \end{bmatrix} \begin{matrix} \left.\vphantom{\begin{matrix}1\\1\\\vdots\\1\end{matrix}}\right\} m_1 \\ \left.\vphantom{\begin{matrix}0\\\vdots\\0\end{matrix}}\right\} m_2 \\ \\ \end{matrix}, \qquad \boldsymbol{\alpha} = \begin{bmatrix} \alpha_1 \\ \alpha_2 \\ \vdots \\ \alpha_l \end{bmatrix}. \tag{5.47}$$

In the notation of (5.40), we then have

$$\boldsymbol{A}_{11} = m, \qquad \boldsymbol{A}_{12} = \boldsymbol{A}'_{21} = (m_1, m_2, \dots, m_l),$$
$$\boldsymbol{A}_{22} = \operatorname{Diag}(m_1, m_2, \dots, m_l).$$

We also have $\boldsymbol{Z}_1 = \boldsymbol{A}_{11}^{-1} = 1/m$. If we insert these expressions into (5.41) we obtain

$$\boldsymbol{Z} = \begin{pmatrix} 1/m & 0 \\ 0 & \operatorname{Diag}(m_1^{-1}, m_2^{-1}, \dots, m_l^{-1}) - (1/m)J_l \end{pmatrix}, \tag{5.48}$$

where $\boldsymbol{J}_l$ is the $l \times l$ matrix each entry of which is 1. If we now use

$$\hat{\boldsymbol{\theta}} = \boldsymbol{Z}\boldsymbol{a}'\mathbf{y},$$

we obtain least squares estimators

$$\hat{\mu} = \sum_i \sum_j y_{ij}/m = \bar{y}_{..}$$

and

$$\hat{\alpha}_i = \frac{1}{m_i} \sum_j y_{ij} - \frac{1}{m} \sum_i \sum_j y_{ij} = \bar{y}_{i.} - \bar{y}_{..} \quad \text{for } i = 1, 2, \ldots, l.$$

The vector $\boldsymbol{P}\mathbf{y}$ becomes

$$\boldsymbol{P}\mathbf{y} = \boldsymbol{aZa}'\mathbf{y} = \begin{bmatrix} \bar{y}_{1.} \\ \vdots \\ \bar{y}_{1.} \\ \bar{y}_{2.} \\ \vdots \\ \bar{y}_{2.} \\ \vdots \end{bmatrix} \begin{matrix} \left.\vphantom{\begin{matrix} a \\ b \\ c \end{matrix}}\right\} m_1 \text{ rows} \\ \left.\vphantom{\begin{matrix} a \\ b \\ c \end{matrix}}\right\} m_2 \text{ rows} \\ \end{matrix}.$$

We therefore obtain $S_{\min}$ from equation (5.46)

$$S_{\min}^{(2)} = \sum_{i=1}^{l} \left\{ \sum_{j=1}^{m_i} (y_{ij} - \bar{y}_{i.})^2 \right\}.$$

If Model 2 is valid, this quantity has a $\sigma^2\chi^2$-distribution on $m - l$ degrees of freedom.

On putting the two models together, we see that

$$\{S^{(1)} - S^{(2)}\} = \sum_{i=1}^{l} m_i(\bar{y}_{i.} - \bar{y}_{..})^2,$$

which has a $\sigma^2\chi^2$-distribution on $l - 1$ degrees of freedom if Model 1 is valid. This is all summed up in the analysis of variance given in Table 5.4.

Formulae for the expected mean squares and the identification of identifiable contrasts follow easily, to complete the picture. A contrast

Table 5.4 *Analysis of variance.*

Source	*CSS*	*d.f.*
Treatments	$\sum m_i(\bar{y}_{i.} - \bar{y}_{..})^2$	$l - 1$
Residual	$\sum\sum(y_{ij} - \bar{y}_{i.})^2$	$m - l$
Total	$\sum\sum(y_{ij} - \bar{y}_{..})^2$	$m - 1$

of parameters is identifiable if the 'arbitrary' parts of estimates resulting from a generalized inverse are identically zero. That is, we have $\mathbf{c}'\boldsymbol{\theta}$ is identifiable if

$$\mathbf{c}'(\boldsymbol{I} - \boldsymbol{Z}\boldsymbol{A}) = 0.$$

If we insert the matrices $\boldsymbol{Z}$ and $\boldsymbol{A}$ for our example we find that any contrast

$$c_0\mu + c_1\alpha_1 + \cdots + c_l\alpha_l$$

is identifiable if

$$c_0 = c_1 + c_2 + \cdots + c_l.$$

5.8.2 Two-way layout

Suppose now we have a two-way layout similar to Example 5.1, so that our model is

$$E(Y_{ij}) = \mu + \alpha_i + \beta_j, \qquad i = 1, 2, \ldots, p; \quad j = 1, 2, \ldots, q.$$

We put $m = pq$, $n = p + q + 1$ and the rank of $\mathbf{a}$ is

$$k = p + q - 1.$$

The first model to fit here is that for a one-way layout, say, by dropping the β_j's. The design matrix is $(\boldsymbol{a}_1, \boldsymbol{a}_2)$ of (5.47) with $m_i = q$, all i, and $l = p$.

We can now redefine $\boldsymbol{A}_{11}$ as the following $(p+1) \times (p+1)$ matrix,

$$\boldsymbol{A}_{11} = (\boldsymbol{a}_1, \boldsymbol{a}_2)'(\boldsymbol{a}_1, \boldsymbol{a}_2)$$
$$= \begin{bmatrix} pq & q & \cdots & q \\ q & q & 0 & \\ \vdots & 0 & \ddots & \\ q & & & q \end{bmatrix}.$$

To add on the β_j's the additional part of the design matrix is

$$\boldsymbol{a}_3 = \begin{bmatrix} 1 & & 0 \\ 0 & \ddots & 1 \\ 1 & & 0 \\ 0 & \ddots & 1 \\ \vdots & & \end{bmatrix} \begin{matrix} \left.\vphantom{\begin{matrix}1\\0\end{matrix}}\right\} q \text{ rows} \\ \left.\vphantom{\begin{matrix}1\\0\end{matrix}}\right\} q \text{ rows} \\ \\ \end{matrix}.$$

This leads to redefining A_{12} and A_{22},

$$A_{12} = \begin{bmatrix} p & p & \cdots & p \\ 1 & 1 & \cdots & 1 \\ \vdots & \vdots & & \vdots \\ 1 & 1 & \cdots & 1 \end{bmatrix} \quad \text{(a } (p+1) \times q \text{ matrix),}$$

$$A_{22} = \begin{bmatrix} p & & 0 \\ & \ddots & \\ 0 & & p \end{bmatrix} \quad \text{(a } q \times q \text{ matrix).}$$

From (5.48) $\boldsymbol{Z}_1$ is now redefined as

$$\boldsymbol{Z}_1 = \begin{bmatrix} \frac{1}{pq} & 0 \\ 0 & \frac{1}{q}\boldsymbol{I} - \frac{1}{pq}\boldsymbol{J}_p \end{bmatrix}.$$

The procedure of Section 5.7 can now be reused to obtain a new $\boldsymbol{Z}$,

$$\boldsymbol{Z} = \begin{bmatrix} \frac{1}{pq} & 0 & 0 \\ 0 & \left(\frac{1}{q}\boldsymbol{I} - \frac{1}{pq}\boldsymbol{J}_p\right) & 0 \\ 0 & 0 & \left(\frac{1}{p}\boldsymbol{I} - \frac{1}{pq}\boldsymbol{J}_q\right) \end{bmatrix}.$$

From here we obtain

$$\hat{\mu} = \bar{y}_{..},$$
$$\hat{\alpha}_i = \bar{y}_{i.} - \bar{y}_{..},$$
$$\hat{\beta}_j = \bar{y}_{.j} - \bar{y}_{..},$$
$$S_{\min} = \sum_i \sum_j (y_{ij} - \bar{y}_{i.} - \bar{y}_{.j} + \bar{y}_{..})^2.$$

The analysis of variance table, expectations of the mean squares, etc., also follow directly.

5.8.3. Comment

The method outlined in this section can be applied to any other form of analysis of variance and, for example, explicit formulae for the analvsis of covariance are readilv obtained.

The conclusion we reach is that redundancy in the model gives no problems in carrying through an analysis. The analysis of variance table, residuals, F-tests, etc., can be carried out by using generalized inverses. Clearly, the parameter estimates contain an arbitrary element but identifiable contrasts are also indicated by the algebra.

References

Boot, J. G. C. (1963) The computation of the generalized inverse of singular or rectangular matrices. *Amer. Math. Monthly*, **70**, 302–303.

Chipman, J. G. C. (1964) On least squares with insufficient information. *J. Amer. Statist. Assoc.*, **59**, 1078–1111.

Cochran, W. G. and Cox, G. M. (1957) *Experimental Designs*, 2nd edn, Wiley, New York.

Dubrulle, A. A. (1975) An extension of the domain of the APL domino function to rank-deficient linear squares systems. *APL 75 Congress Proc.*, ACM, New York, 105–114.

Healy, M. J. R. (1968) Multiple regression with a singular matrix. *Appl. Statist.*, **17**, 110–117.

Seber, G. A. F. (1977) *Linear Regression Analysis*. Wiley, New York.

Strang, G. (1980) *Linear Algebra and its Applications*, 2nd edn, Academic Press, London.

von Hohenbalken, B. and Riddell, W. C. (1979) A compact algorithm for the Moore–Penrose generalized inverse. *APL Quote Quad*, **10**, 30–32.

Wetherill, G. B. (1981) *Intermediate Statistical Methods*, Chapman and Hall, London.

CHAPTER 6

Outliers

6.1 The problem of outliers

We have emphasized in Chapter 2 that outlying observations may occur in a data set due to a variety of causes. There are two quite separate questions which arise and these must be carefully distinguished. One problem is to have some statistical techniques which may indicate outlying observations and so select them for special study. That is the problem which we discuss in the rest of this chapter. The second problem is what to do with these outliers, once they are located, and we make a few remarks on that problem now.

Techniques for locating outliers should *never* be regarded as a means of rejecting observations. We must not go around rejecting the data we do not happen to like or all kinds of curious results can emerge. The best action to take may depend to some extent on the field of application. In some industrial applications, or with industrial plant data, outliers may be fairly common, due to misrecording, etc. In such cases we may be fairly free about deleting a few points. However, in some biological applications, an outlier may indicate that once in a while an entirely different type of response is given. In this case, the outliers may be more informative than the rest of the observations.

As a general guideline on what to do with outliers we offer the following:

(1) Use some statistical technique to see how discrepant they are.
(2) Check back to the owner of the data set to see if any reason can be found for the outliers, such as misrecording or unexpected experimental conditions. If some such reason can be found, the observations may be modified or dropped.
(3) Carry out the analysis with and without the suspect observations, in order to see what effect that has.

The real difficulties arise with multiple outliers, perhaps in clumps. Such multiple outliers so 'pull the regression over' that their presence is masked.

Unfortunately, various theoretical difficulties arise with most of the things we might want to do, but if the suggestions below are followed a fair amount of protection will be given. For an access to the large literature, see Barnett and Lewis (1978), Belsey *et al.* (1980), Cook and Weisberg (1982), Beckman and Cook (1983), or Paul (1983).

6.2 The standardized residuals and an outlier test

The standardized residuals are important, as these are used as a basis of a test of discordancy for an outlier in the response variable.

If we take the regression model in the form (1.1), then the fitted values of the response variable are

$$\begin{aligned}\hat{\mathbf{Y}} &= \boldsymbol{a}(\boldsymbol{a}'\boldsymbol{a})^{-1}\boldsymbol{a}'\mathbf{Y} \\ &= \boldsymbol{H}\mathbf{Y},\end{aligned} \tag{6.1}$$

where

$$\boldsymbol{H} = \boldsymbol{a}(\boldsymbol{a}'\boldsymbol{a})^{-1}\boldsymbol{a}' \tag{6.2}$$

is the hat matrix (see Section 6.3). This is an $n \times n$ matrix which transforms the observations into the fitted values.

We notice that the residuals are

$$\begin{aligned}\mathbf{R} &= \mathbf{Y} - \boldsymbol{H}\mathbf{Y} \\ &= (\boldsymbol{I} - \boldsymbol{H})\mathbf{Y},\end{aligned} \tag{6.3}$$

and, from (1.7), we see that

$$V(\mathbf{R}) = (\boldsymbol{I} - \boldsymbol{H})\sigma^2,$$

so that

$$V(R_i) = (1 - h_{ii})\sigma^2,$$

where $h_{ii} = \mathbf{a}_i(\boldsymbol{a}'\boldsymbol{a})^{-1}\mathbf{a}_i'$ and $\mathbf{a}_i$ is the ith row of the matrix $\boldsymbol{a}$. The error variance σ^2 will be unknown and its unbiased estimate, denoted by s^2, is $R'R/(n-k)$. The estimated variance of R_i (of the ith observation) is, thus,

$$s_i^2 = (1 - h_{ii})s^2.$$

The residual R_i, when standardized by its estimated standard deviation, is

$$t_i = \frac{R_i}{s_i} = \frac{R_i}{s\sqrt{(1 - h_{ii})}}.$$

These are weighted residuals where the weights are inversely proportional to the estimates of the standard errors. The t_i's are not independent but have more or less constant variance. To test whether the response variable contains an outlier, the value of $t = \max_i |t_i|$ is calculated and, if it is sufficiently large, the observation yielding t is declared a discordant outlier.

Let us assume that only the ith observation in the response variable is an outlier. This can be expressed in terms of model (1.1). The null hypothesis is

$$H_0: \boldsymbol{E}(\mathbf{Y}) = \boldsymbol{a\theta} \tag{6.4}$$

and the mean slippage alternative is

$$H_A: \boldsymbol{E}(\mathbf{Y}) = \boldsymbol{a\theta} + \boldsymbol{\delta}, \tag{6.5}$$

where $\boldsymbol{\delta}$ is an $n \times 1$ vector of zeros apart from a possible non-zero value $\boldsymbol{\delta}$ in the ith position. From Section 6.10, we see that the maximum likelihood ratio statistic for testing H_0 versus H_A is a function of $d_i = t_i^2/(n-k)$ and a test based on the maximum likelihood ratio statistic is equivalent to a test based on d_i. Thus, the null hypothesis is rejected when d_i is large. However, the d_i's are not independent and a test of no outliers in the response variable should be based on the statistic

$$d = \max_i (d_i). \tag{6.6}$$

Thus, the null hypothesis of no outliers is rejected at the $100\alpha\%$ significance level in favour of the two-sided alternative $\delta \neq 0$ if

$$d > d_\alpha,$$

where d_α is the critical value of d at the $100\alpha\%$ level of significance. For the test to be of any use we need to find d_α. The exact value of d_α is difficult to find. However, from the first-order Bonferroni inequality (see Section 6.11.1) an upper bound d_0 of the critical value d_α is obtained from

$$\sum_{i=1}^{n} \Pr(d_i > d_0) = \alpha.$$

Since the d_i's are identically distributed as beta (see Section 6.11.2), the above can be written as

$$n \Pr\{x > d_0\} = \alpha$$

$$\Rightarrow n\left[\frac{1}{\beta(\frac{1}{2}, (n-p-1)/2)} \times \int_{d_0}^{1} x^{1/2-1}(1-x)^{(n-p-1)/2-1}\mathrm{d}x\right] = \alpha$$

$$\Rightarrow \beta_{d_0}\left(\frac{1}{2}, \frac{n-p-1}{2}\right) = 1 - \frac{\alpha}{n},$$

where

$$\beta_{d_0}\left(\frac{1}{2}, \frac{n-p-1}{2}\right) = \frac{1}{\beta(\frac{1}{2}, (n-p-1)/2)} \times \int_{0}^{d_0} x^{1/2-1}(1-x)^{(n-p-1)/2-1}\mathrm{d}x$$

is the incomplete beta-function ratio. The value of d_0 for any value of n, p and α can be computed by the computer algorithm of Cran *et al.* (1977). Once the value of d_0 is found, the upper bound of the critical value of t is found as

$$t_0 = [(n-p)d_0]^{1/2}. \tag{6.7}$$

The upper bound of the critical values of t for $\alpha = 0.1, 0.05$ and 0.01, $n = 5(1)\,10(2)\,20(5)\,50(10)\,100$ and $p = 1(1)\,6(2)\,10, 15, 25$ are tabulated by Barnett and Lewis (1978, pp. 335–336). For more details about the derivation and calculation of the approximate critical values of t, see Srikantan (1961), Lund (1975), Prescott (1975) and Paul (1985a).

6.3 The hat matrix

In any discussion of outliers in regression, a fundamental role is played by a matrix called the 'hat matrix' (a term coined by John W. Tukey).

Certain properties of $\boldsymbol{H}$ are important:

(1) It is symmetric (easily seen from (6.2)).

(2) It is idempotent, since

$$\begin{aligned} \boldsymbol{H}^2 &= \boldsymbol{a}(\boldsymbol{a}'\boldsymbol{a})^{-1}\boldsymbol{a}'\boldsymbol{a}(\boldsymbol{a}'\boldsymbol{a})^{-1}\boldsymbol{a}' \\ &= \boldsymbol{a}(\boldsymbol{a}'\boldsymbol{a})^{-1}\boldsymbol{a}' = \boldsymbol{H}. \end{aligned}$$

(3) It follows from (2) that the diagonal elements h_{ii} of $\boldsymbol{H}$ satisfy

$$0 \le h_{ii} \le 1. \tag{6.8}$$

This follows since, by (2),

$$h_{ii} = \sum_{j=1}^{n} h_{ij}^2 = h_{ii}^2 + \sum_{j \ne i} h_{ij}^2.$$

Now consider the effect of centring the explanatory variables. As in (1.18), $\boldsymbol{X}$ denotes the $n \times (k-1)$ matrix of centred explanatory variables, then $\boldsymbol{a}$ becomes

$$\boldsymbol{a} = (\mathbf{1}, \boldsymbol{X}).$$

The residuals are unchanged by this, and $\boldsymbol{H}$ also remains unchanged. Therefore we have

$$\begin{aligned} \boldsymbol{H} = \boldsymbol{a}(\boldsymbol{a}'\boldsymbol{a})^{-1}\boldsymbol{a}' &= \boldsymbol{a}\begin{pmatrix} n^{-1} & \mathbf{0}' \\ \mathbf{0} & (\boldsymbol{X}'\boldsymbol{X})^{-1} \end{pmatrix}\boldsymbol{a}' \\ &= \boldsymbol{X}(\boldsymbol{X}'\boldsymbol{X})^{-1}\boldsymbol{X}' + n^{-1}\mathbf{1}\mathbf{1}'. \end{aligned}$$

Or

$$\boldsymbol{H} = \boldsymbol{H} + n^{-1}\mathbf{1}\mathbf{1}', \tag{6.9}$$

where (6.9) is taken to define $\boldsymbol{H}$. Therefore

$$\tilde{h}_{ii} = h_{ii} - \frac{1}{n},$$

where $\tilde{h}_{ij}$ is the (i, j)th element of $\tilde{\boldsymbol{H}}$. The matrix $\tilde{\boldsymbol{H}}$ is also symmetric and idempotent and therefore

$$\tilde{h}_{ii} = \tilde{h}_{ii}^2 + \sum_{j \ne i} \tilde{h}_{ij}^2.$$

Putting $\tilde{h}_{ii} = h_{ii} - 1/n$ in the above equation it is immediately clear that

$$\frac{1}{n} \le h_{ii} \le 1.$$

(4) As $\boldsymbol{H}$ is symmetric and idempotent it is a projection matrix, and

the eigenvalues are all either zero or one. The number of unit eigenvalues is equal to the rank of $\boldsymbol{H}$ (for proof, see Rao, 1965).

(5) $$\text{Rank}(\boldsymbol{H}) = \text{Rank}(\boldsymbol{a}) = k.$$

From this we also have,

$$tr\,\boldsymbol{H} = \sum_i \boldsymbol{h}_{ii} = (\text{sum of eigenvalues}) = k. \qquad (6.10)$$

Now from equation (6.1) we see that the diagonal elements h_{ii} of $\boldsymbol{H}$ represent the influence of the ith response on the ith fitted value. If any h_{ii} is unduly large, this therefore indicates a point which has an undue weight in the regression, such as an outlying point in the factor space. It is important to see here that $\boldsymbol{H}$ *is a function of the explanatory variables only*. The response variables do not enter into it at all. We say that any point with a large h_{ii} is a *point of high leverage*.

The role of h_{ii} is also important in the outlier test based on t. The power of the maximum likelihood ratio test of the hypothesis that the ith observation is not an outlier is a decreasing function of h_{ii}. For proof, see Gentleman and Wilk (1975).

How large does h_{ii} have to be? Well, we have seen above that

$$\frac{1}{n} \leq h_{ii} \leq 1$$

and it follows from (5) that the average h_{ii} must be equal to k/n. Thus, values of h_{ii} greater than k/n correspond to high leverage points. However, a rule of thumb which has been suggested is to select points for investigation if

$$h_{ii} > \frac{2k}{n}$$

provided this limit is less than unity. This rule can be based on approximate distribution arguments and it seems to work reasonably well in practice. This rule, however, will in general detect too many points as high-leverage points (see Belsley *et al*., 1980). For more details of the hat matrix, see Hoaglin and Welsch (1978).

A more refined way of screening, when the explanatory variables are independently distributed as multivariate normal, is to base a test on $F_i = (n-k)(h_{ii} - 1/n)/(1 - h_{ii})(k-1)$ which has an F-distribution with $k-1$ and $n-k$ d.f. See Belsley *et al.* (1980) for the proof of the distribution of F_i and a justification of the approximate cut-off point

$2k/n$ for large h_{ii}. A point to note here is that the assumption of a multivariate normal distribution for explanatory variables may not be realistic. However, in practice, data analysis in Section 6.7 shows that such a test can pin-point potential high-leverage points. As the F_i's will not be independent we base our test on $F = \max\{F_i\}$. The upper bound F_0 of the critical values of F are obtained by using the first-order Bonferroni inequality from

$$\int_0^{F_0} f(F)\,\mathrm{d}F = 1 - \frac{\alpha}{n}, \tag{6.11}$$

where $f(F)$ is the density of F with $k-1$ and $n-k$ d.f.

6.4 Influence measure

To study how the presence of the ith data point influences the regression analysis, an influence measure is often calculated. One such influence measure is

$$D_i = [(\hat{\boldsymbol{\theta}} - \hat{\boldsymbol{\theta}}_{(i)})'\boldsymbol{a}'\boldsymbol{a}(\hat{\boldsymbol{\theta}} - \hat{\boldsymbol{\theta}}_{(i)})]/ks^2,$$

where $\hat{\boldsymbol{\theta}}$ is the least squares estimate of $\boldsymbol{\theta}$ based on the full data and $\hat{\boldsymbol{\theta}}_{(i)}$ is the least squares estimate of $\boldsymbol{\theta}$ based on the reduced data when the ith data point has been removed from the data. This is the weighted sum of squares of deviation of the regression estimates based on $\hat{\boldsymbol{\theta}}$ from $\hat{\boldsymbol{\theta}}_{(i)}$.

The weights used here are precision of the estimates $\hat{\boldsymbol{\theta}}$. The more precise the estimate, the more weight it gets in the influence measure. Other weightings are possible (see Cook and Weisberg, 1980, p. 497). D_i can be written in an equivalent form, viz.

$$D_i = \frac{t_i^2 w_i}{k}, \tag{6.12}$$

where t_i is the standardized residual and $w_i = h_{ii}/(1 - h_{ii})$. For proof of the above equation, see Section 6.10. Thus D_i is large either when t_i^2 or when w_i (equivalently h_{ii}) is large and when t_i and w_i are both large or moderately large. As has been discussed before, t_i^2 measures the outlying role of the ith value of the response variable and w_i measures the outlying role of the ith row of the explanatory variable matrix.

When the data contains an outlier in the response variable it would be detected by a test based on t, and when the data contains a high leverage point it would be detected by a test based on F. The influence measure is perhaps most important in the situation when the data point is neither a significant outlier nor a significant high leverage point (meaning that neither t_i nor h_{ii} are extreme) but correspond to $\max\{D_i\}$. In such a situation the ith point is most influential and perhaps contains extra information about the data. Cook (1977) discussed an example originally given by Longley (1967), where the observation having the largest D_i does not have the largest t_i or h_{ii} but is influential. In a practical context this situation may arise only very rarely. However, the value of D_i can be used as a check, when the ith data point has been detected as an outlier or as a high leverage point, to see whether the ith point, if removed, causes maximum change in the estimate of $\boldsymbol{\theta}$.

6.5 Illustration of the hat matrix, standardized residual and the influence measure

The data given in Table 6.1 are from Draper and Stoneman (1966). The response is strength and the explanatory variables are specific gravity and moisture content.

Figure 6.1 shows that observation 4 and also, possibly, observation 1 are outliers in the explanatory variables. Figure 6.2 shows that observation 1 is a suspect value and Fig. 6.3 shows that observation 4 is also a suspect value. In this example, when we have only two explanatory variables, it is possible to see some points which either lie

Table 6.1 *Data on wood beams.*

Beam number	*Specific gravity*	*Moisture content*	*Strength*
1	0.499	11.1	11.14
2	0.558	8.9	12.74
3	0.604	8.8	13.13
4	0.441	8.9	11.51
5	0.550	8.8	12.38
6	0.528	9.9	12.60
7	0.418	10.7	11.13
8	0.480	10.5	11.70
9	0.406	10.5	11.02
10	0.467	10.7	11.42

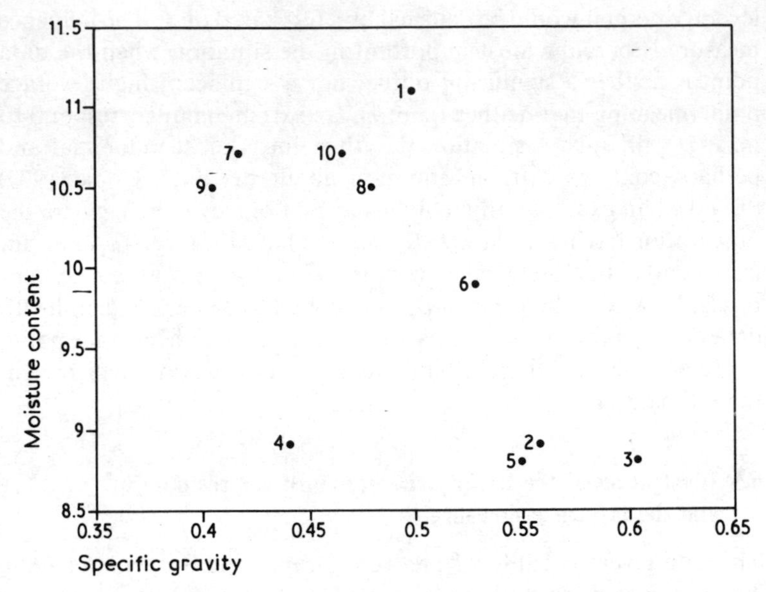

Fig. 6.1 *Scatter graph: moisture content versus specific gravity.*

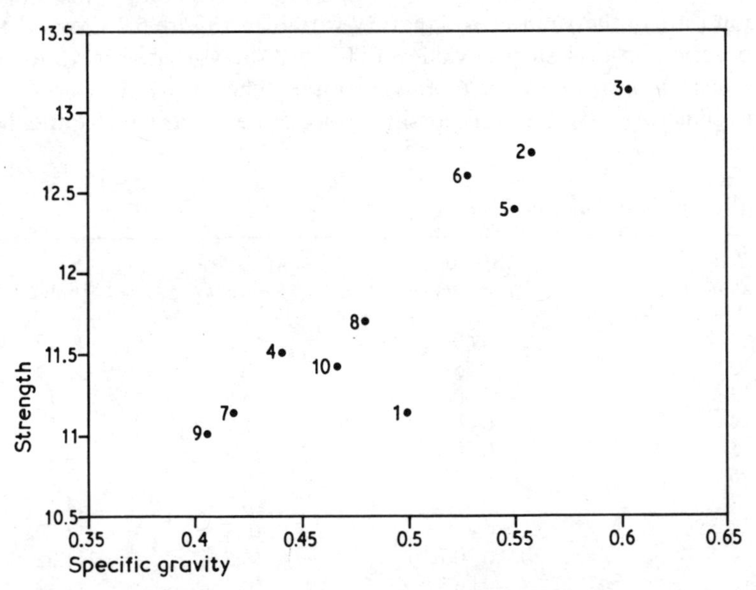

Fig. 6.2 *Scatter graph: strength versus specific gravity.*

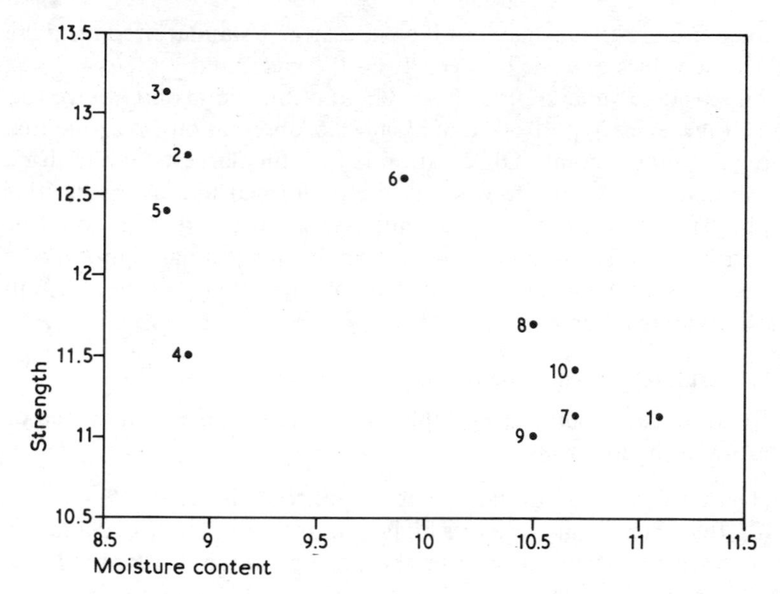

Fig. 6.3 *Scatter graph: strength versus moisture content.*

far out in the factor space or are outliers. However, in a complex situation, where there are many explanatory variables, it will be difficult to carry out an analysis graphically to spot outlying values or high-leverage points. We now give values of R_i, h_{ii}, t_i and D_i (Table 6.2). Several points emerge from Table 6.2. The observation having

Table 6.2 *Values of the residuals, diagonal elements of the 'hat' matrix, the standardized residuals and the influence of measure.*

i	R_i	h_{ii}	t_i	D_i
1	−0.444	0.418	−2.114	1.069
2	0.669	0.242	0.286	0.009
3	0.041	0.417	0.196	0.009
4	−0.167	0.604	−0.966	0.476
5	−0.250	0.252	−1.050	0.124
6	0.450	0.148	1.769	0.181
7	0.127	0.262	0.538	0.034
8	0.117	0.154	0.463	0.013
9	0.066	0.315	0.290	0.013
10	−0.009	0.187	−0.036	0.013

maximum residuals may not have the largest standardized residual. All the values of h_{ii} lie between $1/n = 0.1$ and 1 and $\sum_{i=1}^{10} h_{ii} = 3 = k$. The average value of h_{ii} is $k/n = 0.3$ and four points are above this average. Since $h_{44} = 0.604$ and $2k/n = 0.6$, observation 4 is a potential high-leverage point. Observation 1 has the largest standardized residual. A test based on t could be performed to see whether this point is a significant outlier. Similarly, a test based on F could be performed to see whether observation 4 is a potential high-leverage point. Note that the largest two values of D_i correspond to observations 1 and 4.

6.6 Multiple unusual points and 'How to detect them'

By an unusual point in multiple linear regression we mean one or more of the following:

(1) that there is an outlier in the response variable;
(2) that there is an outlier or high-leverage point in the explanatory variables, which means that there is a point that lies far out in the factor space;
(3) that there is a point which is influential either in respect of the model fit or in respect of the estimation of the parameters.

To detect multiple unusual points two procedures are available and these are:

(1) a block procedure in which all suspected unusual points are tested for discordancy by a single test statistic – reject all or accept all;
(2) a sequential procedure in which the unusual points are tested sequentially.

Thus the test based on standardized residuals by procedure (2) is to test for discordancy of the observation having maximum standardized residual; if this observation is rejected reanalyse the data, after removing this point, to test for a second outlier and continue until no more outliers are detected.

The sequential method may be hampered by the masking effect. This means that if a test statistic is based on the assumption of a single outlier, the presence of more than one outlier will make the spread of the data larger and consequently the test statistic smaller. The opposite of the masking effect is the swamping effect. This occurs if a test statistic is based on the assumption of l outliers when in fact there are less than l of them; this makes the spread of the data smaller and consequently the test statistic larger. For more details of these effects, see Barnett and Lewis (1978). To come back to the sequential method, we note that if a data set contains more than one outlier, because of the masking effect, the very first observation yielding max $\{t_i\}$ may not be declared discordant. Whether there is a masking effect or whether there is no outlier can be detected by reanalysing the data after deleting the observation yielding max $\{t_i\}$ even if it is not found discordant at a particular significance level (say 5%). In any application, to detect multiple outliers by sequential procedure, a useful stopping rule would be to look for two consecutive non-significant values of t.

In comparison to the sequential method, the block method is computationally difficult and expensive. For a discussion of a number of block methods, see Draper and John (1981). We therefore propose that detection of multiple unusual points be based on sequential methods. Thus, to detect mutiple outliers in the response variable, we propose to apply tests on t sequentially. Each time a significant t value is detected, the observation corresponding to t is deleted and the reduced data reanalysed for a further outlier. The sequential procedure is continued until two non-significant t-values can be found. And, to detect multiple high-leverage points, we propose to apply tests on F sequentially. Since the assumption of a multivariate normal distribution for the explanatory variables may not be realistic, a test based on F should be considered only informal and, therefore, we propose to continue applying the test based on F until only one non-significant F value is found.

6.7 Examples

We consider two sets of published data which have been analysed extensively, for possible outliers or high-leverage points or influential points, in the literature.

For each set of data analysed we produce a table giving t, t_0 (equation (6.7)), F and F_0 (equation (6.11)). We also give values of $D = \max\{D_i\}$, to see whether the points found as outliers or high-leverage points are also extremely influential. The values of t_0 and F_0 that are given in the tables are for the 5% level.

The sequential procedure was tried in two orders to see if they would give the same results.

(1) First test for outliers sequentially and then test for high-leverage

points sequentially after deleting the detected outliers. Having deleted the detected outliers and the high-leverage points, repeat the same procedure on the reduced data. Repeat the same procedure on the finally reduced data until no further point is detected as an outlier or high-leverage point.

(2) First test for high-leverage points sequentially and then test for outliers sequentially after deleting the detected high-leverage points. Having deleted the detected high-leverage points and the outliers, repeat the same procedure on the reduced data. Repeat the same procedure on the finally reduced data until no further point is detected as an outlier or high-leverage point.

It is, perhaps, important to carry out the analysis by both the orders (1) and (2) to see whether an observation detected as an outlier in the presence of a high-leverage point is also detected as an outlier when the high-leverage point is removed from the data and vice versa.

EXAMPLE 6.1

(Brownlee's 'stack-loss' data). The data given in Table 6.3 were obtained from 21 days of operation of a plant for the oxidation of ammonia to nitric acid

Table 6.3 *Stack-loss data.* (*Statistical Theory and Methodology in Science and Engineering*, 2nd edn, Brownlee, K. A. (1965); reprinted by permission of John Wiley & Sons, Inc.)

Observation number	x_1	x_2	x_3	y	*Observation number*	x_1	x_2	x_3	y
1	80	27	89	42	11	58	18	89	14
2	80	27	88	37	12	58	17	88	13
3	75	25	90	37	13	58	18	82	11
4	62	24	87	28	14	58	19	93	12
5	62	22	87	18	15	50	18	89	8
6	62	23	87	18	16	50	18	86	7
7	62	24	93	19	17	50	19	72	8
8	62	24	93	20	18	50	19	79	8
9	58	23	87	15	19	50	20	80	9
10	58	18	80	14	20	56	20	82	15
					21	70	20	91	15

x_1 = air flow;
x_2 = cooling water inlet temperature;
x_3 = acid concentration;
y = stack loss.

(Brownlee, 1965, p. 454). The dependent variable is ten times the percentage of the ingoing ammonia to the plant that escapes unabsorbed and so is lost. There are three explanatory variables, namely airflow, cooling water inlet temperature and acid concentration.

Daniel and Wood (1971) decide on a model which includes the linear and quadratic terms of the first variable and the linear term of the second variable, and that four observations (1, 3, 4, 21) were outliers. Andrews (1974) comes to a similar conclusion as the result of a robust analysis. Atkinson (1981) uses diagnostic plots and comes to a similar conclusion. Andrews and Pregibon

Table 6.4 *Values of t, T and D and also of the upper bound of the critical values t_0 and T_0 of t and T for the 'stack-loss' data.*

(1) Values of t and t_0

Observation deleted	t	*Observation number corresponding to t*	t_0
None	2.629	21	2.760
21	3.009	4	2.733
4, 21	2.707	2	2.704
2, 4, 21	2.037	20	2.672
2, 4, 20, 21	2.384	1	2.826
1, 2, 3, 4, 21	2.097	20	2.850
1, 2, 3, 4, 20, 21	2.086	13	2.557

(2) Values of T and T_0

Observation deleted	T	*Observation number corresponding to T*	T_0
None	3.471	1 and 2	7.282
2, 4, 21	11.752	1	7.715
1, 2, 4, 21	236.096	3	7.921
1, 2, 3, 4, 21	5.053	9	8.175

(3) Values of D

Observation deleted	D	*Observation number corresponding to D*
None	0.699	21
21	0.593	2
2, 21	0.645	4
2, 4, 21	2.042	1
1, 2, 4, 21	21.160	3
1, 2, 3, 4, 21	0.708	9

(1978) say 'the same observations are identified for $k=4$ but the summary suggests that the second observation should also be considered as being highly unusual'. In the notation of Andrews and Pregibon k is the number of outliers. Cook (1979) comes to a similar conclusion by looking at values of t_i, h_{ii} and D_i by sequentially deleting observations and also by looking at residual correlations.

We now analyse the 'stack-loss' data by taking the final model of Daniel and Wood. In Table 6.4 we give the results of our analysis.

We first carry out the analysis in order (1) and see from Table 6.4 that the sequential outlier tests detect observations 21, 4 and 2 as outliers and, after deleting observations 21, 4 and 2 from the data, the sequential high-leverage point test detect observations 1 and 3 as high-leverage points. If we now repeat the same procedure on the reduced data, i.e. on the data after deleting observations 1, 2, 3, 4 and 21, we see there is no further point in the data that can be considered an outlier or high-leverage point. Thus by order (1), the conclusion is that the observations 21, 4 and 2 are outliers and the observations 1 and 3 are high-leverage points.

We now analyse the same data by order (2) and see that the sequential high-leverage point tests detect no observation as high-leverage point and the sequential outlier tests detect observations 21, 4 and 2 as outliers. If we repeat the same procedure on the reduced data after deleting the observations 21, 4 and 2, we see that the observations 1 and 3 are, now, detected as high-leverage points and no further point is detected as an outlier. If the same procedure is repeated on the finally reduced data, we see that there is no further point in the data that can be considered an outlier or high-leverage point. Thus we can see that, for this example, the conclusion of the analysis by order (2) is the same as that by order (1).

Table 6.5 *Age at first word and Gessel Adaptive Score.*

Observation number	x	y	*Observation number*	x	y
1	15	95	11	7	113
2	26	71	12	9	96
3	10	83	13	10	83
4	9	91	14	11	84
5	15	102	15	11	102
6	20	87	16	10	100
7	18	93	17	12	105
8	11	100	18	42	57
9	8	104	19	17	121
10	20	94	20	11	86
			21	10	100

x = age in months at first word;
y = Gessel Adaptive Score.

From the values of D, we see that by sequentially deleting the observations 21, 2, 4 and 1, observations 21, 2, 4, 1 and 3 are, in that order, most influential. Thus, we conclude that the observations 2, 4 and 21 are possible outliers and 1 and 3 are possible high-leverage points and all of 1, 2, 3, 4 and 21 are together possible influential points.

EXAMPLE 6.2

(Mickey *et al.*'s (1967) 'first word–Gessel Adaptive Score' data). This is a straight-line regression example. The independent variable is age at first word and the dependent variable is the Gessel Adaptive Score. These data are given in Table 6.5.

Mickey *et al.* (1967) detect observation 19 as an outlier. Andrews and

Table 6.6 *Values of t, T and D and also of the upper bound of the critical values t_0 and T_0 of t and T for the 'first word–Gessel Adaptive Score' data.*

Values of t and t_0

Observation deleted	t	*Observation number corresponding to t*	t_0
None	2.823	19	2.789
19	1.722	3, 13	2.775
19, 3	1.962	13	2.739
19, 13	1.962	3	2.739
18	2.693	19	2.775
18, 19	1.654	3, 13	2.739

Values of T and T_0

Observation deleted	T	*Observation number corresponding to T*	T_0
None	32.940	18	12.268
18	11.271	2	12.321
19	32.776	18	12.321
18, 19	12.001	2	12.385

Values of D

Observation deleted	D	*Observation number corresponding to D*
None	0.678	18
18	1.020	2
19	0.188	18
18, 19	0.797	2

Pregibon (1978) find that observation 18 is the most influential and mention that the nineteenth observation is a significant outlier, but as it is not influential this observation does not need to be discarded. This is an outlier that does not matter. Draper and John (1981) find that observation 19 is an outlier and 18 is an influential point. Dempster and Gasko-Green (1981) find that observations 18 and 2 merely extend the domain of the model but observation 19 is an outlier that does not matter. We now analyse the 'first word–Gessel Adaptive Score' data. We give the results of our analysis in Table 6.6.

We see from Table 6.6 that the sequential outlier tests detect observation 19 as an outlier and, after deleting observation 19 from the data, the sequential high-leverage point tests detect observation 18 as a high-leverage point. If we repeat the same procedure we see that there is no other point in the data that can be considered as an outlier or high-leverage point. Thus, by order (1), the conclusion is that observation 19 is an outlier and observation 18 is a high-leverage point.

We now analyse the same data by first applying the high-leverage point tests and then outlier tests. We see that observation 18 is detected as a high-leverage point (this is also the most influential point), but observation 19 is, now, not detected as an outlier. Thus, by order (2), the conclusion is that there is no outlier and that observation 18 is a high-leverage point.

This illustrates the point that observation 19, detected as an outlier in the presence of a high-leverage point, is not an outlier that matters. We note that after deleting observation 18 (the high-leverage point, as well as the most influential), observation 2, but not 19, is shown to be the most influential. However, comparison with the critical value shows that this is not a significant high-leverage point. We thus conclude that observation 18 is a high-leverage point as well as an influential point and observation 19 is an outlier that does not matter.

6.8* Some identities related to submodel analysis

Bingham (1977) proved a number of useful results related to adding or deleting observations or variables in linear regression analysis. We describe some of his results which will be used in Section 6.9.

Consider model (1.1) and let $\boldsymbol{a}$ and $\mathbf{Y}$ be partitioned as

$$\boldsymbol{a} = \begin{bmatrix} \boldsymbol{a}_1 \\ \boldsymbol{a}_2 \end{bmatrix}, \qquad \mathbf{Y} = \begin{bmatrix} \mathbf{Y}_1 \\ \mathbf{Y}_2 \end{bmatrix},$$

where $\boldsymbol{a}_1$ is $n_1 \times k$, $\boldsymbol{a}_2$ is $n_2 \times k$, $\mathbf{Y}_1$ is $n_1 \times 1$ and $\mathbf{Y}_2$ is $n_2 \times 1$ and where $n_1 + n_2 = n$. Denoting $\boldsymbol{a}'\boldsymbol{a} = \boldsymbol{A}$, $\boldsymbol{a}_1'\boldsymbol{a}_1 = \boldsymbol{A}_1$ and $\boldsymbol{a}_2'\boldsymbol{a}_2 = \boldsymbol{A}_2$, it is immediately apparent that $\boldsymbol{A} = \boldsymbol{A}_1 + \boldsymbol{A}_2$. The model

$$E(\mathbf{Y}_1) = \boldsymbol{a}_1\boldsymbol{\theta} \tag{6.13}$$

is a submodel of model (1.1) with the last n_2 rows of the data removed.

We assume that $\text{rank}(\boldsymbol{a}_1) = \text{rank}(\boldsymbol{A}_1) = \text{rank}(\boldsymbol{A}) = k$. The least squares estimator of $\boldsymbol{\theta}$ by fitting the complete data model is

$$\boldsymbol{\theta} = \boldsymbol{A}^{-1}\boldsymbol{a}'\mathbf{Y}$$

and the residuals are

$$\mathbf{R} = (\boldsymbol{I} - \boldsymbol{H})\mathbf{Y} = \boldsymbol{M}\mathbf{Y},$$

where $\boldsymbol{H} = \boldsymbol{a}\boldsymbol{A}^{-1}\boldsymbol{a}'$ and $\boldsymbol{M} = \boldsymbol{I} - \boldsymbol{H}$. The least squares estimate of $\boldsymbol{\theta}$ by fitting the submodel (6.11) is

$$\tilde{\boldsymbol{\theta}} = \boldsymbol{A}_1^{-1}\boldsymbol{a}_1'\mathbf{Y}_1 \tag{6.14}$$

and the corresponding residuals are

$$\tilde{\mathbf{R}}_1 = (\boldsymbol{I} - \tilde{\boldsymbol{H}}_1)\mathbf{Y}_1 = \tilde{\boldsymbol{M}}_1\mathbf{Y}_1,$$

where

$$\tilde{\boldsymbol{H}}_1 = \mathrm{a}_1\boldsymbol{A}_1^{-1}\boldsymbol{a}_1' \quad \text{and} \quad \tilde{\boldsymbol{M}}_1 - \boldsymbol{I} - \tilde{\boldsymbol{H}}_1.$$

Define $\boldsymbol{C} = \boldsymbol{a}_2\boldsymbol{A}_1^{-1}\boldsymbol{a}_2'$ and $\boldsymbol{D} = \boldsymbol{a}_2\boldsymbol{A}^{-1}\boldsymbol{a}_2'$. We can show that

Result 6.1

$$\boldsymbol{A}_1^{-1} = \boldsymbol{A}^{-1} + \boldsymbol{A}^{-1}\boldsymbol{a}_2'(\boldsymbol{I} - \boldsymbol{D})^{-1}\boldsymbol{a}_2\boldsymbol{A}^{-1}, \tag{6.15}$$

and

$$\boldsymbol{A}^{-1} = \boldsymbol{A}_1^{-1} - \boldsymbol{A}_1^{-1}\boldsymbol{a}_2'(\boldsymbol{I} + \boldsymbol{C})^{-1}\boldsymbol{a}_2\boldsymbol{A}_1^{-1}. \tag{6.16}$$

Proof of Result 6.1
Multiplying the right-hand side of (6.15) on the left by $\boldsymbol{A}_1 = \boldsymbol{A} - \boldsymbol{a}_2'\boldsymbol{a}_2$ we obtain

$$\begin{aligned} &\boldsymbol{I}_k + \boldsymbol{a}_2'(\boldsymbol{I} - \boldsymbol{D})^{-1}\boldsymbol{a}_2\boldsymbol{A}^{-1} - \boldsymbol{a}_2'\boldsymbol{a}_2\boldsymbol{A}^{-1} - \boldsymbol{a}_2'\boldsymbol{a}_2\boldsymbol{A}^{-1}\boldsymbol{a}_2'(\boldsymbol{I} - \boldsymbol{D})^{-1}\boldsymbol{a}_2\boldsymbol{A}^{-1} \\ &\quad = \boldsymbol{I}_k + \boldsymbol{a}_2'[(\boldsymbol{I} - \boldsymbol{D})^{-1} - \boldsymbol{I} - \boldsymbol{D}(\boldsymbol{I} - \boldsymbol{D})^{-1}]\boldsymbol{a}_2\boldsymbol{A}^{-1} \\ &\quad = \boldsymbol{I}_k, \end{aligned}$$

since $\boldsymbol{I} + \boldsymbol{D}(\boldsymbol{I} - \boldsymbol{D})^{-1} = (\boldsymbol{I} - \boldsymbol{D})^{-1}$. Thus (6.15) follows because $\boldsymbol{A}_1$ is of full rank. Equation (6.16) can be proved the same way, multiplying the right-hand side by $\boldsymbol{A} = \boldsymbol{A}_1 + \boldsymbol{a}_2'\boldsymbol{a}_2$.

The relation between $\boldsymbol{C}$ and $\boldsymbol{D}$ can be expressed as

$$\boldsymbol{D} = \boldsymbol{C}(\boldsymbol{I} + \boldsymbol{C})^{-1}, \tag{6.17}$$

$$\boldsymbol{C} = \boldsymbol{D}(\boldsymbol{I} - \boldsymbol{D})^{-1}, \tag{6.18}$$

$$(I+C)^{-1}=I-D, \tag{6.19}$$

$$(I-D)^{-1}=(I+C). \tag{6.20}$$

Equation (6.17) can be checked from (6.16) as

$$a_2A^{-1}a_2'=a_2A_1^{-1}a_2'-a_2A_1^{-1}a_2'(I+C)^{-1}a_2A_1^{-1}a_2'$$
$$\Rightarrow \quad D=C[I-(I+C)^{-1}C]=C(I+C)^{-1}.$$

Equation (6.18) can be obtained the same way from equation (6.15). Equations (6.19) and (6.20) can be obtained from equations (6.17) and (6.18).

Let M be written as the partitioned matrix

$$M=\begin{bmatrix} M_{11} & M_{12} \\ M_{21} & M_{22} \end{bmatrix}=I_n-aA^{-1}a'.$$

Thus $M_{11}=I_{n_1}-a_1A^{-1}a_1'$, $M_{12}=-a_1A^{-1}a_2'$, $M_{21}=-a_2A^{-1}a_1, M_{22}=I_{n_2}-a_2A^{-1}a_2'$. We can establish the following relations:

Result 6.2

$$M_{22}=I-D=(I+C)^{-1}, \tag{6.21}$$

$$M_{12}=M_{21}'=-a_1A_1^{-1}a_2'(I+C)^{-1}, \tag{6.22}$$

$$M_{11}=\tilde{M}_1+M_{12}M_{22}^{-1}M_{21}, \tag{6.23}$$

$$M_{12}M_{22}^{-1}=-a_1A_1^{-1}a_2. \tag{6.24}$$

Proof of Result 6.2

Equation (6.21) is trivial and is obtained from equation (6.19). Equation (6.22) can be proved using (6.16) as

$$a_1A^{-1}a_2'=a_1A_1^{-1}a_2'-a_1A_1^{-1}a_2'(I+C)^{-1}a_2A_1^{-1}a_2'$$
$$\Rightarrow \quad -M_{12}=a_1A_1^{-1}a_2'[I-(I+C)^{-1}C]$$
$$=a_1A_1^{-1}a_2'(I+C)^{-1}.$$

Equation (6.23) can be obtained in the same way from equation (6.15). Equation (6.24) can be obtained from equations (6.21) and (6.22).

We can now express the residuals from the complete data model in terms of the residuals from the submodel. Let **R** be partitioned on

$\mathbf{R} = \begin{bmatrix} \mathbf{R}_1 \\ \mathbf{R}_2 \end{bmatrix}$, where $\mathbf{R}_1$ is $n_1 \times 1$ and $\mathbf{R}_2$ is $n_2 \times 1$, then we can show that

Result 6.3

$$\mathbf{R}_1 = \tilde{\mathbf{R}}_1 - a_1 A_1^{-1} a_2' \mathbf{R}_2. \tag{6.25}$$

Proof of Result 6.3
Equation (6.25) can be proved using equations (6.22) and (6.24) as

$$\begin{aligned} \mathbf{R}_1 &= M_{11}\mathbf{Y}_1 + M_{12}\mathbf{Y}_2 = \tilde{M}_1\mathbf{Y}_1 + M_{12}M_{22}^{-1}M_{21}\mathbf{Y}_1 + M_{12}\mathbf{Y}_2 \\ &= \tilde{\mathbf{R}}_1 - a_1 A_1^{-1} a_2' M_{21}\mathbf{Y}_1 - a_1 A_1^{-1} a_2' M_{22}\mathbf{Y}_2 \\ &= \tilde{\mathbf{R}}_1 - a_1 A_1^{-1} a_2' [M_{21}\mathbf{Y}_1 + M_{22}\mathbf{Y}_2] \\ &= \tilde{\mathbf{R}}_1 - a_1 A_1^{-1} a_2' \mathbf{R}_2. \end{aligned}$$

Again, we can relate the least squares estimate $\hat{\boldsymbol{\theta}}$ obtained from the complete data model to the least squares estimate $\tilde{\boldsymbol{\theta}}$ obtained from the submodel as

Result 6.4

$$\hat{\boldsymbol{\theta}} = \tilde{\boldsymbol{\theta}} + A^{-1} a_2' (I - D)^{-1} \mathbf{R}_2. \tag{6.26}$$

Proof of Result 6.4
Equation (6.26) can be proved using equations (6.16), (6.21) and (6.22) as

$$\begin{aligned} \hat{\boldsymbol{\theta}} &= A^{-1} a' \mathbf{Y} = A^{-1}(a_1'\mathbf{Y}_1 + a_2'\mathbf{Y}_2) \\ &= (A_1^{-1} - A_1^{-1} a_2' (I + C)^{-1} a_2 A_1^{-1})(a_1'\mathbf{Y}_1 + a_2'\mathbf{Y}_2) \\ &= A_1^{-1} a_1' \mathbf{Y}_1 - A_1^{-1} a_2' (I + C)^{-1} a_2 A_1^{-1} a_1' \mathbf{Y}_1 + A_1^{-1} a_2' \mathbf{Y}_2 \\ &\quad - A_1^{-1} a_2' (I + C)^{-1} a_2 A_1^{-1} a_2' \mathbf{Y}_2 \\ &= \tilde{\boldsymbol{\theta}} - A_1^{-1} a_2' [(I + C)^{-1} a_2 A_1^{-1} a_1' \mathbf{Y}_1 - \mathbf{Y}_2 + (I + C)^{-1} C \mathbf{Y}_2] \\ &= \tilde{\boldsymbol{\theta}} - A_1^{-1} a_2' [(I + C)^{-1} a_2 A_1^{-1} a_1' \mathbf{Y}_1 - (I - C)^{-1} \mathbf{Y}_2] \\ &= \tilde{\boldsymbol{\theta}} + A_1^{-1} a_2' [M_{21}\mathbf{Y}_1 + M_{22}\mathbf{Y}_2] \\ &= \tilde{\boldsymbol{\theta}} + A_1^{-1} a_2' \mathbf{R}_2. \end{aligned}$$

Now, from equation (6.15) we see that $A_1^{-1} a_2' = A^{-1} a_2' (I - D)^{-1}$ and hence result (6.4) is proved.

The relationship between the residual sum of squares from the complete data model and that from the submodel is

Result 6.5

$$\mathbf{R}'\mathbf{R} = \tilde{\mathbf{R}}_1'\tilde{\mathbf{R}}_1 + \mathbf{R}_2'(I - D)^{-1}\mathbf{R}_2. \tag{6.27}$$

Proof of result 6.5

Equation (6.27) can be proved using equation (6.25) and the orthogonality relation $\boldsymbol{a}_1'\tilde{\mathbf{R}}_1 = 0$ as

$$\begin{aligned}
\mathbf{R}'\mathbf{R} &= \mathbf{R}_1'\mathbf{R}_1 + \mathbf{R}_2'\mathbf{R}_2 \\
&= (\tilde{\mathbf{R}}_1 - \boldsymbol{a}_1 A_1^{-1}\boldsymbol{a}_2'\mathbf{R}_2)'(\tilde{\mathbf{R}}_1 - \boldsymbol{a}_1 A_1^{-1}\boldsymbol{a}_2'\mathbf{R}_2) + \mathbf{R}_2'\mathbf{R}_2 \\
&= \tilde{\mathbf{R}}_1'\tilde{\mathbf{R}}_1 + \mathbf{R}_2'\boldsymbol{a}_2 A_1^{-1}\boldsymbol{a}_1'\boldsymbol{a}_1 A_1^{-1}\boldsymbol{a}_2'\mathbf{R}_2 + \mathbf{R}_2'\mathbf{R}_2 \\
&= \tilde{\mathbf{R}}_1'\tilde{\mathbf{R}}_1 + \mathbf{R}_2' C\mathbf{R}_2 + \mathbf{R}_2'\mathbf{R}_2 \\
&= \tilde{\mathbf{R}}_1'\tilde{\mathbf{R}}_1 + \mathbf{R}_2'(I + C)\mathbf{R}_2 \\
&= \tilde{\mathbf{R}}_1'\tilde{\mathbf{R}}_1 + \mathbf{R}_2'(I - D)^{-1}\mathbf{R}_2
\end{aligned}$$

6.9* Some identities useful in the analysis of residuals to detect outliers and influential observations in the regression

Suppose that the ith data point has been removed from model (1.1) and let $a_{(i)}$ denote the matrix of the explanatory variables with the ith row deleted and $\mathbf{Y}_{(i)}$ denote the observation vector with the ith value deleted. Then, the least squares estimate of $\boldsymbol{\theta}$ from the submodel based on all the data except the ith data point is

$$\hat{\boldsymbol{\theta}}_{(i)} = (\boldsymbol{a}_{(i)}'\boldsymbol{a}_{(i)})^{-1}\boldsymbol{a}_{(i)}'\mathbf{Y}_{(i)}$$

and the residuals obtained from fitting this submodel are

$$\mathbf{R}_{(i)} = \mathbf{Y}_{(i)} - \boldsymbol{a}_{(i)}\hat{\boldsymbol{\theta}}_{(i)}.$$

Let $S^2 = \mathbf{R}'\mathbf{R}$ denote the residual sum of squares of the complete data model and

$$S_{(i)}^2 = \mathbf{R}_{(i)}'\mathbf{R}_{(i)}$$

denote the residual sum of squares of the submodel.

We can relate the residual sum of squares $S_{(i)}^2$ from the submodel to the residual sum of squares S^2 from the complete model as

$$S_{(i)}^2 = S^2 - R_i^2/(1 - h_{ii}), \tag{6.28}$$

where R_i is the ith residual and $h_{ii} = \mathbf{a}_i A^{-1}\mathbf{a}_i'$ is the ith diagonal element of the hat matrix of the complete data model.

To prove (6.28) we make use of the results of the submodel analysis in the previous section. From the previous section we write $n_1 = n - 1$, $n_2 = 1, \mathbf{a}_1 = \mathbf{a}_{(i)}, \mathbf{a}_2 = \mathbf{a}_i$, the ith row of $\boldsymbol{a}$, $\mathbf{Y}_1 = \mathbf{Y}_{(i)}, Y_2 = y_i$, the ith value of $\mathbf{Y}, \tilde{\boldsymbol{\theta}} = \tilde{\boldsymbol{\theta}}_{(i)}, \tilde{\mathbf{R}}_1 = \mathbf{R}_{(i)}$ and $\mathbf{R}_2 = R_i$. Then equation (6.28) is obtained from equation (6.27).

With the above results, it can be easily verified that equation (6.26) can be written as

$$\hat{\boldsymbol{\theta}} - \hat{\boldsymbol{\theta}}_{(i)} = (\boldsymbol{a}'\boldsymbol{a})^{-1}\mathbf{a}_i' R_i/(1 - h_{ii}),$$

then Cook's distance measure D_i (see Section 6.4) can be written as

$$\begin{aligned} D_i &= (\hat{\boldsymbol{\theta}} - \hat{\boldsymbol{\theta}}_{(i)})'\boldsymbol{a}'\boldsymbol{a}(\hat{\boldsymbol{\theta}} - \hat{\boldsymbol{\theta}}_{(i)})/(ks^2) \\ &= \frac{R_i \mathbf{a}_i (\boldsymbol{a}'\boldsymbol{a})^{-1}(\boldsymbol{a}'\boldsymbol{a})(\boldsymbol{a}'\boldsymbol{a})^{-1}\mathbf{a}_i' R_i}{(1 - h_{ii})^2 ks^2} \\ &= \frac{R_i^2 h_{ii}}{(1 - h_{ii})^2 ks^2} \\ &= \left(\frac{R_i}{s\sqrt{(1 - h_{ii})}}\right)^2 \times \frac{h_{ii}}{1 - h_{ii}} \times \frac{1}{k} \\ &= t_i^2 w_i/k, \end{aligned}$$

where t_i is the ith standardized residual (see Section 6.2) and $w_i = h_{ii}/(1 - h_{ii})$. This result proves equation (6.10) that

$$D_i = t_i^2 w_i/k.$$

6.10* Likelihood ratio test of no outliers in the regression model

In Section 6.2 we mentioned that the likelihood ratio statistic for testing that there is no outlier in the regression model is a function of $d_i = t_i^2/(n - k)$, where t_i is the ith studentized residual. This is a special case of a standard result in least squares theory (e.g. Seber, 1977, p. 100): the likelihood ratio test for comparing the fit of two linear models, one a restricted case of the other, is a function of the ratio of the respective residual sums of squares. In the present case we have the basic model (1.1) with corresponding residual sum of squares $(n - k)s^2 = \mathbf{y}'(\boldsymbol{I} - \boldsymbol{H})\mathbf{y}$. We compare this with the fit of the alternative, augmented model,

$$E(y) = \boldsymbol{a}_* \boldsymbol{\theta}_*,$$

where $\boldsymbol{a}_* = (\boldsymbol{a}, \mathbf{e})$, for $\mathbf{e}$ an $n \times 1$ vector of zeros with a single 1 in the ith position, and $\boldsymbol{\theta}'_* = (\boldsymbol{\theta}', \delta)$. The residual sum of squares from this model is $\mathbf{y}'(\boldsymbol{I} - \boldsymbol{H}_*)\mathbf{y}$ where $\boldsymbol{H}_* = a_*(a'_*a_*)^{-1}a_*$. Now we can partition a'_*a_* and $(a'_*a_*)^{-1}$ in the same way as follows:

$$\boldsymbol{a}'_*\boldsymbol{a}_* = \begin{bmatrix} \boldsymbol{a}'\boldsymbol{a}, & \boldsymbol{a}'\mathbf{e} \\ \mathbf{e}'\boldsymbol{a}, & \mathbf{e}'\mathbf{e} \end{bmatrix},$$

$$(\boldsymbol{a}'_*\boldsymbol{a}_*)^{-1} = \begin{bmatrix} \boldsymbol{b}_{11}, & \boldsymbol{b}_{12} \\ \boldsymbol{b}_{21}, & \boldsymbol{b}_{22} \end{bmatrix}.$$

From standard results on partitioned matrices and their inverses we have

$$\begin{aligned} b_{22} &= (\mathbf{e}'\mathbf{e} - \mathbf{e}'\boldsymbol{a}'(\boldsymbol{a}'\boldsymbol{a})^{-1}\boldsymbol{a}'\mathbf{e})^{-1} = (1 - h_{ii})^{-1}, \\ b_{12} &= -(1 - h_{ii})^{-1}(\boldsymbol{a}'\boldsymbol{a})^{-1}\boldsymbol{a}'\mathbf{e}, \end{aligned}$$

and

$$b_{11} = (\boldsymbol{a}'\boldsymbol{a})^{-1} - (1 - h_{ii})^{-1}(\boldsymbol{a}'\boldsymbol{a})^{-1}\boldsymbol{a}'\mathbf{e}\mathbf{e}'\boldsymbol{a}(\boldsymbol{a}'\boldsymbol{a})^{-1}.$$

Thus

$$\begin{aligned} \boldsymbol{H}_* &= \boldsymbol{a}'(\boldsymbol{a}'\boldsymbol{a})^{-1}\boldsymbol{a}' + (1 - h_{ii})^{-1} \\ &\quad \cdot \{\boldsymbol{a}(\boldsymbol{a}'\boldsymbol{a})^{-1}\boldsymbol{a}'\mathbf{e}\mathbf{e}'\boldsymbol{a}(\boldsymbol{a}'\boldsymbol{a})^{-1}\boldsymbol{a}' - \boldsymbol{a}'(\boldsymbol{a}'\boldsymbol{a})^{-1}\boldsymbol{a}'\mathbf{e}\mathbf{e}' \\ &\quad - \mathbf{e}\mathbf{e}'\boldsymbol{a}(\boldsymbol{a}'\boldsymbol{a})^{-1}\boldsymbol{a}' + \mathbf{e}\mathbf{e}'\} \\ &= \boldsymbol{a}(\boldsymbol{a}'\boldsymbol{a})^{-1}\boldsymbol{a}' + (1 - h_{ii})^{-1}(\boldsymbol{I} - \boldsymbol{H})\mathbf{e}\mathbf{e}'(\boldsymbol{I} - \boldsymbol{H}), \end{aligned}$$

and

$$\begin{aligned} \mathbf{y}'(\boldsymbol{I} - \boldsymbol{H}_*)\mathbf{y} &= (n - k)s^2 + (1 - h_{ii})^{-1}\mathbf{R}'\mathbf{e}\mathbf{e}'\mathbf{R} \\ &= (n - k)s^2 + r_i^2. \end{aligned}$$

The ratio of residual sums of squares is then

$$\begin{aligned} \frac{\mathbf{y}'(\boldsymbol{I} - \boldsymbol{H}_*)\mathbf{y}}{\mathbf{y}'(\boldsymbol{I} - \boldsymbol{H})\mathbf{y}} &= \frac{(n - k)s^2 + r_i^2}{(n - k)s^2} \\ &= 1 + t_i^2/(n - k). \end{aligned}$$

6.11* The Bonferroni inequality and the distribution of $d_i = t_i^2/(n - k)$

6.11.1 First-order Bonferroni inequality

Let $(v_1, v_2, \ldots, v_n)$ be n random variables and further let $v = \max(v_1, v_2, \ldots, v_n)$. Also let

$$P(V) = \Pr(v \geq V),$$

$$P_1(V) = \sum_{j=1}^{n} \Pr(v_j \geq V),$$

and

$$P_2(V) = \sum_{j=2}^{n} \sum_{i=1}^{j-1} \Pr(v_j \geq V, v_i \geq V).$$

Then from the probability law for the joint occurrence of n events we can derive that

$$P_1(V) - P_2(V) \leq P(V) \leq P_1(V).$$

This is the first-order Bonferroni inequality. If V_0 is the exact upper $100\alpha\%$ point of the distribution of v, then $P(V_0) = \alpha$, and the quantity V_1 defined by $P_1(V_1) = \alpha$ is an upper bound of V_0.

6.11.2 Distribution of d_i

In Section 6.2 we mentioned that $d_i = t_i^2/(n-k)$, where t_i is the ith standardized residual, is distributed as beta $(\frac{1}{2}, (n-k-1)/2)$. We now give proof for this.

Theorem 6.1

$$d_i \sim \text{beta}(\tfrac{1}{2}, (n-k-1)/2).$$

[Before proving the above theorem we quote a standard result:
Let $X_{v_1}^2$ and $X_{v_2}^2$ be independently distributed random variables and $X_{v_i}^2 \sim \chi^2(v_i)$ $(i = 1, 2)$. Then

$$\frac{X_{v_1}^2}{X_{v_1}^2 + X_{v_2}^2} \sim \text{beta}\left(\frac{v_1}{2}, \frac{v_2}{2}\right)]$$

Proof
$\mathbf{R} \sim N(0, (I-H)\sigma^2)$ and the marginal distribution of $R_i \sim N(0, (1-h_{ii})\sigma^2)$. Thus,

$$\frac{R_i}{\sigma\sqrt{(1-h_{ii})}} \sim N(0, 1) \quad \text{and} \quad \frac{R_i^2}{\sigma^2(1-h_{ii})} \sim \chi^2(1).$$

Now, since S^2 is the residual sum of squares of the full model and $S_{(i)}^2$ is the residual sum of squares of the submodel after removing the ith point from the data, then

$$S^2/\sigma^2 \sim \chi^2(n-k),$$
$$S_{(i)}^2/\sigma^2 \sim \chi^2(n-k-1).$$

Thus from equation (6.26) it follows that

$$\frac{S^2}{\sigma^2} = \frac{R_i^2}{\sigma^2(1-h_{ii})} + \frac{S_{(i)}^2}{\sigma^2}.$$

Now using Theorem 6.2 below and denoting $\mathbf{R}_1 = \mathbf{R}_{(i)}$ and $R_2 = R_i$, we see that R_i and $\mathbf{R}_{(i)}$ are independently distributed. This also implies that R_i^2 and $\mathbf{R}_{(i)}'\mathbf{R}_{(i)} = S_{(i)}^2$ are independently distributed. If we now denote $R_i^2/(\sigma^2(1-h_{ii}))$ as X_1^2 and $S_{(i)}^2/\sigma^2$ as X_{n-k-1}^2, then from the standard result it follows that

$$\begin{aligned} d_i &= \frac{t_i^2}{n-k} \\ &= \frac{R_i^2/(\sigma^2(1-h_{ii}))}{S^2/\sigma^2} \\ &= \frac{R_i^2/(\sigma^2(1-h_{ii}))}{R_i^2/(\sigma^2(1-h_{ii})) + S_{(i)}^2/\sigma^2} \\ &= \frac{X_1^2}{X_1^2 + X_{n-k-1}^2} \end{aligned}$$

is distributed as beta $(\frac{1}{2}, (n-k-1)/2)$.

Theorem 6.2
The residuals $\tilde{\mathbf{R}}_1$ and $\mathbf{R}_2$ of the submodel analysis in Section 6.8 are independently distributed.

Proof
Using results from Section 6.8, the residuals $\mathbf{R}$ of the full model can be written as

$$\mathbf{R} = \begin{bmatrix} \mathbf{R}_1 \\ \mathbf{R}_2 \end{bmatrix} = \begin{bmatrix} M_{11} & M_{12} \\ M_{21} & M_{22} \end{bmatrix} \begin{bmatrix} \mathbf{Y}_1 \\ \mathbf{Y}_2 \end{bmatrix},$$

from where it can be seen that

$$\mathbf{R}_2 = M_{21}\mathbf{Y}_1 + M_{22}\mathbf{Y}_2.$$

Again from Section 6.8, the residuals $\tilde{\mathbf{R}}_1$ can be written as

$$\tilde{\mathbf{R}}_1 = (I - \boldsymbol{a}_1 A_1^{-1} \boldsymbol{a}_1')Y_1 = \tilde{\boldsymbol{M}}_1 \mathbf{Y}_1.$$

Thus,

$$\begin{aligned} \operatorname{cov}(\tilde{\mathbf{R}}_1, \mathbf{R}_2) &= \tilde{\boldsymbol{M}}_1 \boldsymbol{M}_{21}' \operatorname{var}(\mathbf{Y}_1) \\ &= \tilde{\boldsymbol{M}}_1 \boldsymbol{M}_{21}' \sigma^2. \end{aligned}$$

Now, using (6.20) we see that

$$\begin{aligned} \tilde{\boldsymbol{M}}_1 \boldsymbol{M}_{21}' &= (\boldsymbol{I} - \boldsymbol{a}_1 A_1^{-1} \boldsymbol{a}_1')(-\boldsymbol{a}_1 A_1^{-1} \boldsymbol{a}_2' (\boldsymbol{I} + \boldsymbol{C})^{-1}) \\ &= 0. \end{aligned}$$

Hence $\operatorname{cov}(\tilde{\mathbf{R}}_1, \mathbf{R}_2) = 0$ and $\tilde{\mathbf{R}}_1$ and $\mathbf{R}_2$ are independently distributed.

Excercise 6

1. Use the techniques set out in this chapter to explore the data sets used in the Appendix.

References

Andrews, D. F. (1974) A robust method for multiple linear regression. *Technometrics*, **16**, 523–531.

Andrews, D. F. and Pregibon, D. (1978) Finding the outliers that matter. *J. Roy. Statist. Soc.* B, **40**, 85–93.

Atkinson, A. C. (1981) Two graphical displays for outlying and influential observations in regression. *Biometrika*, **68**, 13–20.

Barnett, V. and Lewis, T. (1978) *Outliers in Statistical Data*. Wiley, New York.

Beckman, R. J. and Cook, R. D. (1983) Outlier...s (with Discussion). *Technometrics*, **25**, 119–163.

Belsley, D. A., Kuh, E. and Welsch, R. E. (1980) *Regression Diagnostics: Identifying Influential Data and Source Collinearity*. Wiley, New York.

Bingham, C. (1977) Some identities useful in the analysis of residuals from linear regression. Technical Report No. 300, School of Statistics, University of Minnesota.

Brownlee, K. A. (1965) *Statistical Theory and Methodology in Science and Engineering*, 2nd edn, Wiley, New York.

Cook R. D. (1977) Detection of influential observations in linear regression. *Technometrics*, **19**, 15–18.

Cook, R. D. (1979) Influential observations in linear regression. *J. Amer. Statist. Assoc.*, **74**, 169–174.

Cook, R. D. and Weisberg, S. (1980) Characterization of an influence function for detecting influential cases in regression. *Technometrics*, **22**, 495–508.

Cook, R. D. and Weisberg, S. (1982) *Residuals and Influence in Regression*, Chapman and Hall, London.

Cran, G. W., Martin, K. J. and Thomas, G. E. (1977) Remark ASR19 and Algorithm AS109: a remark on Algorithm AS63: the incomplete integral, AS64: inverse of the incomplete beta function ratio. *Appl. Statist.*, **26**, 111–114.

Daniel, C. and Wood, F. S. (1971) *Fitting Equations to Data.* Wiley, New York.

Dempster, A. D. and Gasko-Green, M. (1981) New tools for residual analysis. *Ann. Statist.* **9**, 945–959.

Draper, N. R. and John, J. A. (1981) Influential observations and outliers in regression. *Technometrics*, **23**, 21–26.

Draper, N. R. and Stoneman, D. M. (1966). Testing for the inclusion of variables in linear regression by a randomization technique. *Technometrics*, **8**, 695–699.

Gentleman, J. F. and Wilk, M. B. (1975) Detecting outliers, II. Supplementing the direct analysis of residuals. *Biometrics*, **31**, 387–410.

Hoaglin, D. C. and Welsch, R. E. (1978) The hat matrix in regression and ANOVA. *Amer. Statist.*, **32**, 17–22.

Longley, J. W. (1967) An appraisal of least squares program for the electronic computer from the point of view of the user. *J. Amer. Statist. Assoc.*, **62**, 819–841.

Lund, R. E. (1975) Tables for an approximate test for outliers in linear models. *Technometrics*, **17**, 473–476.

Mickey, M. R., Dunn, O. J., and Clark, V. (1967) Note on use of stepwise regression in detecting outliers. *Comput. Biomed. Res.*, **1**, 105–111.

Paul, S. R. (1983) Sequential detection of unusual points in regression. *The Statistician*, **32**, 105–112.

Paul, S. R. (1985a) Critical values of 'maximum studentized residual' statistics in multiple linear regression. *Biom. J.*, **26**, 1–5.

Prescott, P. (1975) An approximate test for outliers in linear models. *Technometrics*, **17**, 127–132.

Rao, C. R. (1965) *Linear Statistical Influence and Its Applications.* Wiley, New York.

Seber, G. A. F. (1977) *Linear Regression Analysis.* Wiley, New York.

Srikantan, K. S. (1961) Testing for the single outlier in the regression model. *Sankhya* A, **23**, 251–260.

Further reading

Paul, S. R. (1985b) A note on maximum likelihood ratio test of no outliers in regression models. *Biom. J.* (to appear).

Wetherill, G. B. (1981) *Intermediate Statistical Methods.* Chapman and Hall, London.

CHAPTER 7

Testing for transformations

7.1 Introduction

In Section 1.8 we outlined briefly the point that difficulties can arise from using multiple regression analysis due to the failure of assumptions made in the ordinary least squares (OLS) analysis. The assumptions can fail because of lack of normality, homoscedasticity, independence, or because the underlying model is not linear in the unknown parameters. In this chapter we examine the use of transformations, either of the response variable or of the explanatory variables, to attempt to satisfy the assumptions of an OLS analysis.

For transformations of the response variable, the Box–Cox transformation has a number of advantages. It is one of the simpler tests which has withstood criticism over the years, and alternatives seem to be in a developmental stage. Basically, the Box–Cox transformation seeks for a simple power transformation of the response variable, which simultaneously leads to a linear model, normally distributed errors and homogeneous variances. The procedure results in a strong indication of the transformations to try.

A number of techniques have been suggested for examining transformations of the explanatory variables and these are briefly reviewed in Section 7.3. One of the tests we found to be very useful was suggested by Ramsey (1969). This is a test for omitted variables or transformations of the explanatory variables. Essentially, the test looks for patterns in the residuals.

This chapter contains some technical discussion of the procedures, together with an examination of problems of implementing them in a general statistical package. A summary of a suggested methodology for implementation is given under each major heading.

7.2 The Box–Cox transformation

In this section we discuss transformations of the response variable, and outline the Box–Cox transformation.

It is assumed that the response variables Y_i are independent whether or not they are transformed; since each observed Y_i is transformed in the same way, this assumption does not enter into the test. It is convenient to restrict the transformations being considered to a family indexed, say, by λ which may be a vector; so Y_i is transformed to $Y_i^{(\lambda)}$ and the column vector $\mathbf{Y}$ to $\mathbf{Y}^{(\lambda)}$, where $\mathbf{Y}^{(\lambda)\prime} = (Y_1^{(\lambda)}, Y_2^{(\lambda)}, \ldots, Y_n^{(\lambda)})$. Box and Cox suggest trying a scalar λ and

$$Y^{(\lambda)} = \begin{cases} (Y^{\lambda} - 1)/\lambda & \text{when } \lambda \neq 0, \\ \log Y, & \text{when } \lambda = 0, \end{cases} \tag{7.1}$$

and tentatively suppose there is one value of λ for which the following assumptions hold:

(1) $E(Y^{(\lambda)})$ may be expressed as a linear combination $\boldsymbol{a\theta}$ of the explanatory variables (where the first column of $\boldsymbol{a}$ consists entirely of 1's as in (1.13));
(2) the variance of $Y_i^{(\lambda)}$ is constant;
(3) the $Y_i^{(\lambda)}$ are normally distributed.

Then the density of $Y_i^{(\lambda)}$ would be $N(\mathbf{a}_{(i)}\boldsymbol{\theta}, \boldsymbol{\sigma}^2)$, where $\mathbf{a}_{(i)}$ denotes the ith row of the matrix of explanatory variables. So the density of Y_i is

$$(2\pi\sigma^2)^{-1/2} \exp\left\{\frac{-(Y_i^{(\lambda)} - \mathbf{a}_{(i)}\boldsymbol{\theta})^2}{(2\sigma^2)}\right\}\left|\frac{\mathrm{d}\, Y_i^{(\lambda)}}{\mathrm{d}\, Y_i}\right|,$$

and the density of $\mathbf{Y}$, in terms of $\mathbf{Y}^{(\lambda)}$, is the product of such terms over i, say, $f(\mathbf{Y})$.

In order to estimate λ (and carry out tests), the method of maximum likelihood is used. Since the observations Y_i are known, the log-likelihood is $\log f(\mathbf{Y})$ as a function of $\lambda, \sigma^2, \boldsymbol{\theta}$, which is

$$L(\lambda, \sigma^2, \boldsymbol{\theta}) = -\frac{n}{2}\log(2\pi\sigma^2) - \frac{(\mathbf{Y}^{(\lambda)} - \boldsymbol{a\theta})'(\mathbf{Y}^{(\lambda)} - \boldsymbol{a\theta})}{2\sigma^2} + \log J(\lambda; Y), \tag{7.2}$$

where

$$\log J(\lambda; \mathbf{Y}) = (\lambda - 1)\sum \log Y_i. \tag{7.3}$$

Considering λ as fixed we maximize first with respect to $\boldsymbol{\theta}$ and σ^2, giving the usual maximum likelihood estimates for regression

$$\hat{\boldsymbol{\theta}} = (\boldsymbol{a}'\boldsymbol{a})^{-1}\boldsymbol{a}'\mathbf{Y}^{(\lambda)}$$

and

$$\hat{\sigma}^2 = n^{-1}\mathbf{Y}^{(\lambda)'}(\boldsymbol{I} - \boldsymbol{H})\mathbf{Y}^{(\lambda)} = \hat{\sigma}^2(\lambda), \quad \text{say,} \tag{7.4}$$

where $\boldsymbol{H}$ is the hat matrix $\boldsymbol{a}(\boldsymbol{a}'\boldsymbol{a})^{-1}\boldsymbol{a}'$; see Section 6.3. After dropping a constant, this gives

$$L(\lambda, \hat{\sigma}^2, \hat{\theta}) = -\frac{n}{2}\log \hat{\sigma}^2(\lambda) + \log J(\lambda; \mathbf{Y}) = L_{\max}(\lambda), \quad \text{say.} \tag{7.5}$$

Then the optimum value of λ, called $\hat{\lambda}$, may be found by plotting $L_{\max}(\lambda)$ against λ by numerical computation for a grid of values of λ. The main work is in calculating $\hat{\sigma}^2(\lambda)$ for each λ. For a confidence interval at the $100(1-\alpha)\%$ level, asymptotic maximum likelihood theory suggests allowing values of λ for which

$$L_{\max}(\hat{\lambda}) - L_{\max}(\lambda) < \tfrac{1}{2}\chi_1^2(\alpha). \tag{7.6}$$

So if $\lambda = 1$ does not lie in this interval we conclude, at this level, that a transformation is needed and that we may select a convenient value in this interval, such as an integer or a half integer.

Box and Cox emphasized that the above assumptions may not be satisfied after doing such a transformation. The transformation represents a compromise and different values of λ may be needed to satisfy different assumptions. A further important point about the Box–Cox transformation is that the result depends upon the explanatory variables included in the model. It is best to carry through the procedure with several different sets of explanatory variables and it is also wise to try the procedure with and without outliers. For numerical illustrations, see Cook and Weisberg (1982).

Atkinson (1983) presents a useful approach to the Box–Cox transformation, using 'constructed variables'. Basically, the approach is as follows. A hypothesis value λ_0 is selected and then the model

$$E(\mathbf{Y}^{(\lambda_0)}) = \boldsymbol{a}\boldsymbol{\theta} + \phi w$$

is fitted, where

$$w = \left(\frac{\mathrm{d}\mathbf{Y}^{(\lambda)}}{\mathrm{d}\lambda}\right)\lambda_0$$

so that a test on the parameter ϕ tests the need for a transformation.

This approach can avoid the calculation of likelihoods, but if several λ's are to be used the transformd variables need adjusting for the Jacobians of the transformation. See Cook and Weisberg (1982) or Atkinson (1983) for details.

In the discussion of the original paper by Box and Cox, Beale raised the point that problems arise with this transformation if some of the Y_i are negative. One way round this, also suggested by Box and Cox, is to let λ have two components and transform

$$Y^{(\lambda)} = \begin{cases} \{(Y+\lambda_2)^{\lambda_1} - 1\}/\lambda_1, & \text{for } \lambda_1 \neq 0, \\ \log(Y+\lambda_2), & \text{for } \lambda_1 = 0, \end{cases} \tag{7.7}$$

and again estimate $\boldsymbol{\lambda}$ by maximum likelihood. The restriction is that $Y_i + \lambda_2$ be positive for each observation. Then equations (7.4) and (7.5) still hold and (7.3) is replaced by

$$\log J(\lambda; \mathbf{Y}) = (\lambda_1 - 1) \sum_{i=1}^{n} \log(Y_i + \lambda_2),$$

where λ_2 is restricted to the range $\lambda_2 > -\min(Y_i)$. Then the right-hand side λ_2 is restricted to the range $\lambda_2 > -\min(Y_i)$. Then the right-hand side of (7.6) becomes $\frac{1}{2}\chi_2^2(\alpha)$. One disadvantage of this method is that, although the resulting value of λ_2 will ensure $\lambda_2 + Y_i > 0$ for $i = 1, 2, \ldots, n$, it may not be positive for all possible observations Y. Another disadvantage is that the grid of λ values is two-dimensional which would greatly increase the computing time. This is already quite considerable for a one-dimensional search.

Another way to overcome the problem of negative Y_i is observations using some prior knowledge of the range of values of Y, such that this corresponds to some natural origin (e.g. some physical criterion such as absolute zero on the temperature scale, $-273\,°\mathrm{C}$). Then λ_2 would not be estimated by maximum likelihood and the right-hand side of (7.6) would again be $\frac{1}{2}\chi_1^2(\alpha)$. In any program, it is easy to allow for replacing Y in (7.1) by $-Y$ or by $\lambda_2 - Y$ for some λ_2 specified by the user.

7.2.1 Some computational points

Consider now the computation of (7.4). One way in which this can be reduced is by centring the variables. As in (1.18) $\boldsymbol{X}$ denotes the $n \times p$ matrix of centred variables. Letting $\mathbf{Y}_c^{(\lambda)}$ denote the centred version of $\mathbf{Y}^{(\lambda)}$, then the residual sum of squares is the same when regressing $\mathbf{Y}_c^{(\lambda)}$

on $\boldsymbol{X}$ as when regressing $\mathbf{Y}^{(\lambda)}$ on $\boldsymbol{a}$; so

$$\hat{\sigma}^2(\lambda) = \mathbf{Y}_{\mathrm{c}}^{(\lambda)\prime}(\boldsymbol{I} - \boldsymbol{H}_X)\mathbf{Y}_{\mathrm{c}}^{(\lambda)}/n, \tag{7.8}$$

where $\boldsymbol{H}_X = \boldsymbol{X}(\boldsymbol{X}'\boldsymbol{X})^{-1}\boldsymbol{X}'$. A suitable grid of values of λ can now be chosen, such as

$$\lambda = -3, -2, -1, -\tfrac{1}{2}, 0, \tfrac{1}{2}, 1, 2, 3.$$

If we assume that $(\boldsymbol{X}'\boldsymbol{X})^{-1}$ has been computed, and $\boldsymbol{X}$ is $n \times p$, then the procedure is to compute $(\boldsymbol{I} - \boldsymbol{H}_X)$ first followed by $\hat{\sigma}^2(\lambda)$ for each value of λ, by pre-multiplying by $\mathbf{Y}_{\mathrm{c}}^{(\lambda)\prime}$ and post-multiplying by $\mathbf{Y}_{\mathrm{c}}^{(\lambda)}$. This leads to

$$N_1 = p^2 n + pn^2 + 9(n^2 + n) \tag{7.9}$$

multiplications, since there are nine values of λ. A disadvantage of this method is that it means storing the matrix $(\boldsymbol{I} - \boldsymbol{H}_X)$, which is $n \times n$, and there may not be room for this.

A second method of calculating (7.4) is to calculate

$$\mathbf{z} = \boldsymbol{X}'\mathbf{Y}_{\mathrm{c}}^{(\lambda)} \tag{7.10}$$

first and then form $\hat{\sigma}^2(\lambda)$ by calculating

$$\hat{\sigma}^2(\lambda) = \mathbf{Y}_{\mathrm{c}}^{(\lambda)\prime}\mathbf{Y}_{\mathrm{c}}^{(\lambda)} - \mathbf{z}'(\boldsymbol{X}'\boldsymbol{X})^{-1}\mathbf{z}.$$

This leads to

$$N_2 = 9(np + p^2 + p + n)$$

multiplications.

To compare the two methods, let $n = 300$ and $p = 8$, then $N_1 =$ 1 551 900 and $N_2 = 24\,948$, so it is much more efficient to use the second method via $\mathbf{z}$. In addition, the first method would involve storing the 90 000 elements of $(\boldsymbol{I} - \boldsymbol{H}_X)$, whereas in contrast, the second method involves storing the eight elements of $\mathbf{z}$.

Since we calculate λ only at a discrete set of values then this makes it impossible to output the confidence interval: there may be only one or two values of λ within it so the values λ_{l} and λ_{u} (lower and upper limits of λ resulting from (7.6)) cannot be read off a smooth graph. One possibility is to interpolate, or else to quote the next value λ_{ll} below λ_{l} out of the discrete set, and the next value λ_{uu} above λ_{u}. Also $\hat{\lambda}$ is not known exactly, so the one λ_{m} giving the maximum out of the discrete set must be quoted. The effect of this also tends to widen the interval so the estimates are conservative, and the message printed out by a computer is not to consider values of $\lambda < \lambda_{\mathrm{ll}}$ or $> \lambda_{\mathrm{uu}}$.

It is really necessary to do some simulations to check the accuracy of the χ^2-level in (7.6), since this is only a first-order approximation. Also the overlap between this test (obtained by testing $L_{\max}(1) < L_{\max}(\hat{\lambda}) - \frac{1}{2}\chi_1^2(\alpha)$) and others needs to be investigated. Other tests which would be likely to overlap strongly are Ramsey's test for wrong functional form of one or more explanatory variables and tests for normality or heteroscedasticity. Also the overlap with tests for outliers is very strong – see, for instance, Atkinson (1982).

7.3 Testing for transformations of the explanatory variables

The Box–Cox transformation, described above, has proved a very useful tool for suggesting transformation of the response variable. Box and Tidwell (1962) suggested that a somewhat similar method be applied to explanatory variables, using a transformation of the same type (7.1). This leads to an iterative procedure and allows powers or logarithms of any or all of the explanatory variables to be entered into the model. A number of problems, however, arise when one tries to implement this, such as ill-conditioning, overfitting and failure of the iterations to converge. Also, it is necessary that the explanatory variables all be positive in order to take their logarithms. If they were not, it it not clear what to do. This method needs considerable development before it could be considered for routine use.

A test for wrong functional form of one or more of the explanatory variables, or for an omitted variable, was proposed by Ramsey (1969). The use of 'wrong functional form' here means that one of the explanatory variables should be replaced by a function of it which is in some sense smooth. Essentially the 'wrong functional form' or the 'omitted variable' is assumed to be approximated by a Taylor series in the powers of the fitted Y's. An approximate description of the test we suggest is that the residuals are regressed against the first three power powers of the fitted Y's. One modification of this is to allow for the residuals not being independent or having constant variance, another modification is to use BLUS (best linear unbiased scalar, see below) residuals instead of OLS ones to give greater power, while a third modification is to produce a simpler but equivalent test which avoids calculating the BLUS residuals.

Some of the details of implementing Ramsey's test are rather complex, so the procedure we adopt is as follows. Sections 7.4 and 7.5 to follow contain the necessary algebra for Ramsey's test and those only interested in using the results should omit these sections. Section 7.6 contains a summary of how to implement Ramsey's test.

7.4* Ramsey's tests

Ramsey (1969) advanced several tests for studying the possibility that a wrong functional relationship was being used in a regression analysis. He considered three cases of which the first and second are most relevant here. They were

(1) that an important explanatory variable had been omitted; and
(2) that one of them x_j, should be replaced by $g(x_j)$, where g is some function.

In both cases, he suggested, an examination of the residuals would provide a clue.

As he pointed out, basing himself on (1.1), if

$$\boldsymbol{Y} = \boldsymbol{a\theta} + \boldsymbol{R},$$

where the covariance matrix of $\boldsymbol{R}$ is $\boldsymbol{I}\sigma^2$, then once $\boldsymbol{\theta}$ has been estimated the estimated residuals $\hat{R}$ have a covariance matrix of $(\boldsymbol{I} - \boldsymbol{H})\sigma^2$ as in (1.7). However, if k parameters are fitted when estimating $\boldsymbol{\theta}$, the rank of $(\boldsymbol{I} - \boldsymbol{H})$ is only $(n - k)$. Theil (1965) set about finding a reduced set of $(n - k)$ residuals $\tilde{R}$ which would have a covariance matrix $\boldsymbol{I}^2$. He proposed the relationship

$$\tilde{\mathbf{R}} = \boldsymbol{A}'\mathbf{Y}, \tag{7.11}$$

where $\boldsymbol{A}$ is an $n \times (n - k)$ matrix such that $\boldsymbol{A}'\boldsymbol{a} = 0$. Accordingly $\boldsymbol{A}'\mathbf{Y} = \boldsymbol{A}'\mathbf{R} = \tilde{\mathbf{R}}$. That makes the covariance matrix of $\tilde{\mathbf{R}}$ to be $\boldsymbol{A}'\boldsymbol{A}\sigma^2$ and adds the constraint, $\boldsymbol{A}'\boldsymbol{A} = \boldsymbol{I}$. The values of $\tilde{\mathbf{R}}$ are BLUS residuals.

To find $\boldsymbol{A}$ it is necessary to select $(n - k)$ data and their residuals from the n available. Ordering so that the first k are to be discarded means selecting $\boldsymbol{J}'\mathbf{R}$, where $\boldsymbol{J}$ has n rows, of which the first k are void and the rest hold an identity matrix. In short, the task is to minimize the sum of squared differences between $\boldsymbol{A}'\mathbf{R}$ and $\boldsymbol{J}'\mathbf{R}$. Theil showed that the sum of squares comes to the trace of $(\boldsymbol{A}' - \boldsymbol{J})(\boldsymbol{A}' - \boldsymbol{J}')$, the minimization of which leads to two further constraints, i.e.

$$\boldsymbol{J}'\boldsymbol{A} = \boldsymbol{A}'\boldsymbol{J} \quad \text{and} \quad \boldsymbol{A}\boldsymbol{A}' = \boldsymbol{I} - \boldsymbol{H}. \tag{7.12}$$

He showed how to find an explicit solution for $\boldsymbol{A}$.

From this point Ramsey (1969) suggested several tests but later work (Ramsey and Gilbert, 1972) indicated that the best was the one called RESET. In essence that involved the regression of $\tilde{\mathbf{R}}$ on

$$\boldsymbol{B} = \boldsymbol{A}'\boldsymbol{Q}, \tag{7.13}$$

where

$$\boldsymbol{Q} = (\hat{\mathbf{Y}}^{(2)} \quad \hat{\mathbf{Y}}^{(3)} \quad \hat{\mathbf{Y}}^{(4)}), \tag{7.14}$$

i.e. alternative or additional explanatory variables. Ramsey comments that in his experience three terms are enough.

Later, Ramsey and Schmidt (1976) were to show that Ramsey's original formulation is nearly equivalent to regressing $\mathbf{Y}$ on $\boldsymbol{X}$ to obtain $\mathbf{Y}$ and $\hat{\mathbf{R}}$, and then regressing $\hat{\mathbf{R}}$ on $(\boldsymbol{I}-\boldsymbol{H})\boldsymbol{Q}$ to apply the usual F-test to examine the hypothesis that in the latter regression analysis all coefficients were zero. It appears then that there is no need to calculate the BLUS residuals. We will now show that the two approaches are in fact equivalent.

The BLUS version of the test is given by (7.13). Suppose instead that the OLS residuals $\mathbf{R}$ were regressed on $(\boldsymbol{I}-\boldsymbol{H})\boldsymbol{Q}$. The test statistic for a significant slope is then

$$F_1 = \{[\mathbf{R}'\boldsymbol{H}_{MQ}\mathbf{R}]/[\mathbf{R}(\boldsymbol{I}-\boldsymbol{H}_{MQ})\mathbf{R}]\}\{(n-k-3)/3\}, \tag{7.15}$$

where $\boldsymbol{H}_{MQ} = (\boldsymbol{I}-\boldsymbol{H})\boldsymbol{Q}\{\boldsymbol{Q}'(\boldsymbol{I}-\boldsymbol{H})\boldsymbol{Q}\}^{-1}\boldsymbol{Q}'(\boldsymbol{I}-\boldsymbol{H})$ since $\boldsymbol{I}-\boldsymbol{H}$ is symmetric and idempotent. The corresponding statistic for the BLUS test is

$$F_2 = \{[\tilde{\mathbf{R}}'\boldsymbol{H}_B\mathbf{R}]/[\tilde{\mathbf{R}}'(\boldsymbol{I}-\boldsymbol{H}_B)\tilde{\mathbf{R}}]\}\{(n-k-3)/3\}, \tag{7.16}$$

where $\boldsymbol{H}_B = \boldsymbol{B}(\boldsymbol{B}'\boldsymbol{B})^{-1}\boldsymbol{B}'$. But (7.13) implies $\boldsymbol{B}'\boldsymbol{B} = \boldsymbol{Q}'\boldsymbol{A}\boldsymbol{A}'\boldsymbol{Q} = \boldsymbol{Q}'(\boldsymbol{I}-\boldsymbol{H})\boldsymbol{Q}$ by (7.12), so that (with (7.11) and (7.12) again)

$$\begin{aligned}\tilde{\mathbf{R}}'\boldsymbol{H}_B\tilde{\mathbf{R}} &= \mathbf{Y}'\boldsymbol{A}\boldsymbol{A}'\boldsymbol{Q}(\boldsymbol{Q}'(\boldsymbol{I}-\boldsymbol{H})\boldsymbol{Q})^{-1}\boldsymbol{Q}'\boldsymbol{A}\boldsymbol{A}'\mathbf{Y}\\ &= \mathbf{Y}'(\boldsymbol{I}-\boldsymbol{H})\boldsymbol{Q}(\boldsymbol{Q}'(\boldsymbol{I}-\boldsymbol{H})\boldsymbol{Q})^{-1}\boldsymbol{Q}'(\boldsymbol{I}-\boldsymbol{H})\mathbf{Y}.\end{aligned} \tag{7.17}$$

But

$$\begin{aligned}(\boldsymbol{I}-\boldsymbol{H})\mathbf{Y} &= (\boldsymbol{I}-\boldsymbol{H})(\boldsymbol{I}-\boldsymbol{H})\mathbf{Y} \quad \text{since } (\boldsymbol{I}-\boldsymbol{H}) \text{ is idempotent,}\\ &= (\boldsymbol{I}-\boldsymbol{H})\mathbf{R} \quad \text{from (1.6).}\end{aligned}$$

Hence (7.17) implies

$$\tilde{\mathbf{R}}'\boldsymbol{H}_B\tilde{\mathbf{R}} = \mathbf{R}'\boldsymbol{H}_{MQ}\mathbf{R}. \tag{7.18}$$

But

$$\mathbf{R}'\mathbf{R} = \mathbf{Y}'\boldsymbol{A}\boldsymbol{A}'\mathbf{Y} = \mathbf{Y}'(\boldsymbol{I}-\boldsymbol{H})\mathbf{Y} = \mathbf{Y}'(\boldsymbol{I}=\boldsymbol{H})'(\boldsymbol{I}-\boldsymbol{H})\mathbf{Y} = \mathbf{R}'\mathbf{R}.$$

With (7.18) this gives

$$\tilde{\mathbf{R}}'(\boldsymbol{I}-\boldsymbol{H}_B)\mathbf{R} = \mathbf{R}'\mathbf{R} - \mathbf{R}'\boldsymbol{H}_{MQ}\mathbf{R} = \mathbf{R}'(\boldsymbol{I}-\boldsymbol{H}_{MQ})\mathbf{R}.$$

Therefore

$$F_1 = F_2, \tag{7.19}$$

where they are distributed as $F(3, n-k-3)$ under the null hypothesis from (7.15) since the residuals are then independent of the fitted values. Thus it is not necessary to calculate the BLUS residuals since (7.15) can be used directly. This is not the same as the test based on OLS residuals investigated by Ramsey and Gilbert (1972), who found that the BLUS residuals' test has more power. They considered the more direct regression of the OLS residuals on $\boldsymbol{Q}$ (as defined above: N.B. this $\boldsymbol{Q}$ defined at (7.13) is as used by Ramsey and Schmidt (1976), but not by Ramsey (1969) who uses it for the matrix $\boldsymbol{B}$ as defined in (7.13) above) rather than on $(\boldsymbol{I}-\boldsymbol{H})\boldsymbol{Q}$. With this version of the test, problems such as which residuals to omit for computing BLUS residuals do not arise.

Then Thursby and Schmidt (1977) showed by more simulation studies that using powers of the explanatory variables (second, third and fourth powers) gives better power results then using powers of the fitted Y's. Thus the $\boldsymbol{Q}$ matrix referred to above should, this suggests, be taken as the one with these powers of explanatory variables. But if there are a lot of explanatory variables and only a limited number of observations this could lead to overfitting. In any case, severe multicollinearity could well arise. For a routine test, especially with response surfaces, it is probably best to use only powers of $\hat{Y}$. This has some similarities with the Box–Cox test, and simulations are really needed to check on how much they overlap.

Ramsey and Schmidt (1976) defines BLUS Variant A of Ramsey's test as regressing $\tilde{\mathbf{R}}$ on $\boldsymbol{B} = \boldsymbol{A}'\boldsymbol{Q}$, where $\boldsymbol{Q}$ is as defined by (7.14) or another observable matrix such as powers of explanantory variables. They also define Variant B as regressing $\tilde{\mathbf{R}}$ on $\boldsymbol{B} = [\mathbf{1}, \boldsymbol{A}'\boldsymbol{Q}]$ which, as they say, 'is the variant of the BLUS test actually described in' Ramsey (1969) and Ramsey and Gilbert (1972). They then describe the above argument to show that Variant A is equivalent to regressing $\mathbf{R}$ on $(\boldsymbol{I}-\boldsymbol{H})\boldsymbol{Q}$ and show by simulations that these two variants give approximately the same results. These simulations are, however, unnecessary and arise out of a confusion by Ramsey (1969). In the latter paper equation (35′) is derived for when $\boldsymbol{X}$ has a column of 1's and equation (36) for when regression is through the origin.

Consider first the case where a constant term is fitted as in (1.13). Then Ramsey's equation (35′) leads to (38) or (40) with $\alpha_0 = 0$: this is (7.13) above or Ramsey and Schmidt's Variant A. Then (7.19) means this is exactly equivalent to their OLS version, to use (7.15) as the test statistic. Now assume the matrix $\boldsymbol{a}$ of explanatory variables does not contain a column of 1's, so the regression is through the origin. Then Ramsey's (1969) equation (36) does not lead to (38) or (40), for the first term in (38) should be $\alpha_0 \boldsymbol{A}'\mathbf{1}$ rather than $\alpha_0 \mathbf{1}$. Thus we should regress $\tilde{\mathbf{R}}$ on $\boldsymbol{B} = [\boldsymbol{A}'\mathbf{1}, \boldsymbol{A}'\boldsymbol{Q}]$ instead of on $\boldsymbol{B} = [\mathbf{1}, \boldsymbol{A}'\boldsymbol{Q}]$. Now if $\boldsymbol{Q}$ is replaced by $(\mathbf{1}, \boldsymbol{Q})$ the same argument applies. So, for regression through the origin, use (7.15) as the test statistic but with

$$\boldsymbol{Q} = (\mathbf{1}, \hat{\mathbf{Y}}^{(2)}, \hat{\mathbf{Y}}^{(3)}, \hat{\mathbf{Y}}^{(4)}), \tag{7.20}$$

and with the degrees of freedom adjusted to $(4, n-p-4)$. Here p denotes the number of explanatory variables.

7.5* Implementing the test

This section concerns the details of the computations and consists mainly of showing how to minimize the amount of calculation. Assume regression is not through the origin and write the sum of squared residuals as S_1,

$$S_1 = \mathbf{R}'\mathbf{R}$$

and write S_2 as the regression sum of squares for regressing $\mathbf{R}$ on $(I - \boldsymbol{H})\boldsymbol{Q}$,

$$S_2 = \mathbf{R}'\boldsymbol{H}_{MQ}\mathbf{R}. \tag{7.21}$$

Then (7.15) may be written in the form

$$F_1 = \frac{(n-k-3)S_2}{3(S_1 - S_2)}. \tag{7.22}$$

Since S_1 will already have been worked out in the program when Ramsey's test is called, being the residual or error sum of squares, it remains only to evaluate S_2.

First consider the effect of centring. The matrix $\boldsymbol{a}$ becomes

$$\boldsymbol{a} = (\mathbf{1}\ \boldsymbol{X}), \tag{7.23}$$

where $\boldsymbol{X}$ is the matrix of centred explanatory variables. From (6.9) we find that

$$\boldsymbol{H} = \boldsymbol{H}_X + n^{-1}\mathbf{1}\mathbf{1}', \tag{7.24}$$

where $\boldsymbol{H}_X$ is given in (7.8). Therefore,we define

$$\boldsymbol{C} = (\boldsymbol{I} - \boldsymbol{H})\boldsymbol{Q} = (\boldsymbol{I} - \boldsymbol{H}_X)\boldsymbol{Q} - \mathbf{1}\bar{\mathbf{q}}, \tag{7.25}$$

where

$$\bar{\mathbf{q}} = n^{-1}\mathbf{1}'\boldsymbol{Q}$$

is the row vector of the means of columns of $\boldsymbol{Q}$.

Now when the dependent variable has a constant value the residuals are zero, since a constant is being fitted; so (1.6) implies

$$(\boldsymbol{I} - \boldsymbol{H})\mathbf{1} = \mathbf{0}. \tag{7.26}$$

so (7.24) implies

$$(\boldsymbol{I} - \boldsymbol{H}_X)\mathbf{1} = \mathbf{1}. \tag{7.27}$$

Therefore (7.25) yields

$$\boldsymbol{C} = (\boldsymbol{I} - \boldsymbol{H}_X)(\boldsymbol{Q} - \mathbf{1}\bar{\mathbf{q}}). \tag{7.28}$$

So we may subtract the column means from $\boldsymbol{Q}$ first to form $\tilde{\boldsymbol{Q}}$,

$$\tilde{\boldsymbol{Q}} = \boldsymbol{Q} - \mathbf{1}\bar{\mathbf{q}} \tag{7.29}$$

and then form

$$\boldsymbol{C} = (\boldsymbol{I} - \boldsymbol{H}_X)\tilde{\boldsymbol{Q}}. \tag{7.30}$$

Since $(\boldsymbol{I} - \boldsymbol{H})$ is symmetric and idempotent, the definitions of $\boldsymbol{C}$ and $\boldsymbol{H}_{MQ}$ in (7.25) and (7.15) imply

$$\boldsymbol{H}_{MQ} = \boldsymbol{C}(\boldsymbol{C}'\boldsymbol{C})^{-1}\boldsymbol{C}'.$$

Then (7.21) gives

$$S_2 = \mathbf{R}'\boldsymbol{C}(\boldsymbol{C}'\boldsymbol{C})^{-1}\boldsymbol{C}'\mathbf{R}. \tag{7.31}$$

Now (7.30) implies

$$\boldsymbol{C}'\boldsymbol{C} = \tilde{\boldsymbol{Q}}'(\boldsymbol{I} - \boldsymbol{H}_X)\tilde{\boldsymbol{Q}},$$

since $(\boldsymbol{I} - \boldsymbol{H}_X)$ is symmetric and idempotent; so the definition of $\boldsymbol{H}_X$ at (7.8) gives

$$\boldsymbol{C}'\boldsymbol{C} = \tilde{\boldsymbol{Q}}'\tilde{\boldsymbol{Q}} - (\tilde{\boldsymbol{Q}}'\boldsymbol{X})(\boldsymbol{X}'\boldsymbol{X})^{-1}(\tilde{\boldsymbol{Q}}'\boldsymbol{X})'. \tag{7.32}$$

As explained around (7.10) it is quicker to form $\tilde{\boldsymbol{Q}}'\boldsymbol{X}$ first, then multiply $(\boldsymbol{X}'\boldsymbol{X})^{-1}$ to the left and right to form the second term in (7.32), then form $\boldsymbol{Q}'\boldsymbol{Q}$ and subtract the results.

It remains, for (7.31), to determine $\boldsymbol{C}'\mathbf{R}$. Since $\mathbf{R} = (\boldsymbol{I} - \boldsymbol{H})\mathbf{Y}$, (7.24) implies

$$\mathbf{R} = (\boldsymbol{I} - \boldsymbol{H}_X)\mathbf{Y} - \mathbf{1}\bar{\mathbf{Y}}. \tag{7.33}$$

Since $(\boldsymbol{I} - \boldsymbol{H}_X)$ is symmetric and idempotent (7.27) implies $(\boldsymbol{I} - \boldsymbol{H}_X)\mathbf{R} = \mathbf{R}$, since both terms on the right of (7.33) remain the same when premultiplied by $(\boldsymbol{I} - \boldsymbol{H}_X)$. It follows from (7.30) that

$$\boldsymbol{C}'\mathbf{R} = \tilde{\boldsymbol{Q}}'\mathbf{R}. \tag{7.34}$$

This again illustrates how efficiency can be improved, since it is quicker to work out $\tilde{\boldsymbol{Q}}'\mathbf{R}$ then $\boldsymbol{C}'\mathbf{R}$. Thus, from (7.31), S_2 may be formed from (7.32) and (7.34) by writing

$$S_2 = (\boldsymbol{C}'\mathbf{R})'(\boldsymbol{C}'\boldsymbol{C})^{-1}(\boldsymbol{C}'\mathbf{R}). \tag{7.35}$$

Suppose now regression is through the origin. Then the matrix $\boldsymbol{a}$ of (1.1) or (1.13) does not contain a column of 1's and $\boldsymbol{X}$ is obtained from $\boldsymbol{a}$ by centring all the columns. It is worth while to work with $\boldsymbol{X}$ rather than $\boldsymbol{a}$ to invert $\boldsymbol{a}'\boldsymbol{a}$ and this is described in Chapter 1. For this section the centring argument above no longer applies and so we work directly with the matrix $\boldsymbol{a}$. Then the degrees of freedom are modified. Here p denotes the number of explanatory variables and (7.22) is replaced by

$$F_1 = \frac{(n-p-4)S_2}{4(S_1 - S_2)}, \tag{7.36}$$

where S_1, S_2 are as in (7.21) and $\boldsymbol{H}_{MQ}$ as at (7.15), but $\boldsymbol{Q}$ is given by (7.20). Again define $\boldsymbol{C}$ by $(\boldsymbol{I} - \boldsymbol{H})\boldsymbol{Q}$. Then S_2 is again as in (7.31). Since $(\boldsymbol{I} - \boldsymbol{H})$ is symmetric and idempotent, the definitions of $\boldsymbol{C}$ and $\boldsymbol{H}$ imply

$$\boldsymbol{C}'\boldsymbol{C} = \boldsymbol{Q}'\boldsymbol{Q} - (\boldsymbol{Q}'\boldsymbol{a})(\boldsymbol{a}'\boldsymbol{a})^{-1}(\boldsymbol{Q}'\boldsymbol{a})'. \tag{7.37}$$

Also (1.6), that $\mathbf{R} = (\boldsymbol{I} - \boldsymbol{H})\mathbf{Y}$, yields

$$\boldsymbol{C}'\mathbf{R} = \boldsymbol{Q}'\mathbf{R}. \tag{7.38}$$

7.6 Summary of Ramsey's test

As in equation (1.18), $\boldsymbol{X}$ denotes the matrix of centred explanatory variables. It is also preferable, from the point of view of numerical analysis, if X has been scaled so that $X'X$ is a correlation matrix, but the following equations hold in either case, provided it has been centred. The procedure for the test as we implemented it is as follows. First form the matrix $\boldsymbol{Q}$ from $\hat{\mathbf{Y}}$, as $\boldsymbol{Q} = (\hat{\mathbf{Y}}^{(2)}, \hat{\mathbf{Y}}^{(3)}, \hat{\mathbf{Y}}^{(4)})$, where $\hat{\mathbf{Y}}^{(i)}$ denotes the column vector consisting of the ith powers of the elements of $\hat{\mathbf{Y}}$, the fitted values after regressing $\mathbf{Y}$ on X with a constant term. Then subtract the column means of $\boldsymbol{Q}$ from each element to form $\tilde{\boldsymbol{Q}}$. Then form $\boldsymbol{C}'\boldsymbol{C}$ using (7.32) and $\boldsymbol{C}'\mathbf{R}$ from (7.34). Substitute these into (7.35) to get S_2 and, finally, use (7.22) to find F, where S_1 is the corrected sum of squares due to deviations in the original regression of $\mathbf{Y}$ against X or the error sum of squares. If this F-value is significant, with $(3, n-p-3)$ d.f., then Ramsey's test is significant.

If the original regression is constrained to be through the origin the procedure is slightly different. Here the matrix $\boldsymbol{a}$ does not contain a column of 1's, and we work with $\boldsymbol{a}$ rather than with the centred matrix X of explanatory variables, and define p to be the number of columns of $\boldsymbol{a}$. S_1 again denotes the error sum of squares after regressing $\mathbf{Y}$ on $\boldsymbol{a}$. Obtain $\boldsymbol{Q}$ from (7.20), where $\mathbf{1}$ denotes a column of n 1's and multiply $\boldsymbol{Q}'\boldsymbol{a}$. Substitute this in (7.37) for $\boldsymbol{C}'\boldsymbol{C}$, and in (7.38) for $\boldsymbol{C}'\mathbf{R}$ (where $\mathbf{R}$ is obtained by regressing $\mathbf{Y}$ on $\boldsymbol{a}$, as is also $\hat{\mathbf{Y}}$). Then substitute these into (7.37) for S_2, and this into (7.36) which is tested for significance on $(4, n-p-4)$ degrees of freedom.

When Ramsey's test (in this form) was tested, it was found that it sometimes failed because $\boldsymbol{C}'\boldsymbol{C}$ was singular. Thus it is necessary to test for singularity of $\boldsymbol{C}'\boldsymbol{C}$ before inverting it for (7.35). In such a case, dropping $\hat{\mathbf{Y}}^{(4)}$ from $\boldsymbol{Q}$ usually produces a non-singular matrix and (7.22) must then be replaced by

$$F_1 = \{(n-p-2)S_2\}/\{2(S_1 - S_2)\}.$$

If $\boldsymbol{C}'\boldsymbol{C}$ is still singular after dropping $\hat{\mathbf{Y}}^{(4)}$ then Ramsey's test must be abandoned.

If $\boldsymbol{C}'\boldsymbol{C}$ is singular when regression is through the origin, the last column of (7.20) is dropped and (7.36) is replaced by

$$F_1 = \{(n-p-3)S_2\}/\{3(S_1 - S_2)\}.$$

After trials of this test, it became clear that some procedure is needed to suggest a variable to transform when the test is significant. As a possible response to this we developed an *ad hoc* procedure which involves no statistical theory, based on the sample correlations between the residuals and the moduli of the centred explanatory

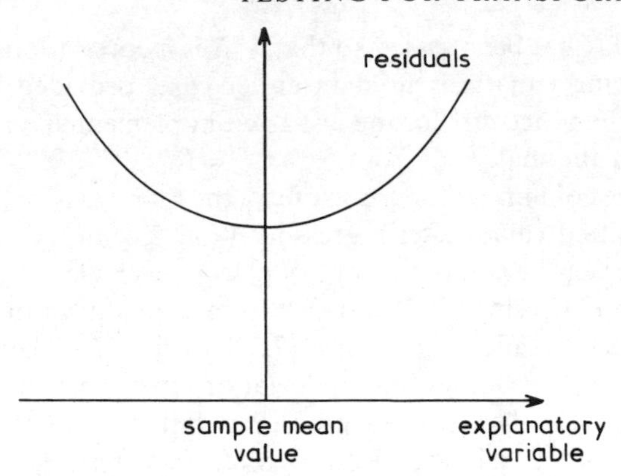

Fig. 7.1 *Finding variables to transform.*

variables. Whenever the modulus of such a correlation is high, we may conclude that the residuals either increase or decrease as the value of that explanatory variable moves away from its sample mean. We pick the largest of the moduli of these sample correlations and, provided it exceeds 0.3, the program outputs a message suggesting the user transforms this one (fig. 7.1). Sometimes it suggests the next best explanatory variable to transform. This would not detect a wrong functional form which has a point of inflection, nor would it be of any assistance when the omitted variable is not a transformation of an existing one. This points to be a weakness in this procedure and there is clearly need for more research on this.

References

Atkinson, A. C. (1982) Regression diagnostics, transformations and constructed variables. *J. Roy. Statist. Soc.* B, **44**, 1–13.

Atkinson, A. C. (1983) Diagnostic regression and shifted power transformations. *Technometrics*, **25**, 23–33.

Box, G. E. P. and Cox, D. R. (1964) An analysis of transformations. *J. Roy. Statist. Soc.* B, **26**, 211–252.

Box, G. E. P. and Tidwell, P. W. (1962) Transformations of the independent variables. *Technometrics*, **4**, 531–550.

Cook, R. D. and Weisberg, S. (1982) *Residuals and Influence in Regression.* Chapman and Hall, London.

Hinkley, D. V. and Runger, G. (1984) The analysis of transformed data. *J. Amer. Statist. Assoc.*, **79**, 302–309.

Ramsey, J. B. (1969) Tests for specification errors in classical linear least squares regression analysis. *J. Roy. Statist. Soc.* B, **31**, 350–371.

Ramsey, J. B. and Gilbert, R. (1972) A Monte Carlo study of some small sample properties of tests for specification error. *J. Amer. Statist. Assoc.*, **67**, 180–186.

Ramsey, J. B. and Schmidt, P. (1976) Some further results on the use of OLS and BLUS residuals in specification errors tests. *J. Amer. Statist. Assoc.*, **71**, 389–390.

Theil, H. (1965) The analysis of disturbances in regression analysis. *J. Amer. Statist. Assoc.*, **60**, 1067–1079.

Thursby, J. G. and Schmidt, P. (1977) Some properties of tests for specification error in a linear regression model. *J. Amer. Statist. Assoc.*, **72**, 635–641.

CHAPTER 8

Testing for normality

8.1 Introduction

Before setting out on a discussion of tests for normality, we must consider briefly the consequences of non-normality. Clearly, if we know that the distribution of the disturbances is, say gamma, then a different model, and different methods, should be used than those discussed in this volume. However, in many situations we may have at least 'near' normality, and we wish to know how serious the effect of deviation from normality might be. Box and Watson (1962) showed that this depends on the distribution of the explanatory variables. If the explanatory variables can be regarded as approximately normal, then the F-test for multiple regression is insensitive to non-normality. If the explanatory variables cannot be so regarded, then there can be a substantial effect on the 'F-test' distribution.

Non-normality can arise for numerous reasons. For example, non-normality could appear in the regression context because of an outlier or groups of outliers. Indeed, because of the masking effect of the presence of groups of outliers, these may be detected by tests for normality rather than by outlier tests. The test of normality can therefore prove to a useful diagnostic to operate routinely. A significant result may often indicate a group of outliers, or perhaps a set of observations taken under different conditions, rather than straightforward variations from the normal distributional form.

There is now a very large literature on tests for normality and many textbooks contain sections on the topic. A standard set of tests, with detailed numerical illustration and tables, is available as a British Standards Institution standard (BS 2846; Part 7; 1982), and this document serves as a reference for those wishing to carry out a test graphically or by a desk calculator.

The objective of this chapter is to examine tests which are suitable for routine incorporation into a computer program, to test whether the disturbances in regression model are normal. The tests were selected mainly for their good power properties, but it is inconvenient to incorporate large tables in such a program without approximating formulae for them; so this was also an important factor. Another consideration was that sometimes papers were unclear on certain points or contained typographical errors, and modifications or improvisations were needed as described below.

We wish to test whether or not the disturbances in regression are normally distributed but the disturbances are not known and we must use residuals. Huang and Bolch (1974) investigated the effects of doing some normality tests on the ordinary least squares (OLS) residuals as well as on best linear unbiased scalar (BLUS) residuals, both unstandardized. For considering the effect on the Type I error level settings when using OLS residuals (as opposed to testing directly on the disturbances), the evidence presented by Huang and Bolch suggests that testing the residuals would act conservatively, but their evidence is a bit conflicting so the present authors did some more work on this point and found that the Type I errors were tripled with some data sets. The Type I error can be put right by omitting certain data points with a high leverage; see below. Huang and Bolch also show that greater power is achieved if OLS residuals are used rather than BLUS residuals. Our conclusion given below is that normality tests be carried out on a subset of the standardized residuals, without any adjustment to the Type I error setting.

We assume here that the disturbances in regression are independently distributed under both the null and alternative hypotheses. The tests described below apply to testing any set of observations Y_1, $Y_2, \ldots, Y_n$ for normality, for which they are assumed to be independent and identically distributed. For the regression situation, as recommended below, these Y_i would be replaced by a subset of the standarized residuals.

8.2 Some tests for normality

In this section we briefly review some tests of normality which might be suitable for our purpose, which is to develop a test suitable to incorporate as a routine diagnostic test in a computer program. For this purpose graphical tests will not be satisfactory, although they would be very useful for communicating results obtained by more precise procedures to the user. In succeeding sections of the chapter we give details of some of the possible tests, so as to facilitate implementation.

8.2.1 The Shapiro–Wilk test

One of the most powerful tests for normality was given by Shapiro and Wilk (1965) and it is suitable for sample sizes ≤ 50. The form of the test statistic is

$$W = \left\{ \sum_{i=1}^{n} a_i Y_{(i)} \right\}^2 \Big/ \left\{ \sum_{i=1}^{n} (Y_i - \bar{Y})^2 \right\},$$

where $Y_{(1)} \leq Y_{(2)} \leq \cdots \leq Y_{(n)}$ and the a_i are constants tabulated by Shapiro and Wilk (1965), together with significance levels for W. The distributions of W for $n \leq 50$ were obtained by simulation and then Johnson curves were fitted. A procedure is given below for approximating to the constants of the Johnson curves, and so of obtaining the significance level of any observed W. Approximations are also given for the a_i.

A separate approach to obtaining significance levels of the W-test is given by Royston (1982a). He uses a transformation to normality, and he carried out separate simulations to approximate the W-distribution for sample sizes up to 2000; for details, see Royston (1982a, b)

8.2.2 Empirical distribution function tests

An empirical distribution function (e.d.f.) test is investigated below which is a variation of the Anderson and Darling (1952, Sec. (2.1) and Example 2) statistic. The statistic is

$$A^2 = -n^{-1} \sum_{i=1}^{n} (2i - 1)[1 + \log\{z_i(1 - z_{n+1-i})\}],$$

where

$$z_i = \Phi\{(Y_{(i)} - \bar{Y})/s\}$$

and s^2 is the sample variance. Large values of A are significant. The use of A^2 is based on work by Stephens (1974, 1976) who obtained a formula to calculate the significance level (see below). Stephens derived modifications of some e.d.f. tests so that their distribution was independent of sample size. The statistic A^2 was chosen because of its superior power properties and it can be used for any $n \geq 10$. However, the Shapiro–Wilk test is more powerful than the A^2-test up to sample size 50.

8.2.3 D'Agostino's test

An alternative test for $n > 50$ was devised by D'Agostino. Some power studies described below indicate that this is on average inferior to the A^2 test; but it is suggested as an alternative for three reasons:

(1) it is simpler to calculate,
(2) it is more powerful for symmetric long-tailed alternatives
(3) it is recommended in the BSI (1982) standard.

If the observations Y_i have already been centred, the formula is more simply

$$D = \left\{ \sum_{i=1}^{n} i Y_{(i)} \right\} \left\{ n^{3/2} \sqrt{\sum_{i=1}^{n} Y_i^2} \right\}.$$

Here no approximations are needed for any coefficients in the test statistic, which is of a simpler form than W. Some Cornish–Fisher expansions enable asymptotic formulae to be obtained for the significance levels of D, see below. This test is particularly useful for large sample sizes, because of the ease of calculation.

8.2.4 Some other tests

The Shapiro and Francia (1972) test was designed to complement the W test for the sample sizes in the range 50 to 99, and is a modification of the Shapiro–Wilk W statistic. They approximate the variance-covariance matrix of normal order statistics by a scalar matrix (a diagonal matrix with equal elements along the diagonal). This is simpler in terms of the coefficients in the numerator and was found to perform about as well as the W-statistic. They published a table of percentage points for n from 50 to 99 but only 1000 trials were used for these, and Pearson *et al.* (1977) show that they are not nearly accurate enough. A separate approach to the distribution of the Shapiro–Francia test is given by Royston (1983). However, as stated below, the A^2-test has advantages for testing normality with larger sample sizes.

Many other tests for normality have been proposed, but mostly they lack appropriate formulae for computer implementation and instead give simulated tables. See, for instance, Pearson *et al.* (1977). Also Green and Hegazy (1976) described some modified e.d.f. tests, but their significance levels depend on n and so lack the simplicity of Stephens' (1974) modified e.d.f. tests, described below.

8.3 The Shapiro–Wilk test

8.3.1 Calculating the test statistic

The Shapiro–Wilk (1965) test was developed to try to summarize formally some of the indications of probability plots. In a probability plot the observations are first ordered, then regressed against the expected values of the standard normal order statistics. A departure from linearity then indicates non-normality, and the fact that such departures are independent of the regression constants makes the approach free of scale and location.

The statistic W used for the Shapiro–Wilk test is proportional to the square of this regression slope (whose expectation under normality is proportional to the variance σ^2) divided by

$$\sum_{i=1}^{n} (Y_i - \bar{Y})^2.$$

Except for a constant, this is the ratio of two estimates of σ^2 (under the normality hypothesis); and a departure from normality has the effect of shrinking W.

Let $Y_{(1)}, Y_{(2)}, \ldots, Y_{(n)}$ be the set of observations written in ascending order, and let m_i denote the expected value of the ith order statistic in n standard normal variates, and let V be the corresponding variance–covariance matrix. Then the squared slope B of that regression line (obtained using generalized least squares (GLS), see Seber (1977), p. 60, or Chapter 9 of this book) is proportional to $\mathbf{m}'V^{-1}(Y_{(1)}, Y_{(2)}, \ldots, Y_{(n)})'$. So W is defined in the form

$$W = \left\{\sum_{i=1}^{n} a_i Y_{(i)}\right\}^2 \Bigg/ \left\{\sum_{i=1}^{n} (Y_i - \bar{Y})^2\right\}, \tag{8.1}$$

where $\mathbf{a}' = (a_1, a_2, \ldots, a_n)$ is chosen so that it is proportional to $\mathbf{m}'V^{-1}$ and normalized for convenience (that is, $\mathbf{a}'\mathbf{a} = 1$). Hence

$$\mathbf{a}' = \frac{\mathbf{m}'V^{-1}}{(\mathbf{m}'V^{-1}V^{-1}\mathbf{m})^{1/2}}. \tag{8.2}$$

Here $W^{1/2}$ is the sample correlation coefficient between $\mathbf{a}'$ and $(Y_{(1)}, Y_{(2)}, \ldots, Y_{(n)})$ so W is always ≤ 1, and $W = 1$ only when $(Y_{(1)}, Y_{(2)}, \ldots, Y_{(n)})$ is a multiple of $\mathbf{a}'$. In fact, $\mathbf{m}'V^{-1}$ is approximately a constant multiple of $\mathbf{m}'$ (see Shapiro and Wilk (1965), p. 596) so that an exactly straight line (in the plot of $Y_{(i)}$ against m_i) would lead to W very close to 1 and vice versa. So if W is significantly less than 1, the hypothesis of normality will be rejected.

The definition of the vector of coefficients is given by (8.2), but the tabulated values (when $n > 20$) were not calculated exactly. The authors used approximations to do so, since V was not known for $n > 20$. They defined

$$C = (\mathbf{m}'V^{-1}V^{-1}\mathbf{m})^{1/2} \tag{8.3}$$

and

$$\mathbf{a}^* = \mathbf{m}^* V^{-1} \tag{8.4}$$

and approximated each separately. Since $\mathbf{a}'\mathbf{a} = 1$, we have

$$\mathbf{a}^{*\prime}\mathbf{a}^* = C^2. \tag{8.5}$$

They wrote down the approximations

$$\hat{a}_i^* = 2m_i \qquad (i = 2, 3, \ldots, n-1), \tag{8.6}$$

$$\hat{a}_1^2 = \hat{a}_n^2 = \frac{\Gamma\{\frac{1}{2}(n+1)\}}{\sqrt{2}\Gamma(\frac{1}{2}n+1)} \quad \text{for } n > 20. \tag{8.7}$$

Now C can be estimated from the approximations (8.6) and (8.7), using the normalizing condition (8.5). Thus, (8.3)–(8.5) imply

$$C^2 = \sum_{i=2}^{n-1} a_i^{*2} + 2C^2 a_1^2 \quad \text{and} \quad \hat{a}_i = \hat{a}_i^*/C \tag{8.8}$$

so C^2 can be approximated by

$$\hat{C}^2 = \left\{\sum_{i=2}^{n-1} \hat{a}_i^{*2}\right\} \Big/ (1 - 2\hat{a}_1^2). \tag{8.9}$$

This is then substituted into (8.8) to get $\hat{a}_i$ for $i = 2, 3, \ldots, n-1$. Then approximations for the right-hand sides of (8.6) and (8.7) were obtained to avoid having to incorporate tables in the computer program, or to having to do repeated divisions and multiplications (in the case of (8.7)). First, some results obtained for D'Agostino's test imply

$$\hat{a}_1^2 = \hat{a}_n^2 \simeq \left[\sqrt{(n+1)}\left\{1 - \frac{0.25}{n+2} - \frac{0.21875}{(n+2)^2} - \frac{0.1484375}{(n+2)^3} - \frac{0.0493164}{(n+2)^4}\right\}\right]^{-1}. \tag{8.10}$$

Now for m_i Harter (1961) investigated some approximations for them. He finds that the approximation

$$m_i = \Phi^{-1}\left(\frac{i-\alpha}{n-2\alpha+1}\right),$$
$$\alpha = 0.327511 + 0.058212X - 0.007909X^2, \tag{8.11}$$

where $X = \log_{10} n$ gives a maximum error not more than 0.001. A FORTRAN algorithm for evaluating Φ^{-1} is given by Beasley and Springer (1977).

These approximations were tried and found to work very well. To summarize, the procedure used is as follows. Obtain m_i $(i = 2, 3, \ldots, n-1)$ from (8.11), then a_i^* from (8.6), followed by $\hat{a}_1$ and $\hat{a}_n$ from (8.10) (here $\hat{a}_1 < 0$ and $\hat{a}_n > 0$). Substitute in (8.9) for C and substitute this into (8.8) for the remaining a_i $(i \neq 1$ or $n)$. This method applies only for $20 < n \leq 50$, since for lower n (8.7) was not used by Shapiro and Wilk and they did not obtain any results for $n > 50$. For $n \leq 20$ corresponding formulae are not available.

8.3.2 Percentage points of the Shapiro–Wilk test

Table 6 of Shapiro and Wilk (1965) was obtained by fitting Johnson curves to some simulated estimates. The method they used is described on pp. 296–297 of Hahn and Shapiro (1967). For each sample size n they find constants $\gamma, \eta, \varepsilon$ such that the W-level is, for a given probability level α, related to the normal deviate Z by

$$Z = \gamma + \eta \log\left(\frac{W-\varepsilon}{1-W}\right). \tag{8.12}$$

For $10 \leq n \leq 50$ we fitted regression equations for these three parameters and obtained

$$\log(-\gamma) = -22.57435 + 0.164518n - 147.753n^{-1} + 255.4251n^{-1/2} - 190.2646n^{-1}\log n - 0.0007575n^2,$$

$$\log \eta = 62.393343n^{-1} + 87.582598n^{-1/2} + 3.0247700 \log n - 38.567346n^{-1} \log n - 19.159578,$$
$$\log \varepsilon = -75.291825n^{-1} + 97.606190n^{-1/2} + 9.4384903 \log n - 1.3012674n^{1/2} - 41.9626. \quad (8.13)$$

(Variable deletion was considered on the basis of the maximum of the moduli of residuals.) Thus for a given α we find Z such that $\alpha = \Phi(Z)$, then

$$W = (\varepsilon + e)/(1 + e) \quad \text{where} \quad e = \exp\{(Z - \gamma)/\eta\}. \quad (8.14)$$

This was found to agree very closely with the tables: the maximum difference was 0.0005.

Some subsequent simulation experiments for $n \leq 50$, however, indicated that these levels were inaccurate (but not significantly so for $n = 20$). It was found that adding

$$\delta = 0.00222 - 0.0317\alpha + 0.126\alpha^2 \quad (8.15)$$

to the tabulated levels of W gave acceptable results, provided α is in the range (0.01–0.1). Equivalently, this δ can be added to the levels obtained by the above approximation formulae. The fact that these tabulated levels are a bit low is also suggested by Table 2 of Huang and Bolch (1974).

8.4 D'Agostino's test

8.4.1 The test statistic

Another test statistic D was proposed by D'Agostino (1971) as

$$D = T/(n^2 S), \quad (8.16)$$

where

$$T = \sum_{i=1}^{n} \{i - \tfrac{1}{2}(n+1)\} Y_{(i)}, \qquad S^2 = \sum_{i=1}^{n} (Y_i - \bar{Y})^2/n,$$

where $Y_{(1)}, Y_{(2)}, \ldots, Y_{(n)}$ denote the observation in ascending order. Then

$$D^2 \left\{ \sum_{i=1}^{n} b_i Y_{(i)} \right\}^2 \Bigg/ \left\{ \sum_{i=1}^{n} (Y_i - \bar{Y})^2 \right\}, \quad \text{where} \quad b_i = [i - \tfrac{1}{2}(n+1)]/n^{3/2}. \quad (8.17)$$

Again this is the ratio of two estimates of σ^2 (under the normality hypothesis), apart from a constant and is similar in form to W. For instance, for $n = 9$ the coefficients are (with a constant modification to a_i)

b_5	b_6	b_7	b_8	b_9
0	0.037	0.074	0.111	0.148
$a_5/3$	$a_6/3$	$a_7/3$	$a_8/3$	$a_9/3$
0	0.032	0.066	0.108	0.196

So D^2 and W are a bit similar with W giving more weight to extreme observations (at least for $n = 9$).

The advantage of D'Agostino's test is that asymptotic formulae for the significance levels can be obtained analytically, but there is no known method for doing this for W. Since D^2 is the ratio of two estimates of σ^2, the distribution of D is independent of σ^2 (under the normality hypothesis) and a sufficiency argument shows that D and S are in fact independent. Hence, from (8.16)

$$E(D^k)E(S^k) = E(D^k S^k) = E(T^k)/n^{2k};$$

so

$$E(D^k) = E(T^k)/\{n^{2k}E(S^k)\}. \tag{8.18}$$

To evaluate the first four moments of D, $E(S^k)$ is given in Kendall and Stuart (1969, vol. I) and $E(T^k)$ by Barnett *et al.* (1967). This enables the following Cornish–Fisher expansion for D_p in terms of Z_p to be used:

$$D_P = E(D) + V_P\surd(\mu_2), \tag{8.19}$$

where

$$V_P = Z_P + \frac{1}{6}\gamma_1(Z_P^2 - 1) + \frac{1}{24}\gamma_2(Z_P^3 - 3Z_P) - \frac{\gamma_1^2}{36}(2Z_P^3 - 5Z_P).$$

Here μ_2 is the variance of D, γ_1 is its skewness, γ_2 its kurtosis $((\mu_4/\mu_2^2) - 3)$, and Z_P is the $100P$ percentile point of the standard normal distribution.

This approach was used by D'Agostino to derive his Table 1 of percentile points, but in his (1971) paper he gives another table for $n \leq 100$ from simulation experiments. The (1971) paper also includes series for μ_2, γ_1 and γ_2 which are not accurate enough for our purpose. The derivation of asymptotic series for γ_1, γ_2 and the moments of D is given in the next section and those prepared to assume the results may skip to Section 8.4.4 for the procedure for testing.

8.4.2* Derivation of asymptotic series

In the derivation, the non-central moments of T and $n^{1/2}S$ were expanded in powers of n^{-1}, using a computer program written to manipulate power series. The same program had subroutines to obtain centred moments and to multiply and divide power series; so equation (8.18) was used to compute the first four non-centred moments of D, and then these were used (within the program) to compute power series for the central moments of D. These, in turn, led to series approximations for γ_1, γ_2 which, with (8.19), enabled these approximations to be checked with D'Agostino's (1971, Table 1) of percentile points.

In order to achieve sufficient accuracy, it was found that five terms were required for all the power series involved, which meant working out the next term in Kendall and Stuart's (1969) series expression for $n^{1/2}S$. Here we have

$$E(n^{1/2}S) = \sqrt{(2)}\frac{\Gamma\{\frac{1}{2}n\}}{\Gamma\{\frac{1}{2}(n-1)\}}, \tag{8.20}$$

with Stirling's approximation

$$\log\Gamma(x+1) \sim \tfrac{1}{2}\log(2\pi) + (x+\tfrac{1}{2})\log x - x + \frac{1}{12x} - \frac{1}{260x^3} + o(x^{-3}). \tag{8.21}$$

Hence we may write

$$\begin{aligned}\Delta &\overset{\text{def}}{=} \log\Gamma(\tfrac{1}{2}n) - \log\Gamma(\tfrac{1}{2}n - \tfrac{1}{2})\\ &= (\tfrac{1}{2}n - \tfrac{1}{2})\log(\tfrac{1}{2}n - 1) - (\tfrac{1}{2}n - 1)\log(\tfrac{1}{2}n - \tfrac{3}{2})\\ &\quad - (\tfrac{1}{2}n - 1) + (\tfrac{1}{2}n - \tfrac{3}{2}) + \frac{1}{12(\tfrac{1}{2}n - 1)} - \frac{1}{12(\tfrac{1}{2}n - \tfrac{3}{2})}\\ &\quad - \frac{1}{360(\tfrac{1}{2}n - 1)^3} + \frac{1}{360(\tfrac{1}{2}n - \tfrac{3}{2})^3} = o(n^{-3})\end{aligned} \tag{8.22}$$

with

$$E(n^{1/2}S) = \sqrt{(2)}\exp(\Delta). \tag{8.23}$$

If (8.22) is substituted into (8.23), a factor of $\sqrt{(n-1)}$ may be extracted from the two logarithm terms on the right of (8.22) giving an expression of the form

$$\frac{E(n^{1/2}S)}{\sqrt{(n-1)}} = \exp\{-\tfrac{1}{2} + (\tfrac{1}{2}n - \tfrac{1}{2})\log\left(\frac{1-2/n}{1-1/n}\right) + (\tfrac{1}{2}n - 1)\log\left(\frac{1-1/n}{1-3/n}\right) + \text{other terms on the right in (8.22)}\}.$$

The logarithms may be expanded and then exp{powers of n^{-1}} expanded to give

$$\frac{E(n^{1/2}S)}{\sqrt{(n-1)}} = 1 - \frac{0.25}{n} - \frac{0.21875}{n^2} - \frac{0.1484375}{n^3} - \frac{0.0493164}{n^4}. \tag{8.24}$$

The computed moments of T (using the results of Barnett *et al.* (1967)) were

$$E(T) = n^2\left\{0.2820948 - \frac{0.2820948}{n}\right\},$$

$$E(T^2) = n^4\left\{0.07957747 - \frac{0.1184670}{n} + \frac{0.04835884}{n^2} - \frac{0.009469267}{n^3}\right\},$$

$$E(T^3) = n^6\left\{0.02244839 - \frac{0.03291164}{n} + \frac{0.01263341}{n^2} - \frac{0.003938010}{n^3} + \frac{0.003249451}{n^4}\right\}$$

and

$$E(T^4) = n^8\left\{0.006332574 - \frac{0.005903257}{n} - \frac{0.003867853}{n^2} + \frac{0.005088538}{n^3} - \frac{0.002414016}{n^4}\right\}. \tag{8.25}$$

Then the first moment of D was found to be

$$E(D) = 0.2820948 - \frac{0.07052370}{n} + \frac{0.008815462}{n^2} + \frac{0.01101933}{n^3} - \frac{0.002892575}{n^4} \tag{8.26}$$

and the three central moments of D

$$\begin{aligned} \mu_2 &= \frac{0.0008991591}{n} - \frac{0.0004779168}{n^2} - \frac{0.004973592}{n^3} + \frac{0.003108496}{n^4}, \\ \mu_3 &= \frac{0.0002314333}{n^2} + \frac{0.001096059}{n^3} - \frac{0.0005389610}{n^4}, \\ \mu_4 &= \frac{0.000002425461}{n^2} + \frac{0.00009018102}{n^3} - \frac{0.0009020809}{n^4}. \end{aligned} \tag{8.27}$$

Then for skewness and kurtosis we obtain

$$\begin{aligned} \gamma_1 &= -\frac{8.5836542}{\sqrt{n}}\left\{1 - \frac{3.938688}{n} + \frac{7.344405}{n^2}\right\}, \\ \gamma_2 &= \frac{114.732}{n}\left\{1 - \frac{8.38004}{n}\right\}. \end{aligned} \tag{8.28}$$

These results enable the Cornish Fisher expansion (8.19) to be used directly

8.4.3* Power properties

The power properties of D'Agostino's test were also examined in the papers of D'Agostino (1971), Green and Hegazy (1976) and Pearson *et al.* (1977). They indicate (see especially p. 243 of Pearson *et al.*) that D'Agostino's test is better than the Shapiro–Wilk test for detecting long-tailed distributions, which is perhaps the most dangerous departure. On p. 208 Green and Hegazy write 'As our introduction indicates, we found the tests of D'Agostino and Shapiro and Wilk to be the two most powerful tests of normality available other than the e.d.f. family of tests.' It depends on what sort of departures are most important, but the Shapiro–Wilk test has been shown to be superior to D'Agostino's for certain alternatives.

8.4.4 Procedure for testing

To summarize the above results, first compute the statistic D from (8.16). Equivalently, first centre the ordered $Y_{(i)}$ then put

$$D = \left\{ \sum_{i=1}^{n} iY_{(i)} \right\} \Big/ \sqrt{\left(n^3 \sum_{i=1}^{n} Y_i \right)}.$$

Now to do a two-sided test on D at level α, put $P = \alpha/2$ and $1 - \alpha/2$ in turn and compute the $100P$ percentile point Z_P of the normal distribution. Then compute $E(D)$ from (8.26), μ_2 from (8.27), γ_1 and γ_2 from (8.28) and substitute them into (8.19) to find D_P. (Of course, the moments need not be calculated from each P as they are independent of P.) Let the values of D_P be denoted by $D_{\alpha/2}$ and $D_{1-\alpha/2}$: then reject the hypothesis of normality whenever D falls outside the interval $(D_{\alpha/2}, D_{1-\alpha/2})$.

For $n \leq 100$ D'Agostino's (1971) Table 1 can be used for testing in place of the Cornish–Fisher expansion.

8.5 Empirical distribution function tests

8.5.1 Discussion

Shapiro and Wilk (1965) and Shapiro *et al.* (1968) compared the performance of the W-test with certain empirical distribution function (e.d.f.) tests. These e.d.f. tests involve estimating, by various weighted integrals, the difference between the null hypothesis function (in this case the normal distribution) and the e.d.f. obtained from observations. These were found to have considerably lower power than the W-test. It has been reported, by Stephens (1974) and by Green and Hegazy (1976), that this was not a fair comparison since the mean and variance were assumed to be known and are not estimated from the data. The use of tests where these are estimated gives much greater power, they claim, which makes them comparable in power with the W-test. (This is confirmed also by Pettit (1977) for the Anderson–Darling statistic.) Two such tests are the Cramér–von

Mises (CVM) and the Anderson–Darling (AD) tests and their application was developed by Stephens (1974). For the asymptotic distribution theory of them and others see, for instance, Stephens (1976). Green and Hegazy (1976) derive some modifications to these which, they show by simulation, give greater power still. The best of these (for a general test for normality) was called A_{22} and is a variant of the AD statistic, giving more weight to variations in the extreme tails. They do not, however, give any formulae or approximations for deriving percentage points, but give estimates from simulations for $n = 5, 10, 20, 40, 80, 160$ at the 5% level only; so in some ways this has disadvantages for implementation in a computer program, since the whole table has to be stored (if other levels were known).

Some power studies by Stephens (1974) showed that AD performed best. So this AD test, using Stephens' method for finding the percentage points, was considered for implementation. This involved modifying the statistic by certain given functions of n and giving a table of significance levels for the modified statistics. These tables do not involve n and each consist of a single row of significance levels (for each statistic), which greatly simplifies the use of the tests. The asymptotic behaviour of the statistics, as $n \to \infty$, is treated theoretically using linear combinations of chi-squared variables.

8.5.2 Procedure for the modified test

Stephens (1974) also explains how the statistics can be evaluated fairly simply for testing either for a normal or for an exponential distribution. Here a modified version is described. For normality the procedure is as follows, when the mean μ and variance σ^2 are unknown. Given a sample $Y_1, Y_2, \ldots, Y_n$, estimate μ by $\bar{Y}$ and σ^2 by

$$s^2 = \frac{1}{n-1} \sum_{i=1}^{n} (Y_i - \bar{Y})^2.$$

Put Y_i ($i = 1, \ldots, n$) into ascending order $Y_{(1)}, Y_{(2)}, \ldots, Y_{(n)}$ and compute

$$z_i = \Phi((Y_{(i)} - \bar{Y})/s). \tag{8.29}$$

Then the AD statistic may be expressed as

$$A^2 = -\left\{ \sum_{i=1}^{n} (2i-1)[\log z_i + \log(1 - z_{n+1-i})] \right\} / n - n, \tag{8.30}$$

then calculate the modified statistic,

$$T^* = \left(1 + \frac{0.75}{n} + \frac{2.25}{n^2}\right) A^2. \tag{8.31}$$

To do a significance test, find the upper tail probability level α, corresponding to the value of T^*, from the formula

$$\alpha = \exp\{1.2937 - 5.709\,T^* + 0.0186\,T^{*2}\}. \tag{8.32}$$

For a given significance level α_0, reject H_0 (the sample comes from a normal population) if $\alpha < \alpha_0$. This relationship (8.32) is only applicable when $\alpha_0 < 0.12$: Stephens has developed alternative equations for $\alpha_0 > 0.12$ (as well as for other e.d.f. statistics). Alternatively (for several values of T^*), given α_0 we can put $\alpha = \alpha_0$ and solve (8.32) to give $T^* = T_0^*$, then reject H_0 whenever $T^* > T_0^*$. If (8.32) is solved by the usual formula for a quadratic equation, roundoff error will lead to the accuracy of the solution being poor. A better method is to use iteration, regarding the squared term as being relatively small. The third iteration then gives

$$T_0^* = A_1(1 + 0.003258\,A_1(1 + 0.006516\,A_1)), \tag{8.33}$$

where

$$A_1 = (1.2937 - \log\alpha)/5.709.$$

For computing purposes an approximation is needed for the normal c.d.f. in (8.29) and one given by Hastings (1955, sheets 43–45) is accurate enough. This uses

$$\Phi(x) \simeq 1 - \left\{\sum_{i=1}^{4} b_i \eta^i\right\} \exp(-\tfrac{1}{2}x^2),$$

where $\eta = (1 + 0.270090x)^{-1}$, $b_1 = 0.0720556$, $b_2 = 0.3052713$, $b_3 = -0.3038699$ and $b_4 = 0.4265428$.

8.6 Tests based on skewness and kurtosis

This test is carried out as follows. Consider a sample $Y_1, Y_2, \ldots, Y_n$ assumed to be independent and identically distributed. Compute the centre moments

$$m_r = \frac{1}{n}\sum_{i=1}^{n}(Y_i - \bar{Y})^r \quad \text{for } r = 2, 3, 4,$$

then the sample skewness is

$$\sqrt{(b_1)} = m_3/m_2^{3/2} \tag{8.34}$$

and the kurtosis is defined here to be

$$b_2 = m_4/m_2^2. \tag{8.35}$$

For an omnibus test of normality we do not want to test b_1 and b_2 separately, and a combined Type I error setting is appropriate. So a large deviation either of b_1 from 0 or of b_2 from 3 should give a significant result, regardless of which one deviates. The distributions of b_1 and b_2 separately are known very accurately but no simple formulae exist for their joint distribution. Asymptotically, however, they are independent and normally distributed; but convergence to this is slow, especially with regard to b_2 (essentially because it involves the fourth moment which is more ill-conditioned). For instance, when $n = 200$ the skewness of b_2 is 0.97 and its kurtosis is 5.25 (see Bowman and Shenton, 1975, p. 243). Gurland and Dahiya (1972) devised a test based on the sample skewness b_1 and kurtosis b_2. The statistic is

$$\hat{Q} = \frac{nb_1}{6} + \frac{3n}{8}\left[\log\left(\frac{b_2}{3}\right)\right]^2, \tag{8.36}$$

and the normality hypothesis is rejected (from asymptotic considerations) whenever $\hat{Q} > \chi_2^2(\alpha)$ where α is the level of the test. Then Bowman and Shenton (1975) suggested, on p. 243, using the statistic

$$S = \frac{nb_1}{6} + \frac{n(b_2 - 3)^2}{24} \tag{8.37}$$

which is asymptotically χ_2^2. The power of these tests is rather low but it has the advantage that it leads to a description of the type of non-normality detected.

8.7 Power studies

Various power studies have been done for normality tests, but since the tests described above were modifications of the published ones some more power studies were needed. So a FORTAN simulation program was written to generate independent sets of observations for various sample sizes.

From the results it was found that W is best overall when $n \leq 50$, followed by A^2, followed by D and $\hat{Q}$. But D and $\hat{Q}$ are good for the scale contaminated example (a symmetric distribution with positive kurtosis) which is an important kind of deviation and for the logistic. It also emerges that D is least good for skewness; for symmetric positive kurtosis D is the best and for negative kurtosis D and A^2 are similar. In view of the weakness of W for symmetric alternatives with positive kurtosis, the best overall test for $n > 50$ is A^2. But D could be important if symmetric positive kurtosis is the main danger. With asymmetry, however, the relative performance of D falls off. So we recommend W for $n \leq 50$ and A^2 for $n > 50$.

8.8 Effect of explanatory variables

The above tests are described in the context of a set of observations $Y_1, Y_2, \ldots, Y_n$, which are assumed to be independent and identically distributed both for the null and for the alternative hypotheses. For the regression problem introduced in Chapter 1 the simplest procedure is to apply the normality tests to the raw residuals R. (We should not apply them to Y since the elements Y_i do not have the same expectations.) This is because in many instances the residuals approximate the disturbances, which are really what we want to test. But the residuals are not independent, nor do they have constant variance; so the Type I error level setting may be inaccurate. The results of Huang and Bolch (1974) suggest that the α-level settings may be conservative or about right. This means that if we apply the usual normality tests to the n residuals, regarding them as independent and identically distributed, with a nominal Type I error level setting, the actual Type I error level will be approximately α or less. In other words, if the Type I error level for testing the disturbances is α, then the level for testing the residuals is approximately α or less. But their results are a bit conflicting. Then White and MacDonald (1980) obtained some similar results; but Weisberg, in a comment on the latter paper, threw some doubt on these conclusions. Pierce and Gray (1982) also suggest considerable caution in carrying out tests of normality on residuals in multiple regression.

We therefore did some more extensive simulations with larger values of k, and found that standardized residuals were better. While most of these results were considered acceptable or conservative, this was not the case with Weisberg's Data Set 3. It follows partially from Weisberg's comments that, with such data sets as his Data Set 3, the observations with large h_{ii} would have small residuals and could be discarded from the test. Now

$$\bar{h}_{ii} = n^{-1} \sum_{i=1}^{n} h_{ii} = k/n \tag{8.38}$$

and one of the criticisms of Weisberg was that White and MacDonald considered only data sets in which the h_{ii} were all approximately equal; so a rule such as the following can be used:

$$\text{Omit observation for which } 1 - h_{ii} < 0.667(n-k)/n, \tag{8.39}$$

and test on the remaining standardized residuals. Clearly, there is a topic here for further research.

The reference list to this chapter is slightly expanded, to contain papers relevant to the above discussion.

References

Anderson, T. W. and Darling, D. A. (1952) Asymptotic theory of certain goodness-of-fit criteria based on stochastic processes. *Ann. Math. Statist.*, **23**, 193–212.

Barnett, F. C. Mullen, K. and Saw, J. G. (1967) Linear estimation of a population scale parameter. *Biometrika*, **54**, 551–554.

Beasley, J. D. and Springer, S. G. (1977) Algorithm 111. The percentage points of the normal distribution. *Appl. Statist.*, **26**, 118–121.

Bowman, K. O. and Shenton, L. R. (1975) Omnibus test contours for departures from normality based on $\sqrt{(b_1)}$ and b_2. *Biometrika*, **62**, 243–250.

Box, G. E. P. and Watson, G. S. (1962) Robustness to non-normality of regression tests. *Biometrika*, **49**, 93–106.

B.S.I. (1982) Draft BS2846: British Standard guide to statistical interpretation of data. Part 7: Tests for departure from normality (ISO/DIS 5479). Obtainable from British Standards Institution, Sales Department, Linfood Wood, Milton Keynes, MK14 6LR.

D'Agostino, R. B. (1971) An omnibus test of normality for moderate and large size samples. *Biometrika*, **58**, 341–348.

Green, J. R. and Hegazy, Y. A. S. (1976) Powerful modified-EDF goodness-of-fit tests. *J. Amer. Statist. Assoc.*, **71**, 204–209.

Gurland, J. and Dahiya, R. C. (1972) Tests of fit for continuous distributions based on generalized minimum chi-squared, in *Statistical Papers in Honour of George W. Snedcor* (ed. T. A. Bancroft). State University Press, Iowa. pp. 115–128.

Hahn, G. J. and Shapiro, S. S. (1967) *Statistical Models in Engineering*. Wiley, New York.

Harter, H. L. (1961) Expected values of normal order statistics. *Biometrika*, **48**, 151–165.

Hastings, C. (1955) *Approximation for Digital Computers*. Princeton University Press, Prinction, NJ.

Huang, C. J. and Bolch, B. W. (1974) On the testing of regression disturbances for normality. *J. Amer. Statist. Assoc.*, **69**, 330–335.

Kendall, M. G. and Stuart, A. (1969) *The Advanced Theory of Statistics*, Vol. I, 3rd edn. Griffin, London.

Pearson, E. S., D'Agostino, R. B. and Bowman, K. O. (1977) Test for departure from normality: comparison of powers. *Biometrika*, **64**, 231–246.

Pettit, A. N. (1977) Testing the normality of several independent samples using the Anderson–Darling statistic. *Appl. Statist.* **26**, 156–161.

Pierce, D. A. and Gray, R. J. (1982) Testing normality of errors in regression models. *Biometrika*, **69**, 233–236.

Royston, J. P. (1982a) An extension of Shapiro and Wilk's test for normality to large samples. *Appl. Statist.*, **31**, 115–124.

Royston, J. P. (1982b) Algorithm AS181. The *W*-test for normality. *Appl. Statist.*, **31**, 176–180.

Royston, J. P. (1983). A simple method for evaluating the Shapiro–Francia *W'*-test for non-normality. *J. Inst. Statist.*, **32**, 297–300.

Seber, G. A. F. (1977) *Linear Regression Analysis*. Wiley, New York.

Shapiro, S. S. and Francia, R. S. (1972) An approximate analysis of variance test for normality. *J. Amer. Statist. Assoc.*, **67**, 215–216.

Shapiro, S. S. and Wilk, M. B. (1965) An analysis of variance test for normality (complete samples). *Biometrika*, **52**, 591–611.

Shapiro, S. S., Wilk, M. B. and Chen, H. J. (1968) A comparative study of various tests for normality. *J. Amer. Statist. Assoc.*, **63**, 1343–1372.

Stephens, M. A. (1974) EDF statistics for goodness of fit and some comparisons. *J. Amer. Statist. Assoc.*, **69**, 730–737.

Stephens, M. A. (1976) Asymptotic results for goodness-of-fit statistics with unknown parameters. *Ann. Statist.*, **4**, 357–369.

White, H. and Macdonald, G. M. (1980) Some large-sample tests for non-normality in the linear regression model. *J. Amer. Statist. Assoc.*, **75**, 16–28.

Further reading

D'Agostino, R. B. (1972) Small sample probability points for the *D* test of normality. *Biometrika*, **59**, 219–221.

Kennard, K. V. C. (1978) A review of the literature on omnibus tests of normality with particular reference to tests based on sample moments. M.Sc. dissertation, University of Kent.

CHAPTER 9

Heteroscedasticity and serial correlation

9.1 Introduction

There are many situations occurring in practice when the simple structure of random variation assumed in (1.1) does not hold; examples will be given below. Suppose that instead of (1.1) we have

$$\left.\begin{aligned} E(\mathbf{Y}) &= \boldsymbol{a\theta}, \\ V(\mathbf{Y}) &= \boldsymbol{V}\sigma^2, \end{aligned}\right\} \tag{9.1}$$

where $\boldsymbol{V}$ is a known $n \times n$ positive definite matrix, then the appropriate method of estimation is generalized least squares (GLS), leading to

$$\hat{\boldsymbol{\theta}}_g = (\boldsymbol{a}'\boldsymbol{V}^{-1}\boldsymbol{a})^{-1}\boldsymbol{a}'\boldsymbol{V}^{-1}\mathbf{Y}. \tag{9.2}$$

If $\boldsymbol{V} = \boldsymbol{I}$, then this reduces to the ordinary least squares (OLS) estimator

$$\hat{\boldsymbol{\theta}} = (\boldsymbol{a}'\boldsymbol{a})^{-1}\boldsymbol{a}'\mathbf{Y}. \tag{9.3}$$

If (9.3) is used when (9.2) is appropriate, then (9.3) is still unbiased, but it is not efficient, and the OLS estimator of σ^2 will in general be biased. All tests will therefore be invalid, Clearly, further research needs to be carried out on how serious the effects are of using OLS when (9.2) ought to be used, but this argument indicates the desirability of testing for departures from the assumption of $\boldsymbol{V} = \boldsymbol{I}$.

Two types of departure are commonly tested for:

(1) The response variables are independent but of unequal variance (heteroscedasticity).
(2) The response variables are of equal variance but serially correlated.

Tests for these points are discussed in this chapter. A brief discussion of problems of estimation of parameters of models with more complex error structure is given in Chapter 13.

9.2 Heteroscedasticity

An excellent survey and discussion of the problems of heteroscedasticity and autocorrelation is given by Judge *et al.* (1980). For some other work on heteroscedasticity see Carroll and Ruppert (1981), Horn (1981) and Cook and Weisberg (1983).

Heteroscedasticity can often be thought of as arising in the following ways.

(1) *The use of averaged data*

Sometimes the 'observations' available to us for use in a regression analysis are averaged data. This happens sometimes in survey analysis, when regressions are performed on data which are themselves estimates based on a complex survey design, see, for example, Holt *et al.* (1980).

(2) *Variances depending on the mean*

In some agricultural experiments on growth, or in economic models, the amount of random variation in a response variable is proportional to some function of the mean.

(3) *Variances depending on the explanatory variables*

Sometimes the situation occurs in which the variance of an observation is some function of the explanatory variables. This can happen in econometric models; for example, if we are relating household expenditure to household income, and various other factors about the household, then the variance of the observations may well depend on a (possibly quadratic) function of income. Other applications arise and an example will be given in Chapter 13 in which a biological problem leads to a similar model.

(4) *Different observers, sites, etc.*

In any large data set, it is very likely that there will have been several observers, sites or machines involved, and these may have different variances.

(5) *Outliers*

Sometimes, herteroscedasticity is due to the presence of a few outliers, and this possibility needs to be examined carefully if heteroscedasticity is found unexpectedly.

If heteroscedasticity is detected, we need to investigate the form that it takes and then some alternative model may be appropriate; see the brief discussion in Chapter 13 and Judge *et al.* (1980). Another course of action is to try the Box–Cox transformation in the hope of obtaining a transformed response variable which fits the OLS model.

Before proceeding to review the tests, it should be noted that a discussion with the client who provides the data is really needed to see whether or not heteroscedasticity is likely to occur, for example, via one of the above five models. Such information could crucially affect the power of the procedures and the interpretation of test results. Furthermore, if we have in mind the possibility of a specific form of heteroscedasticity, then it is easier to test for it.

Clearly, heteroscedasticity is easy to detect if there are (fully) replicated observations in the data set. This occurs very rarely with multiple regression data, and in the following discussion we shall assume that there are no replicated observations.

In Chapters 7 and 8, the primary interest was to develop procedures which can be incorporated routinely into statistical packages to give warning messages, to the effect that certain features of the data need closer examination. The same approach is adopted here and we concentrate on algebraic tests for this purpose. Once heteroscedasticity has been detected, a thorough graphical analysis using residuals plots should be carried out and such plots are now standard techinques. Sections 9.3–9.5 concentrate on algebraic tests for heteroscedasticity.

9.3 Anscombe's test and its variants

Residuals plots from a regression analysis are now commonplace and, following the discussion in Section 9.2, we should plot the residuals against the fitted values, and against any other variables which we consider might be associated with heteroscedasticity. Heteroscedasticity is detected by the range or variance of the residuals varying with the variable or function being plotted. While this form of test for heteroscedasticity is satisfactory for expert statisticians, others may be in doubt about the interpretation of such plots, and a more precise method of testing is required. A more precise test is also required for routine running in a computer program.

Anscombe (1961) introduced a statistic which is a normalized version of $\sum R_i^2 (\hat{Y}_i - \tilde{Y})$, where $\tilde{Y}$ is a weighted average of the fitted values, giving less weight to points with a high influence on the fit, and where R_i are the raw residuals. This tests for a monotonic change in the variance with respect to the fitted values. The test can be looked upon as equivalent to performing a test by plotting residuals against the fitted values.

Anscombe's test statistic is

$$A_\gamma = \frac{\sum_i R_i^2(\hat{Y}_i - \tilde{Y})}{\sum_{i,j} (\delta_{ij} - h_{ij})^2 (Y_i - \bar{Y})(Y_j - \bar{Y}) s^2} \tag{9.4}$$

where $\delta_{ij} = 1$ if $i = j$, otherwise 0, h_{ij} are elements of the hat matrix (or projection matrix)

$$\boldsymbol{H} = \boldsymbol{a}(\boldsymbol{a}'\boldsymbol{a})^{-1}\boldsymbol{a}',$$

and

$$\tilde{Y} = (n-k)^{-1} \sum (1 - h_{ii}) \hat{Y}_i, \qquad s^2 = (n-k)^{-1} \sum_i R_i^2.$$

Then if A_γ differs significantly from zero one assumes heteroscedasticity is present. Anscombe (1961) gives a formula for Var(A_γ) but it is not clear exactly how to carry out a test.

Bickel (1978) suggested the following modifications of Anscombe's statistic:

$$A_{x^2} = \sum_i (\hat{Y}_i - \bar{Y}) R_i^2 / \bar{\sigma}_{x^2}, \tag{9.5}$$

where

$$\bar{\sigma}_{x^2}^2 = \left\{ \sum_i (\hat{Y}_i - \bar{Y})^2 \right\} (n-k)^{-1} \sum_i (R_i^2 - \overline{R^2})^2,$$

$$\overline{R^2} = n^{-1} \sum_i R_i^2.$$

Under the null hypothesis of homoscedasticity, and assuming certain regularity conditions, Bickel showed that A_{x^2} is asymptotically $N(0, 1)$ as $n \to \infty$. Thus large values of $|A_{x^2}|$ indicate the presence of heteroscedasticity. Cook and Weisberg (1983) report some empirical sampling trials on Bickel's test, which show that it can be a very conservative test if the χ^2-levels are used.

A variation of this may be derived formally from the papers by Godfrey (1978) and Koenker (1981) ($\bar{Y}=n^{-1}\sum Y_i$):

$$\phi=\left\{\sum_i(\hat{Y}_i-\bar{Y})R_i^2\right\}^2 \Big/ \left[\sum_i(\hat{Y}_i-\bar{Y})^2 n^{-1}\left\{\sum_i(R_i^2-\overline{R^2})^2\right\}\right], \quad (9.6)$$

and reject H_0 whenever $\phi > \chi_1^2(\alpha)$ for the selected Type I error level $100\alpha\%$. This has the advantage over (9.4) that it is not necessary to compute the matrix $\boldsymbol{H}$, and the advantage over (9.5) that some simulation trials showed the Type I error levels to be closer to asymptotic levels. That is, the simulated significance levels for (9.6) are close to those of χ_1^2, whereas those for A_{x^2} are often a great deal lower. For these reasons the statistic ϕ is recommended.

Some other simulations were done on the statistics ϕ and A_{x^2} using standardized residuals instead of the ordinary residuals. But the Type I error levels for ϕ were sometimes too high (higher than for testing the ordinary residuals); and again the levels for A_{x^2} were too low. So we recommend using ϕ on the ordinary residuals; but when plotting the residuals, the standardized version may be better as it shows up the true departures more, and we are less concerned about error levels for informal plots.

There are various other tests which have been proposed in the literature. For use in a computer package it is very desirable to find a test which does not require the user to interact in any way. Thus the direction should not be a specified explanatory variable. Two possibilities emerge from this:

(1) test on the fitted values;
(2) any unspecified linear combination of the explanatory variables.

Another way would be to test on each explanatory variable separately, reducing the Type I error level by a factor equal to the number of variables p, the number of them (following from the Bonferroni inequality). The problem with the latter approach is that the test would be weak when there are a lot of explanatory variables.

To test for the variance changing monotonically with one of a number of specified variables, or with any linear combination of them, Godfrey (1978) devised a statistic G which is asymptotically distributed as χ^2. Let $\boldsymbol{Z}$ be the matrix with the given variables, on which the changing variance is supposed to depend, including a column of

1's, and let R_i ($i = 1, 2, \ldots, n$) be the residuals obtained from the full regression. The maximum likelihood estimate of error variance from the full OLS regression is

$$\hat{\sigma}^2 = n^{-1} \sum_{i=1}^{n} R_i^2.$$

Then Godfrey defines

$$r_i = (R_i^2/\hat{\sigma}^2) - 1$$

and

$$G = \tfrac{1}{2}\mathbf{r}' \boldsymbol{Z}(\boldsymbol{Z}'\boldsymbol{Z})^{-1}\boldsymbol{Z}'\mathbf{r}; \tag{9.7}$$

so $2G$ is the fitted sum of squares for regression of $\mathbf{r}$ on $\boldsymbol{Z}$. Godfrey shows that this is asymptotically (as $n \to \infty$, under certain regularity conditions) distributed as χ^2 on p_Z degrees of freedom, where p_Z denotes the number of variables included in $\boldsymbol{Z}$ excluding the column of 1's.

If Godfrey's test is used without specifying a subset of the explanatory variables then $\boldsymbol{Z}$ has to be assigned to the entire matrix of explanatory variables. The power of this procedure is weak when there are a moderate number of explanatory variables but Godfrey's test could be effectively used if p_Z is small, say 2 or 3. Otherwise, the test recommended above is to be preferred.

The next two sections of this chapter give a survey of some other tests and procedures for detecting heteroscedasticity. Section 9.4 deals with 'constructive tests' in which estimates of the non-constant variance are obtained. Section 9.5 deals with non-constructive tests. Those not interested in these procedures should pass straight to Section 9.6. Some of these procedures may be useful in special circumstances, as indicated below.

9.4* Constructive tests

If the test statistic (9.6) is significant, we might consider estimating the variance as a function of the fitted value. Anscombe's (1961) statistic A_γ was intended as an estimator for γ in the approximate relationship

$$\sigma_i^2 \propto \exp\{\gamma E(Y_i)\}, \tag{9.8}$$

and then σ^2 could be estimated as $n^{-1}\sum R_i^2$ since σ_i^2 is estimated by R_i^2. Neither Bickel (1978) nor Koenker (1981) mention this point, which perhaps needs looking into for these statistics.

There are some other tests for heteroscedasticity which may be useful in certain circumstances but have disadvantages as described below. The paper by Harvey (1976) also considers a model a bit similar to (9.8), namely

$$\sigma_i^2 = \exp(\mathbf{z}_i'\boldsymbol{\gamma}), \tag{9.9}$$

where the first element of $\mathbf{z}_i$ is 1 and the others are other variables which may be related to the regressors. An iterative method for maximizing the log-likelihood function $\log L$ is described (for estimating $\boldsymbol{\theta}$ and $\boldsymbol{\gamma}$), where he writes

$$\log L = \frac{n}{2}\log(2\pi) - \frac{1}{2}\sum_{i=1}^{n} \mathbf{z}_i'\boldsymbol{\gamma} - \frac{1}{2}\sum_{i=1}^{n} \exp\{-z_i'\gamma\}R_i^2.$$

This is then approximately maximized by two iterations to give an estimator $\boldsymbol{\gamma}^*$. The implication in the paper is that this estimator is substituted into the statistic, called P, say,

$$P = n\log s^2 - \sum_{i=1}^{n} \mathbf{z}_i'\boldsymbol{\gamma}^*,$$

where $s^2 = n^{-1}\sum_{i=1}^{n} R_i^2$. Then if P exceeds the upper χ_{j-1}^2 point, at a chosen level α, the null hypothesis of homoscedasticity is rejected. Here j denotes the number of components of $\boldsymbol{\gamma}$ or $\mathbf{z}_i$. The advantage of this approach, over those described above, is that there is no problem with having negative values of $\mathbf{z}_i'\boldsymbol{\gamma}$ because of the exponential term. One disadvantage is the corresponding increase in complexity and some simulation studies threw doubt on the χ_{j-1}^2 assertion.

Park (1966) tries to fit a relationship of the form $\sigma_i^2 = \sigma^2 \chi_i^\gamma e^v$ for one chosen explanatory variable, 'where v is a well-behaved error term'. Then σ^2 and γ are estimated by regressing $\log R_i^2$ on $\log X_i$, which is applicable only when X_i is always positive. It is this disadvantage (of possibly obtaining negative values of X_i) that Harvey (1976) tries to overcome using exponentials. Park's paper suggests a remedy for the presence of this form of heteroscedasticity but no significance test is given.

Rutemiller and Bowers (1968) describe how to estimate parameters α_i simultaneously with the β_i, using an approximate 'method of scoring', with

$$\sigma_j = \alpha_0 + \sum_{i=1}^{p} \alpha_i X_{ji} \quad \text{for } j = 1, 2, \ldots, n.$$

The solution for derivatives of the log-likelihoods (with respect to the β's and α's) being simultaneously zero is approximated using iteration involving the inverse of the matrix of expected values of second-order derivatives of the log-likelihood. An example is also considered where a quadratic form is assumed for σ_j. The main disadvantage of this method is the number ($2k$ for the linear version) of parameters that need estimating and convergence problems would result. Moreover, the X_{ji} in the formula for σ_j could not be replaced by the fitted values since the β's are estimated simultaneously.

Glejser's (1969) test is similar in that the absolute values of the residuals, as estimates of σ_j, are regressed on one variable which is usually a function of an explanatory variable. For instance, fit $|R_i| = \alpha_0 + \alpha_1 X_i$ or $\alpha_0 + \alpha_1 X_i^{1/2}$ if $X_i > 0$ always. Here the α's are determined after the OLS model has been fitted. Then if α_1 is significant we conclude that heteroscedasticity has been detected.

The above tests are called 'constructive', since although estimators of σ_i^2 are not given in all cases they could be constructed. For instance, with (9.5), A_{x^2} could be taken as an estimate of γ in (9.8) but its effectiveness would depend on the form of the alternative hypothesis.

9.5* Non-constructive tests

Goldfeld and Quandt (1965) describe a 'parametric' test in which the observations are first ordered according to the values of a given explanatory variable. A central portion of j of these observations is omitted and separate regressions are fitted to each of the ends left. Let S_1 and S_2 denote the residual sums of squares thus obtained (with equal numbers $(n-j)/2$ of observations at each end), set $R = S_2/S_1$ and compare with the F-levels on $(\frac{1}{2}\{n-j-2p-2\}, \frac{1}{2}\{n-j-2p-2\})$ d.f. Here the problem of how to order the observations arises, and ordering by the fitted values would imply doing three regressions altogether and R would no longer have this F-distribution. They also describe a nonparametric test in which the observations are ordered as for the parametric version and a regression done on all the observations. A 'peak' in the sequence $|R_1|, |R_2|, \ldots, |R_n|$ is defined as any $|R_j|$ such that $|R_j| > |R_i|$ for $i = 1, 2, \ldots, j$. The number of such peaks is counted, and if significantly large (by comparison with their Table 1 of critical values) it is concluded that the variance increases with the chosen explanatory variable. The table only goes up to $n = 60$ and Hedayat and Robson (1970) write that this test (the peak test) does not take account of the fact that the residuals are not independent and identically distributed.

If we try tests based on BLUS or recursive residuals, there is the question of how to choose which observations to omit for these. The tests developed by Theil (1971) for BLUS residuals and Harvey and Phillips (1974) for recursive residuals, require one of the regressors (or time or observation number) to be singled out, on which the variance is suspected of depending. Then a central portion of observations is omitted and the ratio of error sums of squares for those at either end is compared with F-levels (these are similar to Goldfeld and Quandt's parametric test). Harvey and Phillips also suggest performing their test with different ordering of the recursive residuals, 'to test whether the variances are an increasing function of another explanatory variable', and conjecture that little loss of power would result by omitting the same set of observations (i.e. using the same set of recursive residuals). They argue that this adds to the flexibility of the test, but do not mention the effect such a procedure would have on the Type I error level. Hedayat and Robson (1970) maintain that the particular set of BLUS or recursive residuals chosen will be influenced by what kind of deviation from assumption is being looked for. In this paper the authors considered doing a test, for variance changing monotonically with the mean, on a set of residuals similar in most of their properties to recursive residuals. These are calculated in a stepwise way, adding one observation at a time to produce new regression estimates. They apply them to the Goldfeld and Quandt 'peak' test and show that it gives better results than using ordinary residuals. Here again there is the question of which observations to omit and how to divide the remaining observations into two subsets.

A more general class of tests is suggested by Szroeter (1978), for testing $H_0\colon \sigma_i^2 = \sigma_{i-1}^2$ $(i = 1, 2, 3, \ldots, n)$ against $H_1\colon \sigma_i^2 > \sigma_{i-1}^2$ for some i. Let A be some non-empty subset of $\{1, 2, \ldots, n\}$ and $\{\tilde{\varepsilon}_i; i \in A\}$ denote a set of computable residuals approximating the generally unobservable error terms. Let $\{h_i\colon i \in A\}$ be a prescribed non-decreasing sequence of numbers, and define

$$\tilde{h} = \sum_{i \in A} \tilde{w}_i h_i, \quad \text{where} \quad \tilde{w}_i = \tilde{\varepsilon}_i^2 \Big/ \sum_{j \in A} \tilde{\varepsilon}_j^2. \tag{9.10}$$

The test procedure is to reject H_0 whenever $\tilde{h}$ is significantly larger than $h = n_A^{-1} \sum_{i \in A} h_i$, where n_A denotes the number of elements of A. By omitting a central portion of observations to define A and choosing $h_i = +1$ for one end and $h_i = -1$ for the other, and respectively, ordinary, recursive and BLUS residuals for $\{\tilde{\varepsilon}_i\}$, it is shown that Goldfeld and Quandt's (1965) parametric test, Harvey and Phillips's (1974) test and Theil's (1971) test are all special cases. In Section 3 Szroeter assumes $\{\tilde{\varepsilon}_i\}$ are a set of recursive, BLUS (or similar) residuals, and shows how the above approach can yield any of three statistics with the same distributions as the following: the von Neumann (1941) ratio, the upper or lower bounding ratio for the Durbin–Watson (1950) statistic. The significance levels of these have been tabulated and computer algorithms for those of the Durbin–Watson statistic have been written. In Section 4, the ordinary residuals $\{R_i\}$ are used to construct a statistic for which a bounds test can be applied, again using the Durbin–Watson bounding ratios. But in this case the distribution of the statistic is not obtained, only upper and lower bounds; so there is an inconclusive region which becomes wider as the number of explanatory variables increases. In the last section he showed, asymptotically, that 'the test will tend to be more powerful the higher the degree of correlation between h_i and $\mathrm{Var}(\varepsilon_i)$ $(i = 1, 2, \ldots, n)$', but he does not examine the implications of this for his proposed tests. The precise interpretation of this theorem is not obvious.

Harrison and McCabe (1979) partially overcome this deficiency for a bounds test which uses the ordinary residuals $\{R_i\}$. They propose a statistic of the form

$$b = \left\{\sum_{i \in a} R_1^2\right\} \Big/ \left\{\sum_{i=1}^{n} R_1^2\right\}, \tag{9.11}$$

whose distribution depends on the matrix $\boldsymbol{a}$ of explanatory variables. They find statistics b_L, b_U such that $b_L \leq b \leq b_U$, which are attainable for some matrices $\boldsymbol{a}$, and such that b_L, b_U have beta-distributions independent of $\boldsymbol{a}$. For the inconclusive region they approximate b by the beta-distribution and estimate its parameters by equating the means and variances. But this involves computing eigenvalues with consequent computational burden. Their simulations show that 'the incidence of inconclusiveness is likely to be quite high in small samples', although they tried only $n = 20$. This inconclusive region becomes especially large as the number k of explanatory variables increases towards or beyond $n/2$. The models involve σ_i^2 increasing with some prescribed variable (explanatory or observation number), the tests are all one-sided yet they quote powers for using both b_L and

b_U, which seems a confusion of issues. The powers for using b_L fall short of those for the Goldfeld and Quandt (1965) parametric test GQ, Theil's (1971) BLUS test T and Harvey and Phillips (1974) test HP. The powers for using the beta-approximated significance levels for b (without using the bounds b_L or b_U) are mostly slightly higher than the powers of the three other tests GQ, T and HP.

A modification of the BLUS test was proposed by Ramsey (1969): calculate $(n-k)$ BLUS residuals to correspond with $(n-k)$ of the disturbances (they are intended to estimate a given set of $n-k$ disturbances). These are divided into three groups of sizes $v_i + 1$ $(i = 1, 2, 3)$ and three mean square errors s_1^2, s_2^2, s_3^2 calculated together with the total mean square error s^2. Then Bartlett's M-test, based on likelihood ratio, is reject H_0 (homoscedasticity) whenever $-2\log l^*$ exceeds the upper χ^2 point on three d.f., where

$$l^* = \prod_{i=1}^{3} (s_i^2/s^2)^{v_i/2}.$$

9.6 Serial correlation: The Durbin–Watson test

Another type of departure from the OLS assumption that $\boldsymbol{V} = \boldsymbol{I}$ in (9.1) is that successive observations are correlated. This can occur if our data are time series data, such as some economic time series, or else process data with observations taken sequentially. Serial correlation also occurs sometimes from the use of on-line instruments in industrial processes.

A simple model for correlated errors is that the errors e_i are from the AR (1) process

$$e_i = \rho e_{i-1} + u_i, \qquad (9.12)$$

where the u_i are independent and normally distributed random variables with a variance σ_u^2. For this model the marginal variance of the errors is

$$V(e) = \rho_u^2/(1-\rho^2) \qquad (9.13)$$

and the autocorrelation coefficients are

$$\rho_s = \rho^s,$$

which implies that the effect of individual u_i gradually declines. Before passing on to the test procedure we remark that there could be other forms of correlation in the errors, see Chapter 13.

Durbin and Watson (1950) introduced the following test for AR (1) errors. We calculate

$$D = \sum_{i=2}^{n} (r_i - r_{i-1})^2 \bigg/ \sum_{i=1}^{n} r_i^2. \tag{9.14}$$

The distribution of (9.14) under the null hypothesis $\rho = 0$ can be calculated exactly but it depends on the matrix of explanatory variables $\boldsymbol{a}$. Durbin and Watson were able to obtain bounds which do not depend on $\boldsymbol{a}$ but these leave an inconclusive region of the test. The possible values of D are $(0, 4)$. The procedure is as follows.

For testing for positive serial correlation,

reject the null hypothesis if $D < D_L^*$,
accept the null hypothesis if $D > D_U^*$,

and otherwise regard the test as inconclusive.

For testing for negative serial correlation,

reject the null hypothesis if $D > 4 - D_L^*$,
accept the null hypothesis if $D < 4 - D_U^*$,

and otherwise regard the test as inconclusive.

Clearly, this 'inconclusive region' test is unsatisfactory and Durbin and Watson (1971) gave an approximate procedure which can be used. Otherwise, a method of determining the distribution of D exactly has been given by L'Esperance *et al.* (1976), see also Farebrother (1980).

It is probably well worth incorporating the Durbin–Watson test into a multiple regression package, but it does only test for one type of correlation which may or may not be the most common in any particular field of application.

9.7 Weighted least squares (WLS)

If heteroscedasticity or serial correlation is discovered when analysing residuals, the OLS procedure is not applicable and other methods should be used. A full discussion of the possibilities is outside the scope of this volume, but in the final two sections of this chapter we outline the principle involved.

A particularly simple case is when V of (9.1) is diagonal but with unequal elements. This represents the situation where all the observations are statistically independent but with differing variances. A common case when this arises in the analysis of survey data, when the variables input to regression are all means of differing numbers of observations, and there is no access to the original data. The method of solution for diagonal V is called weighted least squares, (WLS), and there are two cases, known weights and estimated weights.

9.7.1 Known weights

We can write

$$V = D^2, \tag{9.15}$$

where D is a diagonal matrix with standard deviations of the observations on the diagonals. Then in place of

$$\mathbf{Y} = a\boldsymbol{\theta} + \varepsilon, \tag{9.16}$$

we write

$$D^{-1}\mathbf{Y} = D^{-1}a\boldsymbol{\theta} + \boldsymbol{\eta}, \tag{9.17}$$

where $\boldsymbol{\eta} = D^{-1}\varepsilon$ is a vector of statistically independent residuals with equal variances. The procedure is therefore to regress $D^{-1}\mathbf{Y}$ on $D^{-1}a$, using OLS. This simply amounts to minimizing quantities such as

$$\sum_{i=1}^{n} \{y_i - \alpha - \beta_1 x_{i1} - \beta_2 x_{i2}\}^2/\sigma_i^2$$

in the case of regression on two explanatory variables.

The WLS estimator is

$$\hat{\boldsymbol{\theta}} = (a'V^{-1}a)a'V^{-1}Y$$

as in (9.2) but there V is diagonal. The residuals are

$$\mathbf{R} = D^{-1}\mathbf{Y} - D^{-1}a\hat{\boldsymbol{\theta}}, \tag{9.18}$$

that is, using the transformed data. The analysis of variance should also be presented in the transformed scales.

9.7.2 Unknown weights

A model which occurs frequently, for example, in econometric models, is when V depends on the expectation or on $\boldsymbol{\theta}$.

$$E(\mathbf{Y}) = a\boldsymbol{\theta},$$

$$V(\mathbf{Y}) = V(\boldsymbol{\theta}).$$

There are two ways of dealing with this model. One approach is to write out the full likelihood for the model and maximize this directly to estimate $\boldsymbol{\theta}$ and σ^2. For a discussion of suitable optimization methods, see Chambers (1977).

The second approach is to use iterative weighted least squares (IWLS). An unweighted regression is performed first, and the estimates of $\boldsymbol{\theta}$ obtained are used to calculate the weights for a WLS regression. This process can be repeated but Williams (1959) has suggested that one cycle is often sufficient. It has been established under fairly general conditions that the maximum likelihood and IWLS approaches are equivalent (Nelder and Wedderburn, 1972; Bradley, 1973; Charnes *et al.*, 1976).

9.8 Generalized least squares (GLS)

Generalized least squares (GLS) applies when we have a $\boldsymbol{V}$ in (9.1) which is not diagonal. In many ways the analysis proceeds in a similar way to WLS.

9.8.1 Known $\boldsymbol{V}$

Since $\boldsymbol{V}$ is symmetric and positive-definite we can find a matrix $\boldsymbol{P}$ such that

$$\boldsymbol{P}'\boldsymbol{P} = \boldsymbol{V}^{-1}. \tag{9.19}$$

If we then define

$$\mathbf{Z} = \boldsymbol{P}\mathbf{Y}, \qquad \boldsymbol{X} = \boldsymbol{P}\boldsymbol{a}, \tag{9.20}$$

then we have the model

$$\left.\begin{aligned} E(\mathbf{Z}) &= \boldsymbol{X}\boldsymbol{\theta}, \\ V(\mathbf{Z}) &= \boldsymbol{I}\sigma^2. \end{aligned}\right\} \tag{9.21}$$

We then have

$$\hat{\boldsymbol{\theta}} = (\boldsymbol{X}'\boldsymbol{X})^{-1}\boldsymbol{X}'\boldsymbol{Z} = (\boldsymbol{a}'\boldsymbol{V}^{-1}\boldsymbol{a})^{-1}\boldsymbol{a}'\boldsymbol{V}^{-1}\mathbf{Y} \tag{9.22}$$

and

$$V(\hat{\boldsymbol{\theta}}) = (\boldsymbol{X}'\boldsymbol{X})^{-1}\sigma^2 = (\boldsymbol{a}'\boldsymbol{V}^{-1}\boldsymbol{a})^{-1}\sigma^2. \tag{9.23}$$

Residuals can be defined in the transformed scales, as with WLS, but since $\boldsymbol{V}$ is not diagonal, and hence $\boldsymbol{P}$ is not diagonal either, these residuals will not have a clear interpretation. However, it does follow that

$$(\mathbf{Y} - \boldsymbol{a}\hat{\boldsymbol{\theta}})' V^{-1} (\mathbf{Y} - \boldsymbol{a}\hat{\boldsymbol{\theta}})/\sigma^2 \tag{9.24}$$

has a χ^2-distribution with $(n - k)$ degrees of freedom where k is the number of elements in $\boldsymbol{\theta}$.

An application of this method will be given in Chapter 13.

9.8.2 Unknown V

The case of a general unknown V cannot be dealt with since it has too many unknown parameters. However, there may be cases when the form of V is known but is contains unknown parameters, either some of the regression coefficients or other parameters dealing with, for example, the correlation between the ε terms. Here the same two options arise as with WLS. Again, an example will be given in Chapter 13. For access to the literature, see Theil (1971) or Judge *et al.* (1980).

References

Anscombe, J. F. (1961) Examination of residuals. *Proceedings of the Fourth Berkely Symposium on Mathematical Statistics and Probability*, Vol. 1. University of California Press, Berkeley, CA, pp. 1–36.

Bickel, P. J. (1978) Using residuals robustly. I: Tests for heteroscedasticity, nonlinearity. *Ann. Statist.*, **6**, 266–291.

Bradley, E. L. (1973) The equivalence of maximum likelihood and weighted least squares estimates in the exponential family. *J. Amer. Statist. Assoc.*, **68**, 199–200.

Carroll, R. J. and Ruppert, D. (1981) On robust test for heteroscedasticity. *Ann. Statist.*, **9**, 205–209.

Chambers, J. M. (1977) *Computational Methods for Data Analysis.* Wiley, New York.

Charnes, A., Frome, E. L. and Yu, P. L. (1976) The equivalence of generalized least squares and maximum likelihood estimates in the exponential family. *J. Amer. Statist. Assoc.*, **71**, 169–171.

Cook, R. D. and Weisberg, S. (1983) Diagnostics for heteroscedasticity in regression. *Biometrika*, **70**, 1–10.

Durbin, J. and Watson, G. S. (1950) Testing for serial correlation in least squares regression, I. *Biometrika*, **37**, 409–428.

Durbin, J. and Watson, G. S. (1971) Testing for serial correlation in least squares regression, III. *Biometrika*, **58**, 1–42.

Farebrother, R. W. (1980) AS153 Pan's procedure for the tail probabilities of the Durbin–Watson statistic. *Appl. Statist.*, **29**, 224–227.

Glejser, H. (1969) A new test for heteroscedasticity. *J. Amer. Statist. Assoc.*, **64**, 316–323.

Godfrey, L. G. (1978) Testing for multiplicative heteroscedasticity. *J. Econ.*, **8**, 227–236.

Goldfeld, S. M. and Quandt, R. E. (1965) Some tests for homoscedasticity. *J. Amer. Statist. Assoc.*, **60**, 539–547.

Harrison, M. J. and McCabe, B. P. M. (1979) A test for heteroscedasticity based on ordinary least squares residuals. *J. Amer. Statist. Assoc.*, **74**, 494–499.

Harvey, A. C. (1976) Estimating regression models with multiplicative heteroscedasticity. *Econometrica*, **44**, 461–465.

Harvey, A. C. and Phillips, G. D. A. (1974) A comparison of the power of some tests for heteroscedasticity in the general linear model. *J. Econ.* **2**, 307–316.

Hedayat, A. and Robson, D. S. (1970) Independent stepwise residuals for testing homoscedasticity. *J. Amer. Statist. Assoc.*, **65**, 1573–1581.

Holt, D., Smith, T. M. F. and Winter, P. D. (1980) Regression analysis of data from complex surveys. *J. Roy. Statist. Soc.* A, **143**, 474–487.

Horn, P. (1981) Heteroscedasticity: a non-parametric alterative to the Goldfeld–Quandt peak test. *Commun. Statist.* A, **10**, 795–808.

Judge, G. G., Griffiths, W. E., Hill, R. C. and Lee, Tsoung-Chao (1980) *The Theory and Practice of Econometrics.* Wiley, New York.

Koenker, R. (1981) A note on studentizing a test for heteroscedasticity. *J. Econ.*, **17**, 107–112.

L'Esperance, W. L. Chall, D. and Taylor, D. (1976) An algorithm for determining the distribution function of the Durbin–Watson statistic. *Econometrika*, **44**, 1325–1346.

Nelder, J. A. and Wedderburn, R. W. M. (1972) Generalized linear models. *J. Roy. Statist. Soc.* A, **135**, 370–384.

von Neumann, J. (1941) Distribution of the ratio of the mean square successive difference to the variance. *Ann. Math. Statist.*, **12**, 367–395.

Park, R. E. (1966) Estimation with heteroscedastic error terms. *Econometrica*, **34**, 888.

Ramsey, J. B. (1969) Tests for specification errors in classical linear least squares regression analysis. *J. Roy. Statist. Soc.* B, **31**, 350–371.

Rutemiller, H. C. and Bowers, D. A. (1968) Estimation in a heteroscedastic regression model. *J. Amer. Statist. Assoc.*, **63**, 552–557.

Szroeter, J. (1978) A class of parametric tests for heteroscedasticity in linear econometric models. *Econometrica*, **46**, 1311–1327.

Theil, H. (1971) *Principles of Econometrics.* Wiley, New York.

Williams, E. J. (1959) *Regression Analysis.* Wiley, New York.

Further reading

Abdullah, M. (1979) Tests for heteroscedasticity in regression: a review. M.Sc. Dissertation, University of Kent.

Draper, N. R. and Smith, H. (1981) Applied Regression Analysis. Wiley, New York.

Nelder, J. A. (1968) Regression, model building and invariance. *J. Roy. Statist. Soc.* A, **131**, 303–315.

CHAPTER 10

Predictions from regression

10.1 Underfitting

In this chapter we are going to discuss the use of a fitted regression model to make predictions. Clearly, a prediction is obtained by merely inserting the new values of the explanatory variables into the fitted regression equation, and this chapter concentrates mainly on interval estimates. Before going into the question of the variance associated with a prediction, we need to show clearly what the effect is of underfitting and overfitting. Both of these can be described as 'model fitting' errors and both can have a severe effect on predictions.

Firstly, we are going to consider what happens if we fit less than the full model, say only $k_1, k_1 < k$, of the parameters. Let our model be as defined in (1.13)

$$E(\mathbf{Y}) = \boldsymbol{a}\boldsymbol{\theta}, \tag{10.1}$$

but we shall rewrite this as

$$E(\mathbf{Y}) = \boldsymbol{a}_1\boldsymbol{\theta}_1 + \boldsymbol{a}_2\boldsymbol{\theta}_2, \tag{10.2}$$

where $\boldsymbol{\theta}_1$ and $\boldsymbol{\theta}_2$ have k_1 and k_2 components respectively, where $k_1 + k_2 = k$.

The least squares estimator based on the model

$$\mathbf{Y} = \boldsymbol{a}_1\boldsymbol{\theta}_1 + \boldsymbol{\varepsilon} \tag{10.3}$$

is

$$\hat{\boldsymbol{\theta}}_1 = (\boldsymbol{a}'_1\boldsymbol{a}_1)^{-1}\boldsymbol{a}'_1\mathbf{Y}, \tag{10.4}$$

and therefore

$$\begin{aligned} E(\hat{\boldsymbol{\theta}}_1) &= (\boldsymbol{a}'_1\boldsymbol{a}_1)^{-1}\boldsymbol{a}'_1E(\mathbf{Y}) \\ &= \boldsymbol{\theta}_1 + (\boldsymbol{a}'_1\boldsymbol{a}_1)^{-1}\boldsymbol{a}'_1\boldsymbol{a}_2\boldsymbol{\theta}_2 \\ &= \boldsymbol{\gamma}_1 \quad \text{where this defines } \boldsymbol{\gamma}_1. \end{aligned} \tag{10.5}$$

This shows what we would expect – that by fitting less than the full model, biases are introduced unless we happen to have

$$\boldsymbol{a}_1'\boldsymbol{a}_2 = \mathbf{0}$$

or

$$\boldsymbol{a}_1'\boldsymbol{a}_2\boldsymbol{\theta}_2 = \mathbf{0}.$$

For the variance of (10.4) we obtain

$$V(\hat{\boldsymbol{\theta}}_1) = (\boldsymbol{a}_1'\boldsymbol{a}_1)^{-1}\sigma^2 \tag{10.6}$$

as usual. However, it is clear that biases will also be introduced into the residual sum of squares term.

Denote the residual sum of squares based on model (10.3) as RSS_r, then

$$\text{RSS}_r = (\mathbf{Y} - \boldsymbol{a}_1\hat{\boldsymbol{\theta}}_1)'(\mathbf{Y} - \boldsymbol{a}_1\hat{\boldsymbol{\theta}}_1). \tag{10.7}$$

Write

$$\begin{aligned}\mathbf{Y} - \boldsymbol{a}_1\hat{\boldsymbol{\theta}}_1 &= \mathbf{Y} - \boldsymbol{a}_1(\boldsymbol{a}_1'\boldsymbol{a}_1)^{-1}\boldsymbol{a}_1'\mathbf{Y} \\ &= (\boldsymbol{I} - \boldsymbol{H}_1)\mathbf{Y}, \quad \text{say.}\end{aligned}$$

Now $(\boldsymbol{I} - \boldsymbol{H}_1)$ is idempotent,

$$(\boldsymbol{I} - \boldsymbol{H}_1)^2 = (\boldsymbol{I} - \boldsymbol{H}_1)$$

so that

$$\text{RSS}_r = \mathbf{Y}'(\boldsymbol{I} - \boldsymbol{H}_1)\mathbf{Y}.$$

Now

$$\begin{aligned}E(\text{RSS}_r) &= E\{(\mathbf{Y} - \boldsymbol{a\theta})'(\boldsymbol{I} - \boldsymbol{H}_1)(Y - \boldsymbol{a\theta})\} + \boldsymbol{\theta}'\boldsymbol{a}'(\boldsymbol{I} - \boldsymbol{H}_1)\boldsymbol{a\theta} \\ &= \text{tr}\, E\{(\boldsymbol{I} - \boldsymbol{H}_1)(Y - \boldsymbol{a\theta})(Y - \boldsymbol{a\theta})'\} + \boldsymbol{\theta}'\boldsymbol{a}'(\boldsymbol{I} - \boldsymbol{H}_1)\boldsymbol{a\theta} \\ &= \sigma^2\, \text{tr}(\boldsymbol{I} - \boldsymbol{H}_1) + \boldsymbol{\theta}'\boldsymbol{a}'(\boldsymbol{I} - \boldsymbol{H}_1)\boldsymbol{a\theta} \\ &= (n - k_1)\sigma^2 + \boldsymbol{\theta}'\boldsymbol{a}'(\boldsymbol{I} - \boldsymbol{H}_1)\boldsymbol{a\theta} \\ &= (n - k)_1\sigma^2 + \boldsymbol{\theta}_2'\boldsymbol{a}_2'(\boldsymbol{I} - \boldsymbol{a}_1(\boldsymbol{a}_1'\boldsymbol{a}_1)^{-1}\boldsymbol{a}_1')\boldsymbol{a}_2\boldsymbol{\theta}_2\end{aligned}$$

and the last term on the right is the sum of squares bias. This last result shows that the residual mean square will be larger than σ^2, even if the bias in the estimators is zero.

It is clear that any bias in the estimators will severely affect prediction, especially if there is any ‘projection’ beyond the region covered by the data. Underfitting will also inflate the estimate of variance.

EXAMPLE 10.1

Suppose that we fit the model

$$E(Y) = \beta_0 + \beta_1 X_1,$$

when the true model is

$$E(Y) = \beta_0 + \beta_1 X_1 + \beta_2 X_2$$

so that in the notation used above

$$\boldsymbol{\theta}_1 = (\beta_0 \beta_1), \qquad \boldsymbol{\theta}_2 = \beta_2,$$

where the Y's are independently distributed with a variance σ^2. Let the values of the variables be as follows:

$$\begin{array}{rccccc} X_1: & -2 & -1 & 0 & 1 & 2, \\ X_2: & -1 & -2 & 0 & 2 & 1, \\ Y: & Y_1 & Y_2 & Y_3 & Y_4 & Y_5. \end{array}$$

Then we have the following results:

$$\boldsymbol{a}_1 = \begin{bmatrix} 1 & -2 \\ 1 & -1 \\ 1 & 0 \\ 1 & 1 \\ 1 & 2 \end{bmatrix}, \qquad \boldsymbol{a}_2' = (-1 \quad -2 \quad 0 \quad 2 \quad 1)$$

$$\boldsymbol{a}_1' \boldsymbol{a}_1 = \begin{bmatrix} 5 & 0 \\ 0 & 10 \end{bmatrix}.$$

Therefore from (10.4) the bias in estimating $(\beta_0 \quad \beta_1)$ is

$$(\boldsymbol{a}_1' \boldsymbol{a}_1)^{-1} \boldsymbol{a}_1' \boldsymbol{a}_2 \beta_2 = \begin{bmatrix} \frac{1}{5} & 0 \\ 0 & \frac{1}{10} \end{bmatrix} \begin{bmatrix} 0 \\ 8 \end{bmatrix} \beta_2 = \begin{bmatrix} 0 \\ 0.8 \end{bmatrix} \beta_2.$$

This means that $\hat{\boldsymbol{\beta}}_0$ is unbiased but $\hat{\boldsymbol{\beta}}_1$ has bias $0.8\boldsymbol{\beta}_2$.

The practical difficulty here is that we may often omit an important variable simply because we did not know about it. It is wise to inspect the residuals carefully for any special patterns which occur, as this is one way in which the omission of an important variable can be detected. A specific test for an omitted variable was devised by Ramsey (1969), and in effect this test looks for patterns in the residuals; see Section 7.4.

10.2 The effect of overfitting on prediction

We have examined the effect of underfitting on a regression, so we now turn to overfitting.

Suppose we fit the model

$$E(\mathbf{Y}) = \boldsymbol{a\theta}$$

to get predicted (mean) values

$$\hat{\mathbf{Y}} = \boldsymbol{a}(\boldsymbol{a}'\boldsymbol{a})^{-1}\boldsymbol{a}'\mathbf{Y} = \boldsymbol{a}A^{-1}\boldsymbol{a}'\mathbf{Y} \tag{10.8}$$

so that we have

$$V(\hat{\mathbf{Y}}) = \boldsymbol{a}(\boldsymbol{a}'\boldsymbol{a})^{-1}\boldsymbol{a}'\sigma^2 = \boldsymbol{H}\sigma^2. \tag{10.9}$$

Now we add one further variable, denoted by an $n \times 1$ vector $\mathbf{z}$ so that we wish to invert

$$(\boldsymbol{a}\mathbf{z})'(\boldsymbol{a}\mathbf{z}) = \begin{bmatrix} A & \boldsymbol{a}'\mathbf{z} \\ \mathbf{z}'\boldsymbol{a} & \mathbf{z}'\mathbf{z} \end{bmatrix}$$

$$= \begin{bmatrix} A & B \\ B' & C \end{bmatrix}, \quad \text{say.}$$

By using the results in Section 10.7 the inverse is

$$\begin{bmatrix} A^{-1} + mA^{-1}\boldsymbol{BB}'A^{-1} & -mA^{-1}\boldsymbol{B} \\ -m\boldsymbol{B}'A^{-1} & m \end{bmatrix}, \tag{10.10}$$

where

$$m = 1/(\mathbf{z}'\mathbf{z} + \mathbf{z}'\boldsymbol{a}(A^{-1}\boldsymbol{a}'\mathbf{z})).$$

The new predicted values are

$$(\boldsymbol{a}\mathbf{z})\{(\boldsymbol{a}\mathbf{z})'(\boldsymbol{a}\mathbf{z})\}^{-1}(\boldsymbol{a}\mathbf{z})'\mathbf{Y} \tag{10.11}$$

and their variances are

$$\sigma^2(\boldsymbol{a}\mathbf{z})\begin{bmatrix} A^{-1} + mA^{-1}\boldsymbol{BB}'A^{-1} & -mA^{-1}\boldsymbol{B} \\ -m\boldsymbol{B}'A^{-1} & m \end{bmatrix}(\boldsymbol{a}\mathbf{z})'. \tag{10.12}$$

It is readily seen that this is equal to

$$\sigma^2\boldsymbol{H} + m\sigma^2(\boldsymbol{a}A^{-1}\boldsymbol{B} - \mathbf{z})(\boldsymbol{a}A^{-1}\boldsymbol{B} - \mathbf{z})'. \tag{10.13}$$

From this it follows that all the prediction variances cannot decrease and will often increase by adding one variable.

The position is as follows. Increasing the number of variables in a regression equation will reduce the risk of biases which arise from underfitting, but it will increase the prediction variance.

In making the statements above we have assumed a common known value of σ^2. If estimated values are inserted, and if we make the estimated values $\hat{\sigma}^2$ depend on the equation fitted, then this will also affect the result.

EXAMPLE 10.2

(Walls and Weeks, 1969). Suppose we are given the following

y	5	7	7	10	16	20
x	1	2	3	4	5	6

then we assume $V(y) = \sigma^2$ and fit two models.

$$\text{Model 1:} \quad E(Y) = \beta_0 x,$$
$$\text{Model 2:} \quad E(Y) = \beta_1 x + \beta_2 x^2.$$

By applying least squares we get the following results:

$$\text{Model 1:} \quad \hat{y} = 280x/91,$$
$$\text{Model 2:} \quad \hat{y} = (30184x + 1736x^2)/12544.$$

The standardized prediction variances are as given in Table 10.1.

Example 10.2 shows the effect of adding one variable very dramatically. For the first observation, the variance of Model 2 is more than ten times the variance of Model 1. While this is a rather special example, it does demonstrate the error of putting too many variables into a regression model.

A common policy when using multiple regression is to 'throw everything' into the equation. There are a number of reasons against adopting this policy, see Section 11.1.5 but the results of this section provide one powerful argument. To incorporate worthless variables into a regression model will increase prediction variances.

Table 10.1 *Values of* $V(\hat{Y})/\sigma^2$ *for two models.*

x	y	*Model 1*	*Model 2*
1	5	0.0110	0.1183
2	7	0.0440	0.2790
3	7	0.0989	0.3214
4	10	0.1758	0.2589
5	16	0.2747	0.2790
6	20	0.2956	0.7433

10.3 Variances of predicted means and observations

There are two different variances associated with a prediction depending on whether we are predicting a mean or an observation. The choice between these two is determined by the problem on hand.

Let our model be as defined in (10.1) then we have

$$\hat{\boldsymbol{\theta}} = (\boldsymbol{a}'\boldsymbol{a})^{-1}\boldsymbol{a}'\mathbf{y}.$$

The predicted means corresponding to l sets of values of the explanatory variables, which are written '$\boldsymbol{a}_0$' (an $l \times k$ matrix), are given by the vector

$$\hat{Y}(\boldsymbol{a}_0) = \boldsymbol{a}_0(\boldsymbol{a}'\boldsymbol{a})^{-1}\boldsymbol{a}'\mathbf{y}, \qquad (10.14)$$

where $\hat{Y}(\boldsymbol{a}_0)$ is a predicted mean of Y at '$\boldsymbol{a}_0$'. The variance of this is easily seen to be

$$V(\hat{Y}(\boldsymbol{a}_0)) = \boldsymbol{a}_0(\boldsymbol{a}'\boldsymbol{a})^{-1}\boldsymbol{a}_0'\sigma^2. \qquad (10.15)$$

If we use our model in the centred form

$$E(\mathbf{Y}) = \alpha\mathbf{1} + X\boldsymbol{\beta}, \qquad (10.16)$$

where

$$\mathbf{1}'X = \mathbf{0},$$

then we find that

$$V(\hat{Y}(\boldsymbol{a}_0)) = \left(\frac{1}{n}\mathbf{1}\mathbf{1}'X_0(X'X)^{-1}X_0'\right)\sigma^2, \qquad (10.17)$$

where

$$\boldsymbol{a}_0 = (\mathbf{1}X_0).$$

EXAMPLE 10.3

If we fit the model

$$E(Y_i) = \alpha + \beta(x_i - \bar{x}), \qquad i = 1, 2, \ldots, n$$

to data (x_i, y_i), then we have

$$\hat{\alpha} = \bar{y}, \qquad \hat{\beta} = \mathrm{CS}(x, y)/\mathrm{CS}(x, x),$$

where

$$\mathrm{CS}(x, y) = \sum_1^n (x_i - \bar{x})(y_i - \bar{y}).$$

The variance of a predicted mean at a new value x_0 is

$$V(\hat{Y}(a_0)) = \sigma^2\left(\frac{1}{n} + \frac{(x_0 - \bar{x})^2}{\mathrm{CS}(x, x)}\right). \tag{10.18}$$

This shows the way in which the variance increases as x_0 departs from $\bar{x}$.

Having obtained the formulae for the variances, it is necessary to add a word of caution about using it. If we want a prediction at a point in the space of the explanatory variables which is beyond the region covered by the data, or is at a point where observations used to fit the equation are rare, we may have severe doubts about our prediction. Any form of projection is particularly sensitive to the assumption that our fitted model holds near the predicted point, and if there are a few such observational points we may doubt this. A further feature which complicates matters is that correlations within the data may lead us to a model which fits the observed data well but which gives bad predictions. It is important to read Chapter 11 before using a multiple regression model for prediction, and, in particular, to take notice of the section on data splitting and cross-validation.

10.3.1 Variance of a predicted observation

If we want to make a prediction about an actual observation rather than about a mean response to a new set of explanatory variables, the above formulae have to be modified in an obvious way. If we denote a predicted observation as $\hat{y}$, we have

$$V(\hat{y}) = V(\hat{\mu}_y) + V(Y)$$

when the last component represents the random variation of the observation about its expectation so that formulae (10.15) and (10.17) become

$$V(\hat{y}) = (I + \boldsymbol{a}_0(\boldsymbol{a}'\boldsymbol{a})^{-1}\boldsymbol{a}_0)\sigma^2, \tag{10.19}$$

and

$$V(\hat{y}) = \left(I + \frac{1}{n}\mathbf{1}\mathbf{1}'X_0(X'X)^{-1}X_0'\right)\sigma^2. \tag{10.20}$$

Some values of predicted means and variances for the acetylene data are shown in Table 10.2.

The point C used in Table 10.2 is the mean of the explanatory

Table 10.2 *Predicted means and their variances for acetylene data. Predictions at three points, A, B and C.*

			Coef.	*t-value*	*Coef.*	*t-value*	*Coef.*	*t-value*	*Coef.*	*t-value*	*Coef.*	*t-value*
Regression coefficients	x_1		5.32	1.09	0.37	6.04	0.26	6.00	0.22	11.92	0.22	11.44
	x_2		19.24	4.47	19.07	4.43	14.03	3.20	9.99	11.10	10.21	11.08
	x_3		13 766.32	1.32	3200.45	3.10	1104.57	2.55	996.55	2.40	1289.85	
	x_1^2		−0.0019	1.02								
	x_2^2		−0.030	2.60	−0.027	2.40	−0.016	1.30	−0.016	1.36		
	x_3^2		−11 581.68	1.50	−3984.73	2.16						
	x_1x_2		−0.014	4.40	−0.014	4.39	−0.011	3.16	−0.0075	9.06	−0.0081	10.71
	x_1x_3		−10.58	1.28	−2.23	3.22	−0.93	2.45	−0.92	2.24	−1.21	3.23
	x_2x_3		−21.03	2.27	−20.06	2.18	−8.58	0.94				
Predicted Points	*A*	$\hat{Y}$	2 288.9		8.94		25.94		31.45		28.57	
		Var.	5.03E6		105.35		63.53		29.20		26.84	
	B	$\hat{Y}$	2 269.50		−39.17		1.57		32.620		31.97	
		Var.	5.16E6		1097.16		1082.62		0.5970		0.3977	
	C	$\hat{Y}$	42.97		34.31		35.16		35.36		35.24	
		Var.	72.86		0.25		0.14		0.0965		0.0964	

Prediction variables for points A, B and C.

	x_1	x_2	x_3	x_1^2	x_2^2	x_3^2	x_1x_2	x_1x_3	x_2x_3
A:	1 200	23	0.1	240 000	529	0.01	27 600	120	2.3
B:	1 200	10.0	0.04	240 000	100	0.001 6	12 000	48	4
C:	1 212.5	12.4	0.04	1 470 000	184	0.002 56	15 183.7	46.6	0.46

variables, so that we would expect predictions at this point to be relatively good. Both points A and B are 'projections', in various directions, and this shows how the variances of predictions are inflated. Table 10.2 also illustrates the results of Section 10.2, and it shows how the variances of predictions increase due to the inclusion of more variables.

10.4 Confidence intervals

Here we get into the area of simultaneous inference, because the confidence statements we can make differ according to whether we wish to consider each predicted point separately, or whether we want to take some decision or action depending on the set of predicted values simultaneously. A further case is when we wish to put a confidence band round a regression relationship. The difference is simply one of objective, although in any specific case we may be in some doubt as to which to apply.

10.4.1 Confidence intervals for each point separately

If we wish to predict a single point then we use the formula

$$(\text{Predicted mean}) \pm t_\nu(\alpha/2) \times \text{standard deviation}, \qquad (10.21)$$

where ν is the degrees of freedom for error, and α is the significance level. For the predicted mean we use equation (10.14) and for the variance we use either

(10.15) or (10.17) for a predicted mean,
(10.19) or (10.20) for a predicted observation.

These confidence intervals can be made for several points, and for each one, if we use the 90% confidence intervals, there will be a 10% chance of the interval not enclosing the required value. Separate predictions are not independent, but clearly, if we make several such statements, there is a high probability that at least one of our intervals will not contain the required value.

10.4.2 Confidence band for a regression relationship (Working–Hotelling–Scheffé method)

In this case we use the following formula. We shall use the model in the form (10.1) and write the vector

$$\mathbf{x} = (1, x_1, x_2, \ldots, x_{k-1})$$

Table 10.3 *Values of constants for confidence bands* ($\alpha = 5\%$).

k	2	2	4	4	10	10
$n-k$	20	30	20	30	20	30
t	2.09	2.04	2.09	2.04	2.09	2.04
F	2.64	2.58	3.39	3.28	4.85	4.65

as the vector of explanatory variables. The matrix of actual values of $\mathbf{x}$ make up the matrix $\boldsymbol{a}$. The problem is to construct upper and lower bounds which give a band within which the true regression relationship lies, with a confidence $100(1-\alpha)\%$. That is, given the matrix $\boldsymbol{a}$, then for all $\mathbf{x}$ we need a confidence band. The appropriate formula is the set of $\mathbf{y}$ contained in the intervals

$$\mathbf{x}(\boldsymbol{a}'\boldsymbol{a})^{-1}\boldsymbol{a}'\mathbf{y} \pm \{kF_\alpha(k, n-k)\}^{1/2} s\{\mathbf{x}'(\boldsymbol{a}'\boldsymbol{a})^{-1}\mathbf{x}\}^{1/2} \qquad (10.22)$$

for all x, where s^2 is the mean square error. If we wished to consider one point only we would use the multiplying factor

$$t_{n-k}(\alpha/2) \quad \text{instead of} \quad \{kF_\alpha(k, n-k)\}^{1/2}.$$

This shows the price we are paying for covering all values of $\mathbf{x}$. Some values of these constants, referred to as t and F respectively, are given in Table 10.3.

EXAMPLE 10.4

For simple linear regression we denote the corrected sum of squares of the explanatory variable as $\mathrm{CS}(x, x)$, then we have the following formula from (10.22)

$$\bar{y} + \hat{\beta}(x - \bar{x}) \pm (2F_{5\%}(2, n-2))^{1/2} s\left(\frac{1}{n} + \frac{(x-\bar{x})^2}{\mathrm{CS}(x, x)}\right)^{1/2}, \qquad (10.23)$$

where $\hat{\beta}$ is the estimated regression parameter and $\bar{x}$ and $\bar{y}$ are means. This is considerably wider than the corresponding interval for a single observation.

10.4.3 Simultaneous confidence intervals for prediction

We now turn again to the case when we have an $l \times k$ matrix $\boldsymbol{a}_0$ of explanatory variables at which we need l predictions. We now require a simultaneous confidence statement for all predictions and we shall use the model in the form (10.1). There are two methods. Let $\mathbf{x}_i$ denote the ith row of $\boldsymbol{a}_0$.

Scheffé's method gives the following formula:

$$\hat{Y}(\mathbf{x}_i) \pm (lF_\alpha(l, n-k))^{1/2} s(1 + \mathbf{x}_i'(\boldsymbol{a}'\boldsymbol{a})^{-1}\mathbf{x}_i)^{1/2} \qquad (10.24)$$

for $i = 1, 2, \ldots .\ l$. The other method is to apply the Bonferroni inequality to give the result

$$\hat{Y}(\mathbf{x}_i) \pm t_{n-k}(\alpha/2l) s(1 + \mathbf{x}_i'(\boldsymbol{a}'\boldsymbol{a})^{-1}\mathbf{x}_i)^{1/2}. \qquad (10.25)$$

Both methods are conservative and both intervals increase indefinitely in width with l. The statistician is free to examine both and use whichever is narrower, but for values of the parameters usually employed the Bonferroni method is narrower. Some work relevant to this case is given by Dunn (1959).

10.5 Tolerance intervals

Tolerance intervals for future observations are when we make statements of the following kind: '$100P\%$ of future observations will lie in the interval bounded by A and B with confidence $100(1-\alpha)\%$'. In this statement, the words, 'with confidence' refer to the original data on which the regression was calculated, and the statement '$100P\%$ of future observations' refers to the sampling of further observations. Thus, if for the original regression, we make the statement quoted and then do this for many similar regression analyses, then 99% of such statements will be correct.

Tolerance intervals may have to be used in place of confidence statements about future observations if the number of predictions is large or unknown.

EXAMPLE 10.5

Lieberman and Miller (1963). A practical example to illustrate the problem. A new missile is produced and it is known that the speed of the missile is linearly related to the orifice opening which controls the fuel supply. Clearly, the orifice opening can be measured before firing but the speed can only be determined afterwards by measurement. A total of N missiles are fired and based on the data a linear regression is fitted between the speed and orifice opening. On the basis of this one regression equation it is desirable to predict the speed of many new missiles as they are produced. It is not statisfactory to keep on using the results for the prediction intervals for one missile since these prediction intervals are not independent and, in any case, we do not know the number of such statements we require to make.

Lieberman and Miller (1963) give several methods of achieving the tolerance intervals but a satisfactory procedure is to use the

Bonferroni inequality. This results in the following formula.

$$\hat{Y}(\mathbf{x}) \pm \{kF_{\alpha/2}(k, n-k)\}^{1/2} s\sqrt{(x'(\boldsymbol{a}'\boldsymbol{a})^{-1}\mathbf{x})} + sK(P)\{(N-k)/\chi^2_{N-k}(\alpha/2)\}^{1/2}, \tag{10.26}$$

where $K(P)$ is defined by

$$P = \Phi(K(P)) - \Phi(-K(P)).$$

For further details see the source paper, or for tables and explanation, see Odeh and Owen (1980).

10.5.1 Discussion

Some of the methods given above are little known and rarely used. A particular question to consider is when we need *simultaneous* inference rather than inference on each point independently? For further information and references on these methods, see Miller (1966, 1977) and Seber (1977).

10.6 The C_p-statistic

Suppose that in the situation where model (10.1) holds, and model (10.3) is fitted, the fitted model is used to make a prediction at a point $\boldsymbol{a} = (\boldsymbol{a}_1 \boldsymbol{a}_2)'$. The mean square error of the predicted value is

$$E(\text{MSE}_1) = E(\hat{Y}(\boldsymbol{a}_1) - \boldsymbol{a}\boldsymbol{\theta})^2.$$

If we sum this quantity over the data points it gives us a measure of the error of prediction in the region. Therefore, consider the quantity

$$S = (\hat{\mathbf{Y}}(\boldsymbol{a}_1) - \boldsymbol{a}\boldsymbol{\theta})'(\hat{\mathbf{Y}}(\boldsymbol{a}_1) - \boldsymbol{a}\boldsymbol{\theta}), \tag{10.27}$$

where

$$\hat{\mathbf{Y}}(\mathbf{a}_1) = \boldsymbol{a}_1(\boldsymbol{a}_1'\boldsymbol{a}_1)^{-1}\boldsymbol{a}_1\mathbf{Y}. \tag{10.28}$$

This is similar to (10.8), but the predictions are made from model (10.3), whereas the true model is taken to be (10.1). We can rewrite this using γ_1 from (10.5)

$$S = (\hat{\mathbf{Y}}(\boldsymbol{a}_1) - \boldsymbol{a}_1\boldsymbol{\gamma}_1 + \boldsymbol{a}_1\boldsymbol{\gamma}_1 - \boldsymbol{a}\boldsymbol{\theta})'(\hat{\mathbf{Y}}(\boldsymbol{a}_1) - \boldsymbol{a}_1\boldsymbol{\gamma}_1 + \boldsymbol{a}_1\boldsymbol{\gamma}_1 - \boldsymbol{a}\boldsymbol{\theta})$$

$$E(S) = (\hat{\mathbf{Y}}(\boldsymbol{a}_1) - \boldsymbol{a}_1\boldsymbol{\gamma}_1)'(\hat{\mathbf{Y}}(\boldsymbol{a}_1) - \boldsymbol{a}_1\boldsymbol{\gamma}_1) + (\boldsymbol{a}_1\boldsymbol{\gamma}_1 - \boldsymbol{a}\boldsymbol{\theta})'(\boldsymbol{a}_1\boldsymbol{\gamma}_1 - \boldsymbol{a}\boldsymbol{\theta}). \tag{10.29}$$

The first of these terms is the sum of the squares of the variances of the predictions, and the second of these is the sum of the squares of the biases. From (10.28) we obtain

$$V\{\hat{Y}(\boldsymbol{a}_1)\} = \boldsymbol{a}_1(\boldsymbol{a}_1'\boldsymbol{a}_1)^{-1}\boldsymbol{a}_1'\boldsymbol{a}_1(\boldsymbol{a}_1'\boldsymbol{a}_1)^{-1}\boldsymbol{a}_1'\sigma^2$$
$$= \boldsymbol{a}_1(\boldsymbol{a}_1'\boldsymbol{a}_1)^{-1}\boldsymbol{a}_1'\sigma^2.$$

The sum of the variances is the trace of this, so that the first term on the right of (10.29) is

$$\sigma^2 \operatorname{tr}\{\boldsymbol{a}_1(\boldsymbol{a}_1'\boldsymbol{a}_1)^{-1}\boldsymbol{a}_1'\} = \sigma^2 \operatorname{tr}\{(\boldsymbol{a}_1'\boldsymbol{a}_1)^{-1}\boldsymbol{a}_1'\boldsymbol{a}_1\} = k_1\sigma^2. \quad (10.30)$$

Therefore the expectation of S is from (10.29)

$$E(S) = k_1\sigma^2 + (\text{sum of squared bias})$$
$$= k_1\sigma^2 + E(\text{RSS}_{k_1}) - (n - k_1)\sigma^2$$
$$= E(\text{RSS}_{k_1}) - (n - 2k_1)\sigma^2. \quad (10.31)$$

Mallows (1964, 1966, 1973), see also Gorman and Toman (1966), suggested using the statistic C_p, defined below, in choosing a regression equation

$$C_p = \frac{\text{RSS}_p}{\hat{\sigma}^2} - (n - 2_p), \quad (10.32)$$

where p is the number of variables fitted, counting the constant term as one, and where $\hat{\sigma}^2$ is some estimate of σ^2, possibly based on fitting the full model to the data.

Notice that if there is no bias, then

$$E(\text{RSS}_p) = (n - p)\sigma^2,$$

so that

$$E(C_p) = p. \quad (10.33)$$

If, in fact, the estimate of σ^2 from the full model is used in (10.32), then

$$C_p = p, \quad \text{exactly}.$$

If a plot of C_p versus p is made, this will help us in choosing an equation. Broadly, models leading to smaller C_p are preferred, but those points close to the line $C_p = p$ are likely to be for models with a small bias. This will be discussed again in the next chapter.

10.7* Adding a variable to a regression (result for Section 10.2)

Suppose we wish to add one variable to a regression which already has k explanatory variables, then we are faced with inverting a $(k + 1) \times (k + 1)$ matrix, given the inverse of a $k \times k$ matrix.

Therefore let A, A^{-1} be given, and suppose we want the inverse of

$$\begin{bmatrix} A & \mathbf{B} \\ \mathbf{B}' & C \end{bmatrix}, \qquad \begin{array}{l} \mathbf{B} \text{ is } (k \times 1), \\ C \text{ is } (1 \times 1), \end{array}$$

which we write as

$$\begin{bmatrix} D & \mathbf{E} \\ \mathbf{E}' & F \end{bmatrix}, \qquad \begin{array}{l} \mathbf{E} \text{ is } (k \times 1), \\ F \text{ is } (1 \times 1). \end{array}$$

Then we have

$$DA + \mathbf{E}\mathbf{B}' = 1, \tag{10.34}$$

$$D\mathbf{B} + \mathbf{E}C = 0, \tag{10.35}$$

$$\mathbf{E}'A + F\mathbf{B}' = 0, \tag{10.36}$$

$$\mathbf{E}'\mathbf{B} + FC = 1. \tag{10.37}$$

Then from (10.35)

$$\mathbf{E} = -C^{-1}D\mathbf{B}$$

and from (10.36) we have

$$\mathbf{E} = -A^{-1}\mathbf{B}F'.$$

Therefore from (10.34)

$$D + \mathbf{E}\mathbf{B}'A^{-1} = A^{-1},$$

$$D = FA^{-1}\mathbf{B}\mathbf{B}'A^{-1} + A^{-1}.$$

Therefore the required inverse is

$$\begin{bmatrix} A^{-1} + FA^{-1}\mathbf{B}\mathbf{B}'A^{-1} & -FA^{-1}\mathbf{B} \\ -F\mathbf{B}'A^{-1} & F \end{bmatrix},$$

where

$$F = 1/(C + \mathbf{B}'A^{-1}\mathbf{B}). \tag{10.38}$$

Exercises 10

1. Continue Example 10.1 to evaluate the bias in the residual mean square.
2. Show that the average variance of the predicted means at each of the n points of the data set is $(k/n)\sigma^2$, where $k = \text{rank}(a)$.
3. Design and carry out some further trials similar to Example 10.2, to determine the effect on prediction variances of adding un-

necessary variables. In this case, do not assume regression through the origin. How 'special' do you regard the results of Example 10.2?

4. Discuss the practical importance and implications that you attach to an estimated variance of a prediction.

References

Dunn, O. J. (1959) Confidence intervals for the means of dependent normally distributed variables. *J. Amer. Statist.*, **54**, 613–621.

Gorman, J. W. and Toman, R. J. (1966) Selection of variables for fitting equations to data. *Technometrics*, **8**, 27–51.

Lieberman, G. J. and Miller, R. G., Jr. (1963) Simultaneous tolerance intervals in regression. *Biometrika*, **50**, 155–168.

Mallows, C. L. (1964) Choosing variables in a linear regression: a graphical aid. Presented at the Central Regional Meeting of the Institute of Mathematical Statisticians, Manhattan, Kansas.

Mallows, C. L. (1966) Choosing a subset regression. Presented at Joint Statistical Meetings, Los Angles, California.

Mallows, C. L. (1973) Some comments on C_p. *Technometrics*, **15**, 661–675.

Miller, R. G. (1966) *Simultaneous Statistical Inference*. McGraw-Hill, New York.

Miller, R. G. (1977) Developments in multiple comparisons 1966–1976. *J. Amer. Statist. Assoc.*, **72**, 779–788.

Odeh, R. E. and Owen, D. B. (1980) *Tables for Normal Tolerance Limits, Sampling Plans, and Screening* Marcel Dekker, New York.

Ramsey, J. B. (1969) Tests for specification errors in classical linear least squares regression analysis. *J. Roy. Statist.* B, **31**, 350–371.

Seber, G. A. F. (1977) *Linear Regression Analysis*. Wiley, New York.

Walls, R. C. and Weeks, D. L. (1969) A note on the variance of a predicted response in regression. *Amer. Statist.*, **23**, (3), 24–25.

CHAPTER 11

Choosing a regression model

11.1 Preliminaries

There is a very large literature on methods of choosing a regression model but, in spite of this, there is little clear guidance on what to do in a specific case. For access to the literature see Hocking (1976, 1983), Mosteller and Tukey (1977), Seber (1977), Thompson, (1978), Daniel and Wood (1980), and Miller (1984). This chapter presents a review of the main points and for further details these references should be consulted.

Many automated procedures for choosing a regression model are available, but it is important to see these in the right context and never to follow them blindly. Henderson and Vellman (1981) stressed this, and gave examples showing the kind of interactive approach by which regression models should be generated. The general method we suggest is to follow the headings given below.

11.1.1 Objective

It is essential to get the purpose of any statistical analysis clear before setting out on any detailed work. For example, the techniques used may be very different if any analysis is simply for the purpose of summarizing the data, than if it is part of the submission of drug trials results to put before a regulatory authority. We can distinguish three main types of use:

(1) Data summarization. The aim is to present the main features of the data in a reasonably concise way. The use of models in each summarization is common.
(2) Prediction. The aim here is to use the model resulting from our analysis to predict results at values of the explanatory variables which are not observed. This could involve interpolation or extrapolation.

(3) Process studies. Here the objective is to study the behaviour of a system in order to try to control it.

In taking note of the objectives, it is important *not* to confine oneself to the specific objective which might be listed. The statistician should try to find out all the relevant information that the data is giving. However, the stated objectives (if you can get them!) are a clear general guide. For example, they will guide us as to how much effort is needed on the 'model selection' phase.

11.1.2 Basis of the data collection: experiment, survey or observational data

The next important point is to establish the basis of the data collection, as the conclusions we can make depends on this. In an experiment, treatments are allocated to the experimental units and there should be elements of statistical design, such as randomization. In a survey, it is not possible to allocate treatments and we take note of existing affairs, such as people's smoking habits and their health. Other examples of observational data are industrial plant performance data, collected on the plant as it is operating. In no sense are survey or observational data collected under controlled conditions and there may be many important factors which are overlooked.

11.1.3 Nature of the observations

What observations were actually recorded? Were the data 'adjusted' or processed in some way before analysis? In some experiments it may be important to know the order in which the observations were taken.

We should also study the error structure; some variates are known to be fairly precise, but with others there may be some error variance, or considerable scope for subjective judgement. If the data have been coded or transcribed, it is important to know this. Also, if there are missing observations, it may be important to know the reason.

11.1.4 Prior information

Frequently the owner of the data set has prior information about the importance or reliability of observations or variates. Sometimes certain variates are 'required' to be in the final model. All of this prior information needs to be studied (or elicited and studied).

11.1.5 Selection of the set of explanatory variables

The full set of explanatory variables used needs to be chosen with some care. Two frequently made errors are:

(1) to include many variates known to be more or less irrelevant;
(2) to omit variates which could have been taken and which might have increased the precision or made the interpretation easier.

If a whole collection of explanatory variables is thrown in, in the hope that something may be significant, it should not surprise us if the results are meaningless. To put it quite simply, rubbish in means rubbish out! Multiple regression is not an excuse for not thinking.

11.1.6 Data scrutiny

Data scrutiny is a key phase of any analysis and the procedures given in Chapter 2 could be used. For some other possibilities, see Preece (1981). The objective here is to locate wild values, examine the need to transform some variates and to study the reliability of the data.

It should be noted here that considerable care has to be exercised over what is done with any 'wild' values observed. Sometimes they may be safely ignored, especially when reasons connected with the data collection can be discovered. However, sometimes 'wild' values are worthy of special study, and they indicate that our whole theory is inadequate; see Section 6.1.

11.1.7 Preliminary analysis and validation

'Data scrutiny' procedures are not powerful or precise enough to detect all of the important outliers or other problems with data sets. It is therefore necessary to carry out a preliminary regression analysis, and then carry out checks on residuals, such as those discussed in previous chapters. The results of these diagnostic tests may lead to further transformations, deletion of outliers or the introduction of other variables. One major problem is what model to do these diagnostic checks on. Provided there are not too many explanatory variables, it is probably best to regress on them all for this phase of the analysis. However, the introduction of too many explanatory variables can cause a loss of power in the diagnostic tests, and it may be necessary to reduce the number of variates in some suitable way. The strategy used for this 'reduction' of variables is probably not too critical.

If multicollinearity is present, then it should be detected in this phase of the analysis, for example, by estimating the condition number. If multicollinearity is present then ridge regression, or some other biased estimation procedure, could be used as a method of selecting variables. Once the variables are selected ordinary least squares (OLS) estimation could be used; see Chapter 4.

A series of plots which may be of particular value at this stage are partial regression plots. A partial regression plot is obtained as follows. We select any variable, say X_j, and then regress both X_j and the response variable on all remaining variables. The partial regression plot is a plot of the residuals for X_j against the residuals for the response variable. Belsley *et al.* (1980) show that this plot is very useful in detecting leverage points. For other material on this plot, see Mosteller and Tukey (1977).

11.1.8 Model selection

Finally, we are ready with a (hopefully) 'clean' data set to use a model selection strategy. It is important here to use such prior information as was gleaned in Section 11.1.4 above. Diagnostic checks following this analysis may lead to a further revision of the data or the model.

11.1.9 Data splitting and cross-validation

Even after we have carefully followed through all of the stages listed above, a fundamental problem arises in that no regression model can be properly validated from the sample used to estimate it. Suggestions have therefore been made to arbitrarily split the data, and use one part for estimation and fitting and the other part for validation. For example, if the estimation sample results in several almost equal models, we may well prefer to use one which achieved good results on the validation sample. Unfortunately, there is no established methodology for doing this data splitting and there would appear to be some inefficiency in using it. However, if our objective is prediction or model building, then some validation from separate data is very desirable.

One proposal is to use a validation sample of one and to do this in all n possible ways. With regression this can be done easily without a lot of recalculation. However, since we are only deleting one point in n we may well expect that, in general, the effects of doing this are rather marginal. Severe danger can arise in regression due to clumps of points with special features, or due to the presence of some unknown factor, present throughout most of the data set but which changes with a new data set. The 'one at a time' procedure would miss these features.

For access to some of the literature on this, see Stone (1974) and Copas (1983).

11.1.10 Discussion

In this discussion we see the place that subset selection strategies takes in a regression analysis. It is important to recognize the limited, though important, area of the whole model-fitting process that they cover and to note the essential steps prerequisite to using them.

11.2 Computational techniques: stepwise procedures

Hocking (1976) pointed out that three distinct ingredients of techniques for the model selection process in multiple regression analysis can be identified:

(1) the computational techniques used to identify a set of possible models to be considered;
(2) the criterion function used to select a particular model;
(3) the estimation of parameters in the chosen model.

Usually all three of these are put together but we shall discuss them separately. On estimation, several authors (Rencher and Pun, 1980; Copas, 1983) have pointed out that if least squares estimation is used in any situation where the same data is used for selecting the model as for estimating the parameters, and there is competition between models, then the least squares estimators are biased. According to Miller (1984) and Berk (1978) this bias can be substantial. Some comments on how to cope with the bias are made later. For the rest of this section we consider simply the computational techniques of the stepwise type, and the results of these are sets of possible models, to which we need to apply a criterion.

11.2.1 Forward selection

With this strategy we start with no variables in the model, and add the one which gives the largest increase in sum of squares and so on. Usually there is a stopping criterion related to some F-value. We denote by S_r and $\hat{\sigma}_r^2$ the residual sum of squares and estimate of variance from a model with a constant and r explanatory variables, and denote by $S_{r+(j)}$ and $\sigma^2{}_{r+(j)}$ the same quantities calculated on the $(r+1)$ variables including the original r and variable j, then variable i is added if the F-ratio F_i satisfies

$$F_i = \operatorname*{Max}_j \{(S_r - S_{r+j})/\hat{\sigma}^2_{r+j}\} > F_{\text{IN}}, \tag{11.1}$$

where i is the maximum over j. The choice of F_{IN} is discussed below.

11.2.2 Backward selection

With this strategy we start with all variables in the equation and delete variable i if

$$F_i = \operatorname*{Min}_j \{(S_{r-j} - S_r)/\hat{\sigma}_r^2\} < F_{\text{OUT}} \tag{11.2}$$

where i is the minimum over j and where the choice of F_{OUT} is discussed below.

11.2.3 Forward stepwise selection

This strategy is as in the forward selection strategy, but at each stage the possibility of deleting a variable, or several variables, is considered.

11.2.4 Backward stepwise selection

This strategy is as in the backward selection strategy, but at each stage the possibility of adding a variable is considered.

11.2.5 Choice of stopping rules

A number of stopping rules have been suggested for use in stepwise procedures and a review is given by Hocking (1976). If the aim is to let the strategies work through most of the 'reasonable' subsets, constant values of F_{IN} and F_{OUT} can be used. Clearly, we can consider using

$$F_{IN} = F(\alpha;1, n - r - 1)$$

and

$$F_{OUT} = F(\alpha;\ 1, n - r),$$

but Pope and Webster (1972) show that use of the central F in this context is wrong, because the process of selection invalidates the basic conditions for it to hold. Unfortunately, there is as yet no practicable substitute and, clearly, if F-distribution points are used, they must be regarded as an approximation.

Wilkinson and Dallal (1981) obtained by simulation some percentage points of the multiple correlation coefficient under 'forward selection'. In one example quoted, the authors showed that a final regression said by the F-procedure to be significant at 0.1%, was in fact only significant at 5%. Rencher and Pun (1980) give some related results on the inflation of R^2.

It should be recognized that in the forward selection procedures, using F-levels, the choice of a small α will limit the number of models that would be explored. Therefore, it would be better to choose α large, say even 0.25. Similarly, a much larger value of α could be used for F_{OUT} in order to increase the number of models explored.

Another point about the stopping rule is whether or not to keep updating the estimate of σ^2, particularly with 'backward' procedures. There would be good grounds for retaining an estimate of σ^2 based on the 'full' model, but this would lead to inconsistencies in presenting results if the usual 'analysis of variance' table is given.

11.2.6 Discussion

None of the stepwise procedures given above can claim any kind of optimality, and it is not guaranteed that the best subsets of any given size will be covered. The forward selection strategies can be seriously misleading because of the restriction to adding one variable at a time. Also, forward selection strategies may be working with badly inflated estimates of σ^2, whereas a watch can be kept on this with 'backward' strategies. Preference should be given to the 'backward' stepwise strategy for problems with a moderate number of variables.

If we are dealing with a problem with, say 30 variables, of which only a few are expected to be important, then the 'backward' strategy is very expensive in computing and the forward stepwise strategy could be tried.

It is worth noting that in computer applications the time taken to evaluate F percentage points may be considerable, and this is one reason why constant values are sometimes used. Further, it should be noted from this discussion that standard stepwise strategies, with F-levels used, are open to very serious criticisms if any meaning is given to 'significance'. Unfortunately, there is little proposed to replace them and they will probably remain in popular use for some time.

There are really two separate problems. One is the question of a suitable stopping rule for forward/backward or stepwise rules, bringing in the question of when estimated regression coefficients are significantly different from zero. Forsythe *et al.* (1973) attacked this problem and proposed a permutation test.

The second problem is one of testing the difference between alternative regression models. This is a 'multiple comparisons' type problem and Spjøtvoll (1972a) discussed this when it can be assumed that all variables have a multivariate normal distribution. The test is of the Scheffé type, and conservative, so that it may not be felt to be satisfactory.

In any case, these approaches on the 'distribution theory' side of the problem do not deal with selection bias in the estimators; Copas (1983) and references should be studied in this regard. We shall suggest some approaches later.

11.3 Computational techniques: other procedures

11.3.1 All possible regressions

If the number of variables p is not too large, say about a dozen, then all $\binom{n}{p}$ regressions can be calculated and algorithms have been written to perform the necessary calculations efficiently. These algorithms are based on the idea of generating subsets which differ by only one variable and then using the SWEEP operator (see Chapter 3). For good algorithms see Furnival (1971) or Morgan and Tatar (1972); see also Clarke (1981, 1982). Another possibility is to use the 'element analysis' method of Newton and Spurrell (1967a, b). No comparison of the various algorithms has been made, but Hocking (1976) suggests the second Furnival algorithm, pending further results.

11.3.2 Search for favourable solutions

If our aim is to explore the best subsets with a given number of explanatory variables, then it is possible to limit our search to favourable subsets by the following technique. First, the full model is fitted and then we order the explanatory variables according to the modulus of their t-statistics. We denote by $S(1, 2, 3)$ the residual sum of squares for the full model less variables 1, 2 and 3. If we find

$$S(1, 2, 3) \leq S(4)$$

then the residual sum of squares, after dropping any other set of three variables, is at least as great as $S(1, 2, 3)$. If the full model has p variables, then the full model, less variables (1, 2, 3), is the best subset of size $(p - 3)$.

If we find

$$S(1, 2, 3) > S(4)$$

then the calculations must be extended in order to calculate the best subset of size $(p - 3)$. We incorporate variable 4 and search among all $\binom{4}{3}$ subsets of variables (1, 2, 3, 4). If we find that the minimum residual sum of squares among these subsets is less than or equal to $S(5)$, then the procedure stops. If this is not true we must incorporate variable 5 and search among all $\binom{5}{2}$ subsets of variables (1, 2, 3, 4, 5) and so on. This procedure could end up with a lot of computing but this is not usual. Algorithms of this type have been given by Beale *et al.* (1967), Hocking and Leslie (1967) and a number of papers have followed suggesting modifications and improvements. One of the most efficient of these algorithms has been given by Furnival and Wilson (1974), but the algorithm by Hocking and Leslie (1967) has the advantage of revealing more of the 'nearly best' sets.

For regressions with a moderate number of explanatory variables these procedures must be attractive. Some criterion will have to be used to select the number of explanatory variables included and to choose a particular model. A good practical procedure would be for the statistician to choose a model subjectively, bearing in mind the results of the calculations, the residual sums of squares and the prior information (if any) that he has. It may be that a model that is not quite best in terms of sums of squares makes much more sense physically.

11.3.3 Factorial exploration

Daniel and Wood (1980) bring in the use of fractional replicate factorial designs to explore regressions models. If there are p explanatory variables, then we have 2^p possible models and the p-variables $X_1, \ldots, X_p$ can be regarded as 'factors'. The procedure is then to fix on some variables to be definitely included and then carry out a suitable fractional replicate of the remaining models. In this procedure we would usually disregard three factor interactions, but we would seek for fractional replicates in which all main effects and two factor interactions are aliased with higher-order interactions. Daniel and Wood (1980) used the C_p-criterion, and the usual Yates algorithm for 2^n factorial experiments to estimate the 'average' effects on C_p of introducing any combination of variables into the model. This technique could be very useful for exploring regressions with a moderate number of explanatory variables.

11.3.4 Swapping strategies

A variety of 'swapping' strategies can be used, and while they are not guaranteed to find the best subsets of any given size, they are often effective and fairly cheap in computing. The basic idea is that we fix on the size of subset we are considering, say 5, and we suppose that there are many more possible explanatory variables. We start with a given subset, say $(X_1, X_2, X_3, X_4, X_5)$, and we then search for a replacement of X_1 which gives the smallest residual sum of squares. After this we proceed to X_2, X_3, etc. and then repeat until no more improvements can be made. Since there is a finite number of possibilities the process is obviously convergent, but it may converge on a different result if a different starting set is used.

Variations on this theme are possible, including replacements two at a time, etc. In spite of being non-optimal, it may be a useful technique in some circumstances.

11.3.5 Discussion

Berk (1978) gave some interesting comparisons of stepwise strategies and the 'all possible' regressions approach. He showed that if 'forward selection' agrees with the 'all possible' regressions approach at every subset size, then so does 'backward selection'. He also showed that both 'forward selection' and 'backward selection' can agree at every subset size and still be a long way short of the optimal selections. However, he also showed that in 'fairer' comparisons there is not much difference between the strategies.

11.4 Criteria for selecting models

In our discussion so far we have assumed that we are selecting models on the basis of the residual mean square. Many other criteria have been suggested, and we shall outline a few of them here. As we stated earlier, we are distinguishing here between a computational method chosen to reveal 'good' subsets and a criterion used to choose one particular model. Whatever the criterion chosen, it is better not to follow it blindly, but to use it, together with prior knowledge, etc.

11.4.1 The residual mean square

As the number of variables is reduced from the full model, the residual mean square either increases or else it goes through a minimum. There is usually a point at which the residual mean square increases sharply for a further reduction in the number of variables. A choice can be made at this critical value, or at the minimum if there is one.

One possibility is to calculate the minimum residual mean square for each of a range of numbers of explanatory variables, say 4, 5, 6... variables, and then plot this.

11.4.2 The multiple correlation coefficient

The multiple correlation coefficient is

$$R^2 = 1 - \{(\text{residual sum of squares})/(\text{total corrected sum of squares})\} \tag{11.3}$$

and since this is directly related to the residual sum of squares, its use is closely related to Section 11.4.1.

Several points about the multiple correlation coefficient need to be noted carefully:

(1) Increasing the number of explanatory variables will increase the multiple correlation.
(2) For simple linear regression, with a given slope, the multiple

correlation can be increased or decreased by increasing or decreasing the variation of the explanatory variable.

(3) For a given residual mean square, the size of the multiple correlation depends on the size of the regression coefficient.

If the explanatory variables are randomly chosen from some distribution, say normal, then use of the multiple correlation coefficient may be satisfactory, bearing in mind the points made above. However, if the explanatory variables are fixed and controllable, then the multiple correlation simply reflects this (controlled) variation in the explanatory variables.

11.4.3 The adjusted multiple correlation coefficient

An adjusted multiple correlation coefficient has been suggested because of the criticisms noted in Section 11.4.2 above. It is defined as

$$\bar{R}^2 = 1 - (n-1)\frac{\text{RMS}}{\text{Total SS}}. \tag{11.4}$$

With this definition, an extra variable included only increases $\bar{R}$ if its F-value is greater than unity.

As far as criteria go, this does not carry us any further, since maximizing $\bar{R}^2$ is equivalent to minimizing the residual mean square. A criticism of both R^2 and $\bar{R}^2$ is that neither incorporates into the decision criterion any consideration of the effects of incorrect specification of the model.

11.4.4 C_p

It is clear from the derivation of C_p in the previous chapter that C_p is a standardized mean squared error of predicting the current data. Several authors have suggested using C_p as a criterion for choosing a model and we look for models with a small C_p, and preferably we look for a C_p close to p which means a small bias. The statistic C_p has a minimum and we could choose the subset of variables which leads to this. Sometimes a C_p plot will show an obvious choice, but on other occasions it may show that no one choice is clear. Some practice is required in interpreting C_p plots.

Figure 11.1 shows a C_p plot for the acetylene data (see also Table 11.1). For other illustrations and comments on interpretation,

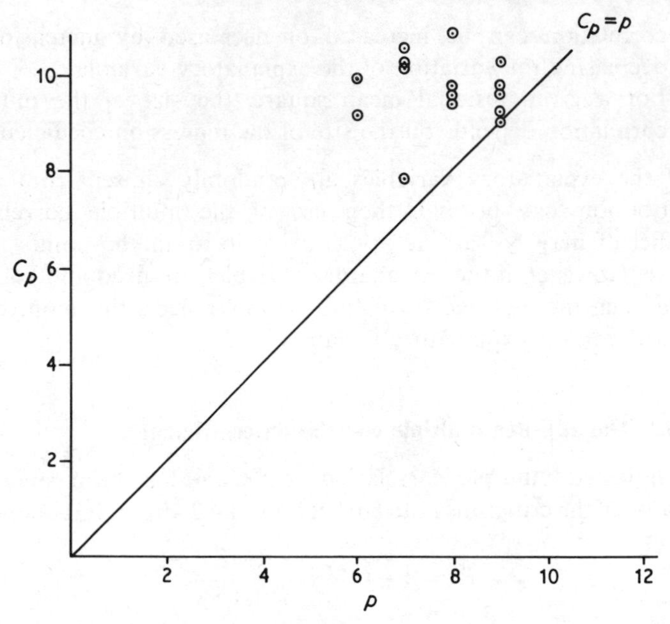

Fig. 11.1 *C_p plot for acetylene data.*

Table 11.1 *Values of p and C_p where $C_p \leq 11$.*

p	C_p	*Variables in equation*
10	10	1, 2, 3, 5, 6, 7, 8, 9, 10
9	9.03	1, 2, 3, 6, 7, 8, 9, 10
9	9.19	2, 3, 5, 6, 7, 8, 9, 10
9	9.65	1, 2, 3, 5, 6, 7, 8, 10
9	9.74	1, 2, 5, 6, 7, 8, 9, 10
9	10.26	1, 2, 3, 5, 6, 8, 9, 10
8	9.51	1, 2, 5, 6, 8, 9, 10
8	9.59	1, 2, 3, 5, 6, 8, 10
8	9.85	1, 2, 5, 6, 7, 8, 10
8	10.91	2, 3, 5, 6, 8, 9, 10
7	7.86	1, 2, 5, 6, 8, 10
7	10.18	1, 2, 3, 5, 6, 8
7	10.18	1, 2, 5, 6, 7, 8
7	10.20	1, 2, 5, 6, 8, 9
7	10.65	2, 3, 5, 6, 8, 9
6	9.14	1, 2, 5, 8, 10
6	9.99	1, 2, 5, 6, 8

Variables X_1 to X_4 as in Table 1.1. $X_5 = X_1^2$, $X_6 = X_2^2$, $X_7 = X_3^2$, $X_8 = X_1X_2$, $X_9 = X_1X_3$, $X_{10} = X_2X_3$.

see Gorman and Toman (1966), Mallows (1973) or Daniel and Wood (1980).

One disadvantage of C_p is that it seems to be necessary to evaluate C_p for all, or most, of the possible subsets, to allow interpretation. A further point to note is that the derivation of C_p assumed unbiased estimators and, as we have seen, the selection and stopping rules will introduce bias.

11.4.5 PRESS

The prediction sum of squares was introduced by Allen (1974) as a criterion and was mentioned by Hocking (1976). If $Y(i)$ is the predicted value of the response variable at the ith set of explanatory variables, using all of the data except the response at the ith value, then PRESS is defined

$$\text{PRESS}_r = \sum_{i=1}^{n} (Y_i - \hat{Y}(i))^2, \tag{11.5}$$

where r is the number of explanatory variables included. It can be shown that this sum of squares can be restated as follows,

$$\text{PRESS}_r = \mathbf{R}'\boldsymbol{D}^{-2}\mathbf{R}, \tag{11.6}$$

where $\mathbf{R}$ is the vector of residuals obtained by fitting all of the data, and $\boldsymbol{D}$ is a diagonal matrix whose diagonal elements are equal to those of

$$\{\boldsymbol{I} - \boldsymbol{X}(\boldsymbol{X}'\boldsymbol{X})^{-1}\boldsymbol{X}'\}.$$

This criterion does have an intuitive appeal if the purpose is prediction.

11.4.6 Tukey's rule

Tukey's rule can be stated quite simply as follows.

Tukey's rule
Choose the set of explanatory variables which leads to a minimum of s^2/v, where s^2 is the mean square error and v is the residual degrees of freedom.

For a derivation see Anscombe (1981); there are two approximate arguments.

The first approach is to consider the error in fitting each of the n data points as being made up of lack of fit (due to omitted variables) and the basic random error, these combined errors can be termed 'noise'. Anscombe (1981) shows that under certain assumptions and approximations the main component of the mean square noise is a term s^2/v. This leads to Tukey's rule.

The second approach is to note that from (10.9)

$$V(\hat{\mathbf{Y}}) = \boldsymbol{H}\sigma^2,$$

so that

$$\operatorname{tr} V(\hat{\mathbf{Y}}) = \sigma^2 \operatorname{tr}(\boldsymbol{H}) = k\sigma^2.$$

Thus the average variance of a predicted observation is

$$(1 + k/n)\sigma^2.$$

If k is small compared with n, then this term is approximately ns^2/v, which again leads to Tukey's rule. This latter argument shows that Tukey's rule might be particularly appropriate to prediction.

Anscombe suggests that when using Tukey's rule, it may be better to limit the sets of explanatory variables considered by some *a priori* arguments, rather than to use Tukey's rule in an unrestricted search. (This may be a good general advice anyway.)

Anscombe also shows that Tukey's rule is very similar in behaviour to C_p, and that when there is a difference in outcome between two rules, use of C_p leads to a larger number of explanatory variables.

11.4.7 Mean squared error of $\hat{\boldsymbol{\beta}}$

All of the fitting criteria considered so far fit the data and, instead of this, we could form a criterion based on how close the regression parameters are to their true values. This may be particularly appropriate if prediction is in view, and the mean squared error of $\hat{\boldsymbol{\beta}}$ as a criterion was mentioned in Chapter 4. The point is that biased estimators of $\boldsymbol{\beta}$, such as ridge, can lead to a smaller mean squared error of $\hat{\boldsymbol{\beta}}$. When we consider that most estimators in common use are biased, due to selection and stopping rules, this may lead to biased estimation being considered more favourably.

We cannot use this criterion to choose a particular equation, but it does lead us to consider biased estimation, such as ridge regression. This was discussed in Chapter 4.

11.4.8 Other criteria

Many other criteria have been suggested. For surveys and reviews of criteria, see Hocking (1976), Seber (1977), Thompson (1978) and Judge *et al.* (1980).

Unfortunately, there is insufficient knowledge at present to give firm advice on the choice of a criterion, and possibly because of this, the residual mean square will remain a favourite for some time. For problems involving prediction, however, Tukey's rule, C_p or PRESS should be considered.

11.5 Freedman's simulation

Freedman (1983) carried out a simulation experiment which brings out clearly the dangers of applying some of the foregoing strategies blindly.

A matrix of 100 observations by 51 variables was created in which all entries were independently drawn from a standard normal distribution. The regression was carried out in two stages:

(1) firstly, the final variable was regressed on the remaining 50 variables;
(2) in the second run only those variables reaching a 25% significance level by the usual t (or F) procedures were included.

The result of the second run was that the regression was highly significant and six of the regression coefficients were significant at the 5% level. Freedman (1983) reports the results of ten such experiments, all of which led to somewhat similar results, and he gives an asymptotic argument to show why this happens.

Results of this kind are obviously very disturbing, but perhaps such an experiment should be made a part of all regression courses!

11.6 Discussion

In spite of all of the work which has been done in this area recently, the choice and estimation of subset regressions remains a very difficult problem.

The statistician will no doubt want to try two or three criteria and selection strategies, and compare the results, taking any prior knowledge into account.

For the non-statistician who wants a precise rule to follow, we cannot oblige. Indeed, it would be better if he studied the results of two or three strategies before making a decision.

The importance of going through the procedures outlined in Section 11.1 cannot be overstressed. Many statistical packages will permit users to go ahead and 'fit' regression models to dirty data, or to data with strong multicollinearities, etc., without printing any comment. There is no excuse for this with the computing power we have available today. A package can be written so that users are encouraged (or forced!) to look at their data before fitting a model, and also encouraged to run diagnostic checks.

With regard to selection bias, one possibility is to use ridge regression, since this technique will shrink some biases. Another possibility is to divide the data into two, use one part to identify the explanatory variables, and the other part to estimate the parameters of the model. Unless there is a superabundance of a data there will be resistance to using this technique.

Exercises 11

1. Repeat Freedman's experiment (Section 11.5), with a smaller sample size if necessary. In a similar way try out the various subset regression strategies suggested in this chapter. Write a report on your conclusions.
2. Use the techniques given in this chapter to analyse the data sets given in the Appendix.

References

Allen, D. M. (1974) The relationship between variable selection and data augmentation and a method for prediction. *Technometrics*, **16**, 125–127.

Anscombe, R. J. (1981) *Computing in Statistical Science Through APL*. Springer-Verlag, New York.

Beale, E. M. L., Kendall, M. G. and Mann, D. W. (1967) The discarding of variables in multivariate analysis. *Biometrika*, **54**, 357–366.

Belsley, D., Kuh, E. and Welsch, R. E. (1980) *Regression Diagnostics: Identifying Influential Data and Sources of Collinearity*. Wiley, New York.

Berk, K. N., (1978) Comparing subset regression procedures. *Technometrics*, **20**, 1–6.

Clarke, M. R. B. (1981) A given algorithm for moving from one linear model to another without going back to the data. Algorithm AS153. *Appl. Statist.*, **30**, 198–203.

Clarke, M. R. B. (1982) The Gauss–Jordan sweep operator with detection of collinearity. Algorithm AS178. *Appl. Statist.*, **31**, 166–168.

Copas, J. B. (1983) Regression, prediction and shrinkage. *J. Roy. Statist. Soc.* B, **45**, 311–354.

Daniel, C. and Wood, F. S. (1980) *Fitting Equations to Data.* Wiley, New York.

Forsythe, A. B., Engelman, L., Jennrich, R. and May, P. R. A. (1973) A stopping rule for variable selection in multiple regression. *J. Amer. Statist. Assoc.*, **68**, 75–77.

Freedman, D. A. (1983) A note on screening regression equations. *Amer. Statist.*, **37**, 147–151.

Furnival, G. M. (1971) All possible regressions with less computation. *Technometrics*, **13**, 403–408.

Furnival, G. M. and Wilson, R. W., Jr. (1974) Regression by leaps and bounds. *Technometrics*, **16**, 499–512.

Gorman, J. W. and Toman, R. J. (1966) Selection of variables for fitting equations to data. *Technometrics*, **8**, 27–51.

Henderson, H. V. and Vellman, P. F. (1981) Building multiple regression models interactively. *Biometrics*, **37**, 391–411.

Hocking, R. R. (1976) The analysis and selection of variables in linear regression. *Biometrics*, **32**, 1–49.

Hocking R. R. (1983) Developments of linear regression methodology: 1959–1983. *Technometrics*, **25**, 219–249.

Hocking, R. R. and Leslie, R. N. (1967) Selection of the best subset in regression analysis. *Technometrics*, **9**, 531–540.

Judge, G. G., Griffiths, W. E., Hill, R. C. and Lee, Tsoung Chao (1980) *The Theory and Practice of Econometrics.* Wiley, New York.

Mallows, C. L. (1973) Some comments on C_p. *Technometrics*, **15**, 661–675.

Miller, A. J. (1984) Selection of subsets of regression variables. *J. Roy. Statist. Soc.* A, **147**, 389–425.

Morgan, J. A. and Tatar, J. F. (1972) Calculation of the residual sum of squares for all possible regressions. *Technometrics*, **14**, 317–325.

Mosteller, F. and Tukey, J. W. (1977) *Data Analysis and Regression*, Addison-Wesley, Reading, MA.

Newton, R. G. and Spurrell, D. J. (1967a) A development of multiple regression for the analysis of routine data. *Appl. Statist.*, **16**, 51–65.

Newton, R. G. and Spurrell, D. J. (1967b) Examples of the use of elements for clarifying regression analysis. *Appl. Statist.*, **16**, 165–171.

Pope, P. T. and Webster, J. T. (1972) The use of an F-statistic in stepwise regression procedures. *Technometrics*, **14**, 327–340.

Preece, D. A. (1981) Distribution of final digits in data. *The Statistician*, **30**, 31–60.

Rencher, A. C. and Pun, F. C. (1980) Inflation of R^2 in best subset regression. *Technometrics*, **22**, 49–54.

Seber, G. A. F. (1977) *Linear Regression Analysis.* Wiley, New York.

Spjøtvoll, E. (1972a) Multiple comparison of regression functions. *Ann. Math. Statist.*, **43**, 1076–1088.

Spjøtvoll, E. (1972b) A note on a theorem of Forsythe and Golub. *SIAM J. Appl. Math.*, **23**, 307–311.

Stone, M. (1974) Cross-validatory choice and assessment of statistical predictions. *J. Roy. Statist. Soc.* B, **36**, 111–147.

Thompson, M. L. (1978) Selection of variables in multiple regression. Part I. A review and evaluation. Part II. Chosen procedures, computations and examples. *Int. Statist. Rev.*, **46**, 1–19 and 129–146.

Wilkinson, L. and Dallal, G. E. (1981) Tests of significance in forward selection regression with an F-to enter stopping rule. *Technometrics*, **23**, 377–380.

CHAPTER 12

The analysis of response surface data

12.1 Introduction

The fitting of quadratic response surfaces involves an important and rather special application of multiple regression in which a second degree polynomial in the original independent variables is fitted to the response under analysis. This can be called multiple quadratic regression. For a very simple example refer back to Data Set 1.1 in Chapter 4. Much of the research on response surface analysis has concentrated on the design of suitable experiments and good discussions of this and of general methodology can be found in Box and Wilson (1951), Box (1954), Myers (1971), Box *et al.* (1978) and Davies (1978, Chapter 11). We shall concentrate here on the fitting of the polynomial and subsequent interpretation assuming that the data have already been collected. Obviously consideration of the design will enter at certain points in the discussion but the present chapter is not intended as an exposition of this aspect of the subject.

We have the following situation. A response, for example total cost per unit or purity of a product of a chemical process, is known to depend on a number of variables or factors. Examples of such factors are the concentration of the reactants and the temperatures and pressure at which the reaction is conducted. A series of experiments have been run with different combinations of factor levels. Using the resulting yields it is required to determine if and where in the region of the factor space investigated the optimum response is produced and, if this does not appear to be in this region, in what direction it is most likely to be. It may be that the response is an optimum throughout a region of the factor space and again, if this is the case, the shape and position of this region needs estimating. In all the situations it is necessary to estimate the shape of the response surface in the experimental region or factor space.

The true relationship between the response, y say, and the factors, of which we assume that there are $p, x_1, x_2, \ldots, x_p$, say, is assumed to be unknown and probably complex. However, throughout any relatively small region it may be possible to approximate this relationship by a polynomial of small degree, in particular, by a quadratic polynomial

$$y = \beta_0 + \sum_{i=1}^{p} \beta_i x_i + \sum_{i=1}^{p} \sum_{j=i}^{p} \beta_{ij} x_i x_j, \tag{12.1}$$

with the first degree equation as a special case of this:

$$y = \beta_0 + \sum_{i=1}^{p} \beta_i x_i. \tag{12.2}$$

The observed response will differ from the model because of the presence of experimental or random errors and also because the polynomial is inevitably an imperfect representation of the true response surface. If the random errors are assumed to be independent with a common variance, least squares estimation is appropriate and the techniques discussed in previous chapters are applicable. Note that we are always underfitting however, since the quadratic polynomial is only an approximation to the true relationship, and the least squares estimation will always be biased to a certain extent. How serious this is for any particular analysis is hard to say, although experimental designs exist which reduce the seriousness of this for certain omitted higher-order terms, assuming that the higher degree model is appropriate. Although this underfitting will not be dwelt on, it should be kept in mind that the results below assume that the effect is negligible. In the following sections we shall consider some problems that arise in the fitting and subsequent interpretation of quadratic response surfaces and a particular example will be used as an illustration.

12.2 Example

We consider the following set of data, supplied by Mr Gordon Thackray of ICI plc. The data are set out in Table 12.1. They were obtained from experiments on degassing an aqueous liquid stream, which was pumped through a nozzle into a vessel in which a specified

Table 12.1

y	x_1	x_2	x_3	x_4	x_5	x_6	x_7
234	0.99	6.0	12.7	497	70	2.0	16.2
191	0.99	5.0	12.5	468	71	2.0	16.1
234	0.99	3.0	11.2	497	68	2.0	15.4
234	0.43	6.2	12.7	582	69	1.8	15.5
235	0.43	5.0	12.5	617	68	1.8	15.5
191	0.41	4.0	12.0	580	69	1.8	15.6
85	0.40	2.9	11.0	475	68	1.8	15.5
85	0.32	2.0	9.9	497	67	1.8	15.5
85	0.30	2.0	9.9	624	66	1.8	16.4
149	0.99	6.0	12.7	504	68	1.8	16.5
184	0.99	5.0	12.5	416	69	1.8	16.5
161	0.99	3.9	11.9	444	75	1.8	15.9
150	0.99	3.0	11.2	475	74	1.8	16.0
156	0.99	1.0	8.0	504	71	1.8	15.9
89	0.38	2.9	11.0	427	75	1.8	15.9
92	0.37	2.5	10.5	355	75	1.8	15.9
106	0.33	1.0	8.0	532	70	1.8	15.8
149	0.99	6.0	12.7	497	70	1.8	15.9
177	0.99	4.0	12.0	497	69	1.8	15.8
134	0.45	3.0	11.2	419	71	1.8	15.7
92	0.34	3.0	11.2	458	68	1.8	15.7
170	0.99	6.0	12.7	451	71	1.8	15.6
156	0.99	6.0	12.7	490	71	1.8	15.6
227	0.46	1.6	9.2	568	67	1.8	14.8
177	0.99	3.0	11.2	454	68	1.8	14.6
142	0.99	2.5	10.5	355	74	1.8	15.2
64	0.39	3.0	11.2	340	73	1.8	15.1
57	0.39	2.5	10.5	369	73	1.8	15.1
50	0.38	1.0	8.0	410	73	1.8	15.2
206	0.99	1.0	8.0	639	67	1.5	14.9
185	0.99	1.5	9.0	589	66	1.5	14.9
234	0.99	2.0	9.9	625	66	1.5	14.8
128	0.34	1.0	8.0	582	66	1.5	15.3
163	0.35	1.5	9.0	596	66	1.5	15.2
184	0.35	2.0	9.9	532	66	1.5	15.2
199	0.99	3.0	11.2	575	66	1.8	15.3
209	0.99	3.0	11.2	582	66	1.8	15.4
220	0.99	3.0	11.2	624	66	1.8	15.3
205	0.99	3.0	11.2	532	64	1.8	15.3
149	0.36	3.0	11.2	639	65	1.8	15.4
135	0.36	3.0	11.2	653	66	1.8	15.3
135	0.35	3.0	11.2	639	66	1.8	15.4
138	0.36	3.0	11.2	646	66	1.8	15.4
127	0.39	6.0	12.7	397	75	1.8	15.5

(Contd.)

Table 12.1 (*Contd.*)

y	x_1	x_2	x_3	x_4	x_5	x_6	x_7
92	0.38	6.0	12.7	424	75	1.8	15.5
85	0.38	6.0	12.7	446	75	1.8	15.6
85	0.39	6.0	12.7	424	74	1.8	15.6
126	0.41	5.0	12.5	550	75	1.8	16.2
57	0.40	5.0	12.5	448	76	1.8	16.2
126	0.99	6.0	12.7	416	76	1.8	16.1
119	0.99	5.0	12.5	416	76	1.8	16.1
142	0.47	6.0	12.7	482	72	1.8	15.7
128	0.47	6.0	12.7	482	72	1.8	15.7
113	0.47	5.0	12.5	523	72	1.8	15.8
134	0.44	4.0	12.0	423	72	1.8	15.7
129	0.99	4.0	12.0	404	74	1.8	16.8
134	0.99	3.0	11.2	404	74	1.8	16.9
134	0.99	2.0	9.9	389	74	1.8	16.9
64	0.43	4.0	12.0	404	74	1.8	16.9

y: outlet gas concentration, ppm.
x_1: vessel pressure, bar.
x_2: nozzle pressure, bar.
x_3: flow rate, $m^3 hr^{-1}$.
x_4: inlet gas concentration, ppm.
x_5: liquid temperature, °C.
x_6: pH.
x_7: liquid concentration.

liquid level was maintained. Note that in this example we are not solely concerned with finding the optimum settings of the x variables; one variable x_4, the inlet gas concentration, is not fixed in advance, and the experimenter wishes to know how the shape of the response surface and position of the optimum for the other x variables, depend on this variable. The optimum in this case corresponds to a minimum concentration of gas as output for a given input concentration.

Before attempting to fit the response surface, the data are first subject to some simple transformations in order to reduce the likelihood of numerical problems, and to remove correlations between the original x variables and the generated squared and cross-product variables. These correlations can be very high for the original data. In the present example the correlations between the original x variables and their squares are all greater than 0.98, and for two, $\{x_5\}$ and $\{x_7\}$, the correlations could be a potential source of multicollinearity. The correlations between the original variables and the cross-products are not usually as high but can still be large enough to complicate model selection. For example, the correlations between $\{x_1\}$ and $\{x_1x_2\}, \{x_1x_3\}, \ldots, \{x_1x_7\}$ are respectively 0.713, 0.962, 0.930, 0.992, 0.988, 0.996. These correlations are, of course, a function of the design and for certain standard response surface designs this problem does not arise. We circumvent the problem here by transforming the data as follows:

(1) The variables are centred and normalized, i.e. we calculate new variables

$$Z_{i(j)} = (x_{i(j)} - \bar{x}_i)/\text{s.e.}\{x_i\}, \qquad j = 1, \ldots, n,$$

where

$$\text{s.e.}\{x_i\} = \left\{ \sum_{j=1}^{n} (x_{i(j)} - \bar{x}_i)^2/(n-1) \right\}^{1/2}$$

is the standard error of $\{x_i\}$, and $x_{i(j)}$ is the jth observation or setting of the ith variable.

(2) The squared variables are calculated as

$$Z_{ii(j)} = (Z_{i(j)} - C_i)^2, \qquad j = 1, \ldots, n,$$

where C_i is chosen to produce a zero correlation between $\{Z_i\}$ and $\{Z_{ii}\}$, i.e. $C_i = \frac{1}{2}\sum Z_{i(j)}^3$. The $\{Z_{ii}\}$ are then normalized.

(3) The cross-product variables are calculated as

$$Z_{ij(k)} = (Z_{i(k)} - C_{i(j)})(Z_{j(k)} - C_{j(i)}), \qquad k = 1, \ldots, n,$$

where $C_{i(j)}$ and $C_{j(i)}$ are chosen to produce zero correlations between $\{Z_i\}$ and $\{Z_{ij}\}$ and $\{Z_j\}$ and $\{Z_{ij}\}$, i.e.

$$C_{i(j)} = \frac{Z_{j(k)}Z_{i(k)} - \sum Z_{i(k)}Z_{j(k)} \sum Z_{i(k)}Z_{j(k)}}{1 - \left(\sum Z_{i(k)}Z_{j(k)}\right)^2}.$$

The $\{Z_{ij}\}$ are then normalized.

We work below with the new variables $Z_1, \ldots, Z_7, Z_{11}, Z_{12}, \ldots, Z_{77}$. It is a straightforward matter to transform back to the original variables when an appropriate model has been selected. This procedure does not necessarily reduce the correlations between pairs of linear variables or between pairs of second-order terms and, if these correlations are high, the usefulness of the full procedure is somewhat diminished. The normalization of the variables only, step 1, may be thought to be adequate. This will generally reduce considerably the correlation between a regressor variable and its square and so reduce the likelihood of numerical problems.

12.3 Model selection

The problem of selecting a model in multiple regression has been discussed in Chapter 11. The comments made there apply equally well in the present context. There are, however, some additional points which are special to model selection for multiple quadratic regression and in this section we shall consider these. It should be noted first, however, that we are discussing the statistical aspects only of model selection. This is not to say that in any given situation there will be no prior knowledge about the relationships between the regressor variables and the response, nor that the use of such information is unimportant. For example, physical considerations may point to a linear relationship between the response and a certain function of a regressor variable. We might then be inclined to use the function of the variable in the regression equation rather than the variable itself. The statistical selection of variables in a model should complement such prior knowledge rather than be used in its place.

The first point we note is that with the addition of generated cross-product and quadratic regression variables, the total number of variables from which we have to select a model will typically be large. In our present example there are seven original predictor variables, from which we generate seven squared and twenty-one cross-product variables, making a total of 35 in all. For the reasons set out in Chpaters 10 and 11, a model which includes all terms, regardless of the contribution of individual terms, will rarely be a satisfactory choice. This is especially true in the present context when we generate such a large number of new variables from the originals.

The second point to be made is that unlike the variables in standard multiple regression, those in polynomial regression have a structure, e.g. x_i^2 and $x_i x_j$ are related to x_i in a way that they are not related to x_k. As a consequence there are certain logical considerations, which precede statistical ones in the choice of model. These are summarized in the following two requirements:

(1) A quadratic term should only be included in the model if accompanied by the corresponding linear term.
If a quadratic term stands alone in the model it implies that the turning point of the surface is at the origin with respect to this variable. This origin is determined by the design of the experiment and is to some extent arbitrary. We are not justified therefore in making such constraints on the fitted surface.

(2) A cross-product term should only be included in the model if accompanied by both corresponding linear terms.
The presence of the cross-product term indicates that the effect on the response of changing one of the variables depends on the setting of the other and vice versa. Having accepted that the settings of both variables affect the response we are rarely, if ever, justified in assuming that the linear components are negligible. The cross-product term is generally interpreted as accounting for variation in the response over and above the variation accounted for by the two linear terms. Its interpretation in the absence of either or both these terms is problematical. For example, the relationship between the response and two x variables, x_1 and x_2 say, including the cross-product term but omitting one linear term is

$$y = \beta_1 x_1 + \beta_{12} x_1 x_2 + \text{function of the other variables.}$$

For fixed values of x_1 and $x_3, x_4, \ldots$, the response is then a linear function of x_2, with the slope coefficient proportional to x_1. Such a relationship depends on the choice of origin for x_1 and, as in point (1) above, we are rarely justified in constraining the fitted surface in this way.

We could summarize both these points in the requirement that the structure of the fitted surface (i.e. the terms that are included) should be invariant under changes of location of the x variables. Both points refer to the inclusion of linear terms of variables which enter the model in quadratic or cross-product terms and we see that the requirement is a sensible one when we consider the transformations discussed in the previous section. If we work with transformed variables, and omit some of these linear terms, the transformation back to the original variables will normally reintroduce the linear terms anyway. For this reason alone it is best that these linear terms be included in the original model.

It can also be argued that the model ought to be 'balanced'. By this is meant the inclusion of all quadratic terms whose corresponding cross-product terms are in the model and vice versa. The justification for this does not rest on logical arguments like (1) and (2) above, and, since the requirement can lead to inclusion of many non-significant terms in the model, which in turn can produce much less precise estimates of the optimum, the approach is not recommended here.

Although we now describe one particular approach to model selection for multiple quadratic regression there are, of course, a variety of possible alternatives. In each case a combination is used of one of the selection procedures discussed in Chapter 11 and the logical constraints described above. The present approach, which was developed by the late Dr J. Köllerström, has two main distinguishing features. It is based on the forward stepwise technique and it sets an upper limit of six on the number of quadratic terms which are allowed in the model, regardless of the number of original regressor variables. The procedure consists of three stages:

Stage 1. First only the linear terms are considered. Variables are included using forward stepwise selection with a very small critical F-value, for example, rejecting those with an F-statistic less than 0.001. The same process is then repeated with the quadratic terms, keeping the linear terms already selected in the model. Finally, any linear term which does not accompany its corresponding quadratic term in the model is added.

This first stage eliminates all variables whose linear and quadratic terms have an additional contribution so small as to make it very unlikely that a corresponding cross-product term will have a non-negligible contribution. Only the variables selected in this stage are included in the following two stages.

Stage 2. The cross-products corresponding to the variables in the Stage 1 model are included in the pool of regression variables to be considered and a forward stepwise procedure is again used to select a model, with the critical F-value increased to 1. At this stage terms can be included and dropped to conform to the 'balance' criterion given above, if this is required. The other two requirements (1) and (2) are necessarily adhered to. In this stage an attempt is made to select all potentially important cross-product terms in addition to the linear and quadratic terms from Stage 1.

Stage 3. The final model is selected using the Anscombe–Tukey approach (Section 11.4) from the terms included in the model output from Stage 2, i.e. the subset is chosen which minimizes the ratio of the mean square error and residual degrees of freedom, with the additional condition, for present purposes, that the subset corresponds to a model satisfying requirements (1) and (2) above, and if wanted, the condition of 'balance'. The justification for the Anscombe–Tukey method is essentially that it gives the most precise predictions in standard multiple regression and will therefore give good, if not best, predictions for models subject to the logical requirements of multiple quadratic regression.

For the example data set we get the following:

Stage 1. All the linear terms are included and all the quadratic except variable 4.

Stage 2. Two cross-product terms are entered, Z_{16} and Z_{27}, and the linear variable Z_3 and quadratic variables Z_{22}, Z_{33}, Z_{66} and Z_{77} are dropped.

Stage 3. The model output from the previous stage contains ten predictor variables $Z_1, Z_2, Z_4, Z_5, Z_6, Z_7, Z_{11}, Z_{55}, Z_{16}$ and Z_{27}. The corresponding residual mean square is 836.0 on 50 degrees of freedom giving a value of 16.72 for the Anscombe–Tukey (A–T) statistic. Within our constraints there are several other models which could be considered from the set of ten variables. Two examples are:

(1) omitting Z_2, Z_7, Z_{27}, A–T = 17.24,
(2) omitting Z_6, Z_{16}, A–T = 18.45.

None of these possible models, however, produces a smaller A–T statistic than the original ten-variable model and, consequently, we keep this model as the final choice. Note that if we also required the final equation to be 'balanced' we would need to introduce a further eleven variables, more than doubling the number of terms in the model.

Before proceeding to the examination and interpretation of the fitted surface it is important to consider the adequacy of the fit. Some of the methods described in Chpater 6 are appropriate here. In particular, an examination of the standardized residuals can point to a systematic lack of fit due to underfitting, which may occur in just a part, or thoughout most, of the factor space. Also if the experiment has not been designed with a view to providing an even distribution of points throughout the factor space, some points which are relatively isolated may have an excessively high leverage. Their influence on the fitted surface can be checked by omitting these points from the data set and comparing the resulting and original parameter estimates. Should the difference be excessive we could argue that either more observations should be obtained in neighbouring regions of the factor space, or attention be restricted to the region covered when these points are excluded.

12.4 Interpretation of the fitted surface

Once the model has been selected and fitted it is necessary to interpret the resulting quadratic polynomial equation. Precisely what we do must depend on the purposes for which the data were collected but usually we shall want some interpretable representation of the surface, remembering that a direct visualization is impossible when the factor space has dimension greater than two. We begin by transforming back to the original variables, if a transformation was used in the first place. For the example data set the final fitted model in the transformed variables $\{Z_i\}$ and $\{Z_{ij}\}$ is

$$\begin{aligned} y = 233.00 &+ 259.11Z_1 + 38.34Z_2 + 151.13Z_4 - 155.54Z_5 \\ &+ 31.07Z_6 - 71.39Z_7 - 719.10Z_{11} + 82.71Z_{55} \\ &+ 92.65Z_{16} + 61.85Z_{27}. \end{aligned}$$

Transforming to the $\{Z_i\}$, using

$$Z_{ii} = W_{ii}(Z_i^2 - 2C_iZ_i + C_i^2),$$

and

$$Z_{ij} = W_{ij}(Z_iZ_j - C_{i(j)}Z_j - C_{j(i)}Z_i + C_{i(j)}C_{j(i)}),$$

where W_{ii} and W_{ij} are normalizing coefficients, we get

$$\begin{aligned} y = 232.45 &+ 321.61Z_1 + 17.77Z_2 + 151.13Z_4 - 156.93Z_5 \\ &+ 36.54Z_6 - 72.53Z_7 - 5470.97Z_1^2 + 497.17Z_5^2 \\ &+ 706.72Z_1Z_6 - 570.51Z_2Z_7. \end{aligned}$$

Since the coefficients in the second equation are linear combinations of those in the first, we can also directly transform the covariance matrix for the latter coefficients $\sigma^2(\boldsymbol{a}'\boldsymbol{a})^{-1}$ in the general multiple regression notation from Section 1.3) to obtain the matrix for the former. The unknown variance σ^2 is estimated in the normal way from the residual mean square. The equation can be transformed a second time to get the expression in terms of the original observations, the $\{x_i\}$. For simplicity, we shall work in terms of the $\{Z_i\}$ here noting that there are also some numerical advantage in keeping the variance of the predictor variables equal to one.

In general, we can write the fitted model, possibly after transformation, as

$$y = b_0 + \sum_{i=1}^{p} b_i x_i + \sum_{i=1}^{p} \sum_{j=i}^{p} b_{ij} x_i x_j, \tag{12.3}$$

where the $\{x_i\}$ may be the original predictor variables or some other transformation of these as required. Before considering specific methods for interpreting the model (12.3) we note that some important information is to be obtained by examining the model directly. Some of the coefficients may be zero and those variables which have a zero linear coefficient, and consequently zero second-order coefficients as well, have a negligible influence on the response in the experimental region. An impression of the relative effects of the other variables can be gained from the relative sizes of their coefficients, taking into account the different scales on which they are measured. Some variables may only enter the equation through the linear part and these have a simple proportional effect on the response, independent of the other variables. These variables will be separated out when we investigate the shape of the surface. It is also useful to note which pairs of variables have a non-zero cross-product coefficient, i.e. those variables which are interacting and those with a non-zero quadratic coefficient, i.e. those for which the surface is curved in the experimental region.

Unfortunately, if there are more than a very few non-zero second-order coefficients in the model it is very difficult to appreciate the shape of the surface from a simple inspection of these. Using a computer plotting routine it is not difficult to construct contour maps, or even two-dimensional pictures of the three-dimensional surface, for pairs of variables with the remaining variables fixed. A judicious choice of such pairs and levels at which to fix the remainder can produce a useful set of maps from which a visualization of the surface can be made. Another method for examining the surface is ridge analysis. Details will not be given here, an exposition can be found in Draper (1963). Perhaps the most important aid to the interpretation of a response surface is the canonical representation and the remainder of this section will be devoted to this approach.

We begin by writing (12.3) in matrix notation, separating out those variables with zero second-order coefficients and dropping altogether those with all coefficients equal to zero:

$$y = b_0 + \mathbf{x}_1'\mathbf{b}_1 + \mathbf{x}_2'\mathbf{b}_2 + \mathbf{x}_2'\boldsymbol{B}\mathbf{x}_2,$$

where $\mathbf{x}_1$ consists of those variables with zero second-order coefficients, $\mathbf{b}_1$ of the corresponding linear coefficients, $\mathbf{x}_2$ consists of those variables with at least one non-zero second-order coefficient, $\mathbf{b}_2$ of the corresponding linear coefficients and

$$(\boldsymbol{B})_{ij} = \begin{cases} \frac{1}{2}b_{ij}, & i < j, \\ b_{ii}, & i = j, \\ \frac{1}{2}b_{ji}, & i > j, \end{cases}$$

again corresponding to the variables in $\mathbf{x}_2$. It is assumed that $\mathbf{x}_1$ and $\mathbf{x}_2$ have dimensions v and w respectively where $v + w \leq p$. Note that there are $p - w - v$ variables with all coefficients equal to zero which have been dropped. In order to investigate the shape of the surface we can hold the value of $\mathbf{x}_1$ constant, noting that changes in any element of $\mathbf{x}_1$ simply increases or decreases the value of y by a proportionate amount. Fixing $\mathbf{x}_1$ at $\tilde{\mathbf{x}}_1$ and writing $\tilde{b}_0 = b_0 + \tilde{\mathbf{x}}_1'\mathbf{b}_1$ we get

$$y - \tilde{b}_0 = \mathbf{x}_2'\mathbf{b}_2 + \mathbf{x}_2'\boldsymbol{B}\mathbf{x}_2. \tag{12.4}$$

Before proceeding we note that should $v > 0$, then, if the surface has an extremum in the experimental region, this must be on the boundary of the region. This follows from the fact that whatever value y may take in the interior of the region this value can be increased or decreased as required by increasing or decreasing the elements of $\mathbf{x}_1$ until the boundary is reached. For the example data set variable 3 is dropped and $\mathbf{x}_1$ consists of one variable, Z_4, the inlet concentration. This is rather convenient since it tells us that in the experimental region the outlet concentration is approximately linearly related to the inlet concentration independently of the values of the remaining variables. Consequently $\mathbf{x}_2$ consists of the remaining five predictor variables $(Z_1, Z_2, Z_5, Z_6, Z_7)'$,

$$\mathbf{b}_2 = (321.61, 17.77, -156.93, 36.54, -72.53)'$$

and

$$B = \begin{bmatrix} -5470.97 & 0 & 0 & 353.36 & 0 \\ 0 & 0 & 0 & 0 & -285.26 \\ 0 & 0 & 497.17 & 0 & 0 \\ 353.36 & 0 & 0 & 0 & 0 \\ 0 & -285.26 & 0 & 0 & 0 \end{bmatrix}.$$

The first step towards canonical representation of the surface is to relocate the origin at the centre of the quadratic surface (12.4), i.e. at the point $\mathbf{k} = -\frac{1}{2}\boldsymbol{B}^{-1}\mathbf{b}_2$. For this we assume that $\boldsymbol{B}$ is non-singular. This will be true in theory since all the rows are non-null and a linear dependence among them has probability zero, but may be false in practice due to the approximate way in which real numbers are stored in the computer. The problem is related to that of multicollinearity discussed in Chapter 4. Should $\boldsymbol{B}$ be singular the surface is a degenerate one and there is no unique centre; any solution of $2\boldsymbol{B}\mathbf{k} = -\mathbf{b}_2$ will then serve. The equation of the relocated surface is

$$y - b_0^* = \mathbf{t}'\boldsymbol{B}\mathbf{t}, \tag{12.5}$$

where $\mathbf{t} = \mathbf{x}_2 - \mathbf{k}$ and $b_0^* = \tilde{b}_0 + \frac{1}{2}\mathbf{k}'\mathbf{b}_2$. In the second step the axes are rotated until they lie along the Principal Axes of the surface. For this we need to calculate the eigenvalues and eigenvectors of the matrix $\boldsymbol{B}$ (the canonical decomposition of $\boldsymbol{B}$), as explained in Section 3.6. By definition $\boldsymbol{B}$ is symmetric, so all of its eigenvalues are real, and its eigenvectors are mutually orthogonal. These eigenvectors define the Principal Axes of the surface. We can write

$$\boldsymbol{B} = \boldsymbol{Q}'\boldsymbol{\Lambda}\boldsymbol{Q}$$

where $\boldsymbol{Q} = [\mathbf{q}_1, \mathbf{q}_2, \ldots, \mathbf{q}_w]$ and $\boldsymbol{\Lambda} = \mathrm{diag}(\lambda_1, \lambda_2, \ldots, \lambda_w)$ for $\mathbf{q}_i$ the eigenvector of $\boldsymbol{B}$ corresponding to the ith largest eigenvalue, λ_i. Substituting for $\boldsymbol{B}$ in (12.5) above we get

$$\begin{aligned} y - b_0^* &= \mathbf{t}'\boldsymbol{Q}'\boldsymbol{\Lambda}\boldsymbol{Q}\mathbf{t} \\ &= \sum_{i=1}^{w} u_i^2 \lambda_i, \end{aligned} \tag{12.6}$$

where $u_i = \mathbf{q}_i'\mathbf{t}$. This is the canonical representation of the fitted surface. The shape of this surface is determined by the signs and relative sizes of the eigenvalues and its orientation in terms of the original variables is determined by the eigenvectors. There are many possible surface types that may arise and a full discussion of the interpretation of the canonical representation is given in Davies (1978, Chap. 11). We shall concentrate here on a few main points.

We shall assume from now on that the matrix $\boldsymbol{B}$ is non-singular. Should this not be the case it means that $\boldsymbol{B}$ has one or more zero eigenvalues, the corresponding eigenvectors of which define linear combinations of the original variables which have no effect on the response. In geometrical terms the subspace spanned by these eigenvectors corresponds to a horizontal ridge (one zero eigenvalue) or plateau (more than one zero eigenvalue) in the surface. The interpretation of the surface in the remaining, or complementary subspace, is the same as for a non-singular $\boldsymbol{B}$, all of the eigenvalues of which are non-zero. We therefore proceed under this latter assumption, bearing in mind the effect on the interpretation of zero eigenvalues.

In the simplest case all the eigenvalues have the same sign. The contours of the surface are then w-dimensional ellipsoids and there is a unique extremum at the centre of the surface $\mathbf{k} = \frac{1}{2}\boldsymbol{B}^{-1}\mathbf{b}_2$. If the eigenvalues are negative this is a maximum and vice versa. It is advisable at this point to recall that we are dealing here only with the shape of the surface for the variables in $\mathbf{x}_2$. This extremum may lie outside the experimental region in which case it should be confirmed by further experimentation in its neighbourhood before much confidence is placed in it. Using the methodology in Box and Hunter (1954) it is also possible to calculate a confidence region for the point $\mathbf{k}$. Let $\sigma^2 \boldsymbol{V}$ be the variance–covariance matrix of $\mathbf{h} = \mathbf{b}_2 + \boldsymbol{B}x_2$ for a given value of $\mathbf{x}_2$, then a $100\alpha\%$ confidence region for $\mathbf{k}$ is defined by the set of points $\mathbf{x}_2$ satisfying

$$\mathbf{h}' \boldsymbol{V}^{-1} \mathbf{h} \leq S^2 w F_{w,m}(\alpha),$$

where S^2 is the residual mean square estimator of σ^2 on m degrees of freedom and $F_{w,m}(\alpha)$ is the upper α point of the F-distribution with w and m degrees of freedom. Note that both $\mathbf{h}$ and $\boldsymbol{V}$ are functions of $\mathbf{x}_2$ and so the shape and size of the region depends both on the fitted surface and the variance–covariance matrix of the parameter estimates, which is a function of the experimental design and residual variance. This can be very informative and the region can be updated as more observations are added.

It cannot be assumed, however, that a surface will take this particularly simple form and examples in which the eigenvalues $\{\lambda_i\}$

take both positive and negative signs are common. This indicates that the surface contains a saddle point. Maxima can be found when moving across the surface in certain directions, minima in other directions; the directions being determined by the eigenvectors $\{v_i\}$. Consequently no unique extremum exists, which is obvious from equation (12.6) when some of the $\{\lambda_i\}$ have opposite signs. In those directions corresponding to relatively small eigenvalues the surface is extended with little curvature and in the experimental region in the directions of the corresponding eigenvectors the surface is approximately ridge-shaped. Such a ridge will be rising or falling depending on the sign of the eigenvalue and the position of **k**. If the purpose of the analysis is to find an optimum it may then be necessary to conduct further experiments to extend the experimental region in the direction of a rising (or falling) ridge.

We now return to the example. The matrix $\boldsymbol{B}$ was given earlier, its eigenvalues are

$$-5493.70, \quad 497.17, \quad 285.26, \quad -285.26 \quad 22.73$$

with corresponding eigenvectors,

$$\begin{bmatrix} 0.998 \\ 0 \\ 0 \\ -0.064 \\ 0 \end{bmatrix} \begin{bmatrix} 0 \\ 0 \\ 1.000 \\ 0 \\ 0 \end{bmatrix} \begin{bmatrix} 0 \\ 0.707 \\ 0 \\ 0 \\ -0.707 \end{bmatrix} \begin{bmatrix} 0 \\ 0.707 \\ 0 \\ 0 \\ 0.707 \end{bmatrix} \begin{bmatrix} 0.064 \\ 0 \\ 0 \\ 0.998 \\ 0 \end{bmatrix}$$

Note the simple form of the eigenvectors. This is the consequence of the small number of non-zero elements of $\boldsymbol{B}$ and will aid interpretation of the surface. Had the equation been 'balanced', all of the elements of $\boldsymbol{B}$ would have been non-zero, complicating, but not substantially altering, the canonical decomposition. We see from the first eigenvector that the dominant eigenvalue corresponds almost entirely to variable Z_1, the vessel pressure. This eigenvalue has a negative sign which indicates that there is a turning-point, which is a maximum, for this variable at the centre of the surface. The centre for all five variables $(Z_1, Z_2, Z_5, Z_6, Z_7)'$ is

$$\tfrac{1}{2}\boldsymbol{B}^{-1}\mathbf{b}_2 = (0.052,\ 0.127,\ -0.158,\ 1.756,\ 0.031)'.$$

Since the experimental range of $\{Z_1\}$ is $(-0.142, 0.122)$ we see that the centre for this variable is well within the experimental region. In fact, the only variable for which the centre lies outside the experimental region is $\{Z_6\}$ for which the range is $(-0.788, 0.216)$. The second eigenvalue corresponds exactly to variable Z_5, the liquid temperature. This is to be expected since there was no evidence of interaction between $\{Z_5\}$ and the other variables in the experimental region. Note the positive sign of the eigenvalue indicating a minimum with respect to this variable.

The next two eigenvalues are of equal and opposite sign and correspond to the sum and difference respectively of variables Z_2 and Z_7, the nozzle pressure and liquid concentration. In the direction of their difference there is a minimum, in the direction of their sum a maximum. Such behaviour is a consequence of two variables having zero quadratic coefficients but a non-zero cross-product coefficient. The final eigenvalue which corresponds almost wholly to variable Z_6, the pH of the liquid, is considerably smaller than the others. The centre is also well outside the experimental region for this variable. In the direction of Z_6 therefore the surface approximates a ridge which decreases as Z_6 is increased across the experimental region.

For our example then the canonical representation has a relatively simple form; the interpretation of all the eigenvectors being straightforward. This need not be the case when B contains more non-zero elements. There is no simple answer, however, to the question of locating an optimum value of x to minimize the output gas concentration, but this may not be unexpected and the information gained on the response surface in the experimental region provides a basis for the next step in the investigation.

12.5 Further reading

The literature on response surface methodology is vast, for example, the review of Hill and Hunter (1966) contains 165 references and much has been written since. However, as mentioned in the introduction, much of this work has been on the design of response surface experiments, rather than on the analysis of response surface data. Some more recent developments which are related to the present chapter can be found in Bradley and Srivastava (1979) on correlation in polynomial regression, Park (1977), Driscoll and Anderson (1980) on variable selection in multiple quadratic regression, and Box and Draper (1982) on lack of fit for response surface designs.

References

Box, G. E. P. (1954) The exploration and exploitation of response surfaces: some general considerations and examples. *Biometrics*, **10**, 16–60.

Box, G. E. P. and Draper, N. R. (1982) Measures of lack of fit for response surface designs and predictor variable transformations. *Technometrics*, **24**, 1–8.

Box, G. E. P. and Hunter, J. S. (1954) A confidence region for the solution of a set of simultaneous equations with an application to experimental design. *Biometrika*, **41**, 190–199.

Box, G. E. P., Hunter, W. G. and Hunter, J. S. (1978) *Statistics for Experimenters*. Wiley, New York.

Box, G. E. P. and Wilson, K. B. (1951) On the experimental attainment of optimal conditions (with discussion). *J. Roy. Statist. Soc.* B, **13**, 1–45.

Bradley, R. A. and Srivastava, S. S. (1979) Correlation in polynomial regression. *Amer. Statist.*, **33**, 11–14.

Davies, O. L. (1978) *The Design and Analysis of Industrial Experiments*, 2nd edn. Longman, London.

Draper, N. R. (1963) 'Ridge analysis' of response surfaces. *Technometrics*, **5**, 469–479.

Driscoll, M. R. and Anderson, D. J. (1980) Point-of-expansion, structure, and selection in multivariable polynomial regression. *Commun. Statist. Theor. Methods* A **9** (8), 821–836.

Hill, W. J. and Hunter, W. G. (1966) A review of response surface methodology: a literature survey. *Technometrics*, **8**, 571–590.

Myers, R. H. (1971) *Response Surface Methodology*. Allyn and Bacon, Boston.

Park, S. H. (1977) Selection of polynomial terms for response surface experiments. *Biometrics*, **33**, 225–229.

CHAPTER 13

Alternative error structures

13.1 Introduction

In everything done so far it has been assumed that only the response variables are subject to error or random variation, and that these errors are statistically independent. In practice, examples where very different error structures are present are very common, as the examples below illustrate. This chapter presents only a brief introduction to the problems presented by these alternative error structures.

EXAMPLE 13.1

In order to test the validity of a theoretical relationship between diameter and velocity of flow in the blood vessels of animals, Mayrovitz and Roy (1983) measured the internal diameters d_i and average blood velocities (across the vessel) v_i in individual vessels in the muscles of anaesthetized rats. The intention was to test whether the velocities were proportional to a power of the diameters, with particular interest in a possible proportional relationship.

The experiment involved observing tiny blood vessels (in the range 6 to 108 μm) which were presumably not perfect cylinders, so neither diameter nor velocity measurements can be regarded as exact. Even if precise measurements were possible it would be too much to expect that a biological system would conform exactly to such a theoretical relationship.

To explain how this relationship is derived and to describe the departures from it, it is convenient to think of the observed quantities d_i and v_i as being approximately equal to hypothetical "true" values D_i and V_i which satisfy the relationship exactly. The equation satisfied by the D_i and V_i may be written as

$$V_i = \gamma D_i^{\beta} \tag{13.1}$$

for some unknown parameters β and γ.

The relationship is derived by supposing that species have evolved in such a way as to minimize the power required to maintain the flow in each vessel of their vascular systems. Ignoring the integration of the system into the whole body, this power is required for two main purposes, overcoming friction and maintaining the blood in the vessel. For a fixed length of artery with internal diameter D containing blood undergoing laminar flow at a fixed rate F, these two components are proportional to F^2D^{-4} and D^2 respectively. The power is therefore minimized with respect to D when flow is proportional to D^3, a result known as Murray's law (Murray, 1926). If the blood has average velocity V and the vessel has circular cross-section, then V is proportional to FD^{-2} and hence, by Murray's law, to D. However, some turbulence can be expected to occur in the blood vessels; for completely turbulent flow the two components of power are proportional to F^3D^{-5} and D^2, leading to an optimum when flow and average velocity are proportional to $D^{7/3}$ and $D^{1/3}$ respectively. For a flow pattern intermediate between these extremes we might propose a relationship between diameter and velocity of the form $V = \gamma D^{\beta}$, as in (13.1).

Writing X, Y and α for $\log D$, $\log V$ and $\log \gamma$ respectively, equation (13.1) is seen to be equivalent to the linear relationship

$$Y_i = \alpha + \beta X_i. \tag{13.2}$$

If x_i, y_i are now taken to represent the logarithms of the corresponding observed values, that is, $\log d_i$ and $\log v_i$, then we may regard $x_i - X_i$ and $y_i - Y_i$ as departures from the true values and write

$$x_i = X_i + \delta_i,$$

$$y_i = Y_i + \varepsilon_i.$$

Here δ_i and ε_i might be taken to be independently distributed random variables with zero expectations and variances σ_δ^2 and σ_ε^2 respectively. If normality is also assumed then we have a model with unknown parameters α, β, σ_δ^2, σ_ε^2 and X_1, X_2, for $i = 1, 2, \ldots$.

In Chapter 1 we defined a regression relationship as one between the expected value of a random variable and the observed values of other variables. By this definition, the situation described in Example 13.1 is not regression, since the X_i are not observed. The method of least squares is not really appropriate to this type of model, although on occasions we use it as a close approximation, see below. Example 13.1 represents what we call a functional relationship model.

EXAMPLE 13.2

Nelder (1968) introduced a model he called a regression model of the second kind with the following situation. We suppose that we have an agricultural experiment with fertilizers, and denote by x the amount of fertilizer applied to a plot; the corresponding yield is denoted by y. For the ith plot the relationship between y and x is taken to be

$$y_i = a_i + b_i x_i, \tag{13.3}$$

where (a_i, b_i) have a joint distribution over the plots used, which we suppose to be bivariate normal with expectation (α, β) and a variance matrix

$$\begin{bmatrix} V_{00} & V_{01} \\ V_{01} & V_{11} \end{bmatrix}.$$

In Example 13.2 it is the parameters of the relationship (13.3) which are random variables. This is equivalent to a regression model in which we have

$$\left.\begin{aligned} E(Y) &= \alpha + \beta x, \\ \text{and} \qquad V(Y) &= V_{00} + 2xV_{01} + x^2V_{11}. \end{aligned}\right\} \tag{13.4}$$

This is therefore case (3) of Section 9.2.

EXAMPLE 13.3
In a certain chemical process, manufacturing a continuously flowing material, measurements are made at regular intervals of a quality characteristic, denoted y. Measurements are also made of three explanatory variables, x_1, x_2 and x_3. The relationship

$$Y = \alpha + \beta_1 x_1 + \beta_2 x_2 + \beta_3 x_3 + \varepsilon \tag{13.5}$$

is thought to hold. However, the errors ε_i, for $i = 1, 2, \ldots$, are not independent, but autocorrelated, due to the nature of the sampling of the chemical process.

By our definition Example 13.3 is a regression relationship, but ordinary least squares is not appropriate because of the autocorrelation in the errors. This situation applies frequently in economic and other time series problems.

In the subsequent sections of this chapter we shall discuss some of the situations illustrated by these examples. In Sections 13.2–13.5 we discuss what might be called the 'errors in both variables' model, and Section 13.6 discusses models with autocorrelated and similar errors.

The topics mentioned in this chapter cover a vast area, and the models considered represent commonly occurring situations. It is therefore unfortunate the techniques are not more widely known.

The material in Sections 13.4 and 13.5 covers the estimation of functional and structural relationships, and these methods apply when we do have to take errors in the explanatory variables into account. This is a rather difficult area and one which statisticians have tended to neglect.

13.2 The effect of errors in the explanatory variables

Suppose we have simple linear regression with a model

$$Y = \alpha + \beta X + t, \tag{13.6}$$

where Y and X are underlying true values and t is distributed independently $N(0, \sigma_t^2)$. Let Y be measured with measurement error ε, which we assume to be distributed independently $N(0, \sigma^2)$. The distinction between t and ε here is that t is a variable representing the conditional distribution of Y given X, whereas ε is a measurement error which may be reduced by improved technique. We have then, say

$$y = Y + \varepsilon,$$

and

$$y = \alpha + \beta X + t + \varepsilon.$$

We now suppose that there is also measurement error in X, so that we do not observe X directly, but instead observe the random variable

$$x = X + \delta,$$

where δ is distributed independently $N(0, \sigma_\delta^2)$. Our relationship now becomes

$$y = \alpha + \beta x - \beta\delta + t + \varepsilon. \tag{13.7}$$

A regression relationship here would represent $E(y|x)$, since it is x which is observed. Three questions now arise:

(1) Under what conditions is the relation between $E(y|x)$ and x still linear?
(2) When can we use ordinary least squares to estimate this relationship?
(3) What techniques do we use when ordinary least squares cannot be used?

The following example demonstrates that the relationship between $E(Y|x)$ and x is, in general, not linear.

EXAMPLE 13.4

Consider the situation where there exists an exact relation

$$Y = \alpha + \beta X, \tag{13.8}$$

but where X is uniformly and continuously distributed on the range (b, c). Let δ have the distribution

$$\delta = \begin{cases} -d & \text{with probability } 0.05 \\ 0 & \text{with probability } 0.90, \\ +d & \text{with probability } 0.05, \end{cases}$$

then Fig. 13.1(a) shows the situation defined. Fig. 13.1(b) shows that $E(y|x)$ is very non-linear. With a continuous distribution of errors δ, we can see that, in

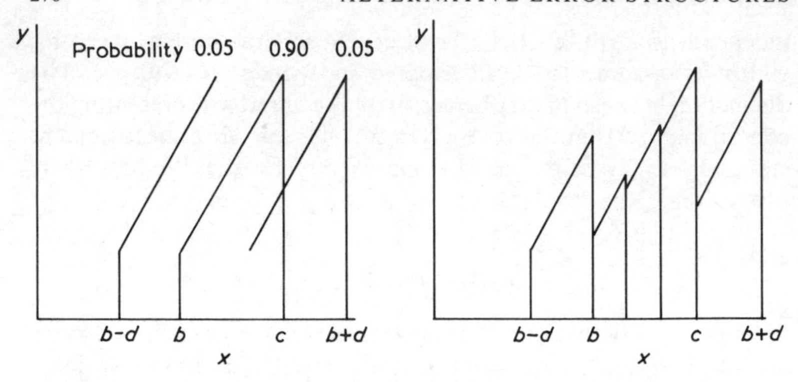

Fig. 13.1 *Effect of errors on a linear relationship. (a) Data for Example 13.4. (b) Expectations conditional on x.*

general, a linear relationship between Y and X will be transformed into a curved relationship between $E(y|x)$ and x. Lindley (1947) studied this problem and examined the precise conditions under which a regression relation remains linear. The result is complicated and we refer readers to the original article, but if the distribution of X is normal, then only a normal distribution for δ satisfies the conditions. (Lindley's result applied to all multiple regression models.)

Now suppose that X is a random variable and our distributions are such that the regression relation from (13.7) is still linear, what is the new regression coefficient? We are going to write (13.7) (with $\varepsilon = 0$) in the form

$$y = \alpha + \gamma x + t$$

from which we obtain

$$C(y, x) = \gamma V(x),$$
$$V(x) = V(X) + V(\delta),$$

so that

$$\gamma = C(y, X)/\{V(X) + V(\delta)\}.$$

Now

$$C(y, X) = C(y, X + \delta) = C(y, X) = C(Y, X)$$
$$= \beta V(X).$$

Therefore

$$\gamma = \beta\{1 + V(\delta)/V(X)\}^{-1} \tag{13.9}$$

and $\gamma \neq \beta$ unless $\delta = 0$, that is, unless X is observed without error.

We see that the effect of the errors in the regression is to flatten out the line, as we might expect. If the error variance is not small relative to $V(X)$, this flattening out effect could be quite serious. Clearly, the application of standard regression techniques in such a situation will estimate γ, not β, and β cannot be obtained from the estimate unless $V(\delta)$ is known. There is a similar attenuation effect when values of X are not random, but are selected in some other way.

If we assume now that there is still a linear relationship between $E(y|x)$ and x, we can ask under what circumstances it is appropriate to use ordinary least squares (OLS), even though it estimates $\gamma \neq \beta$. In discussing this, it is helpful to use an example introduced by Lindley (1947). An experimenter has lumps of metal of different shapes and sizes, and he wishes to determine the mass and volume of each piece, and it is reasonable to suppose that each observation is subject to error. One possible problem is to estimate the density of the metal, and this reduces to estimating a functional relationship, and methods appropriate to this situation are described in Section 13.4. Lindley then considers a different problem. Suppose the experimenter does not wish to carry out the weighings for reasons of convenience, but wished to predict the masses of the lumps of metal from the measured volumes. Suppose further that the distributions are such that the regression of weight on measured volumes is linear, then a common view is that this second problem is answered by using the ordinary regression of weight on measured volume, and, for example, Hald (1952) quotes a similar example concerning the prediction of starch content of potatoes from measurements of specific gravity. Madansky (1959) says,

> One should note that there is another problem in this context which is also called the prediction problem. This is the situation in which one can never hope to observe X or Y without error, and hence is only interested in predicting $E(Y) = Y$ for a new observed $x = X + \delta$. But this is just the case in which the least squares regression of y on x works, for our independent variable is no longer X but instead x, which is observed without error.

Undoubtedly, the kind of argument just quoted is used frequently by practising statisticians to justify ignoring the rather difficult problem of estimating functional and structural relationships, but we must beware, because there are severe dangers in it. Suppose we have the model

$$Y = \alpha + \beta X$$

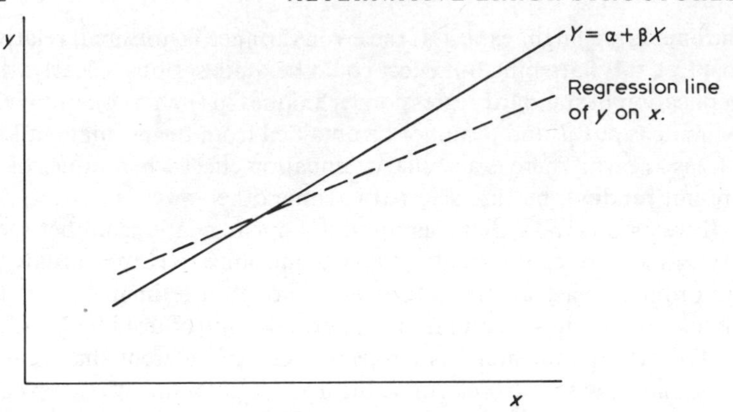

Fig. 13.2 *Functional relationships and regression lines.*

but we observe $y = Y + \varepsilon$, $x = X + \delta$. We have seen that the effect of errors in x is to flatten out the linear relationships, as shown in Fig. 13.2. The position and slope of this regression line is largely determined by

(1) the relationship between Y and X;
(2) the distribution of X values;
(3) the error variance σ_δ^2.

If the distribution of X-values is altered, a different regression line would be produced. It is easy to see that if an X outside the range used in the regression relation occurs, a large prediction error can result unless the functional relationship, and not the regression line, is used in prediction. If the X-values in Fig. 13.2 were translated, a different regression line should be used, whereas it is the same underlying functional relation. This point has been stressed by Ehrenberg (1968).

Let us apply this to Lindley's example. If the lumps of metal used in estimating the regression line are a random sample of those to be used in practice, the use of the ordinary regression line is correct. If, however, the distribution of volumes of the metal lumps changes for some reason, serious errors could result from using the regression line. Similar reasoning applies to Hald's starch content example.

A further point here is that the regression line is also affected by the error variance σ_δ^2, so that if different observers have different error variances, different regression lines should be used. This effect is likely to be much less important than the one mentioned above.

Bearing these points in mind, there are conditions under which it is legitimate to use a regression line when both variables are in error. Some guidance on which of the two regression lines to use in such cases is given, for example, by Cox (1968). We need to take care that the distribution of X-values remains constant in use, but since these values are not observed, this can be difficult to check. In many cases of the kind quoted, it may be safer to use the functional relationship provided this is feasible; see the discussion in Sections 13.4 and 13.5 below. Although much of this discussion has been for one explanatory variable only, the extension of the arguments to many explanatory variables is clear.

Exercises 13.2

1. In a medical study, patients diagnosed as hypertensive are given a course of treatment intended to lower blood pressure, and have their blood pressure recorded before and after the treatment, giving values denoted by x and y respectively. The reduction in pressure expected from the treatment is thought to depend on the pressure before treatment; to investigate this dependence, the regression is calculated of $x - y$ on x. By treating each measurement as the sum of a 'true' value and an error of measurement, explain why this procedure can be expected to be misleading. (Consider first the case in which the expected reduction does not depend on the initial blood pressure.) Investigate the published suggestion that a test of correlation between $x - y$ and $x + y$ provides a test of whether the expected reduction in pressure depends on the initial pressure.
2. An explanatory variable x is called a *controlled variable* if it can be controlled or adjusted to specified values, for example, a particular set of lengths, weights or voltages. The specified value x_i may still differ from the true value X_i; this may be expressed by the relation

$$x_i = X_i + \delta_i,$$

where, since x_i is fixed, δ_i is independent of x_i. Show that OLS is appropriate for simple linear regression on a controlled variable but not for quadratic regression.

13.3 Working rules for bias due to errors in explanatory variables

In the above discussion we have seen that the presence of errors in the explanatory variables introduces a bias into the estimates of the regression coefficients. The purpose of this section is to present some working rules with can be used to detect when such biases may be large.

We consider our usual model (1.13), where

$$E(\mathbf{Y}) = \boldsymbol{a}\boldsymbol{\theta}$$

and $\boldsymbol{a}$ is $(n \times k)$. Suppose we observe

$$\boldsymbol{W} = \boldsymbol{a} + \boldsymbol{\Delta}, \tag{13.10}$$

where $\boldsymbol{\Delta}$ represents error, we then use

$$\hat{\boldsymbol{\theta}} = (\boldsymbol{W}'\boldsymbol{W})^{-1}\boldsymbol{W}'\mathbf{Y}$$

to estimate $\boldsymbol{\theta}$. If we take expectations conditional on $\boldsymbol{\Delta}$, we get

$$\begin{aligned} E_Y(\hat{\boldsymbol{\theta}}|\boldsymbol{\Delta}) &= (\boldsymbol{W}'\boldsymbol{W})^{-1}\boldsymbol{W}'\boldsymbol{a}\boldsymbol{\theta} = (\boldsymbol{W}'\boldsymbol{W})^{-1}\boldsymbol{W}'(\boldsymbol{W} - \boldsymbol{\Delta})\boldsymbol{\theta}, \\ &= \boldsymbol{\theta} - \mathbf{b} \end{aligned}$$

where $\mathbf{b}$ is the bias,

$$\mathbf{b} = (\boldsymbol{W}'\boldsymbol{W})^{-1}\boldsymbol{W}'\boldsymbol{\Delta}\boldsymbol{\theta}. \tag{13.11}$$

Now write the matrix $\boldsymbol{\Delta}$ in the form

$$\boldsymbol{\Delta}' = (\boldsymbol{\delta}_1, \boldsymbol{\delta}_2, \ldots, \boldsymbol{\delta}_n),$$

then we shall assume that the $\boldsymbol{\delta}_i$ are random and independently distributed with

$$\begin{aligned} E(\boldsymbol{\Delta}) &= \mathbf{0}, \\ V(\boldsymbol{\delta}_i) &= \boldsymbol{U}, \end{aligned}$$

where $\boldsymbol{U}$ is a $k \times k$ matrix. In this case Davies and Hutton (1975) showed that the bias $\mathbf{b}$ is approximately

$$\mathbf{b} \approx n(\boldsymbol{a}'\boldsymbol{a} + n\boldsymbol{U})^{-1}\boldsymbol{U}\boldsymbol{\theta}, \tag{13.12}$$

so that we could estimate the bias by using

$$\hat{\mathbf{b}} = n(\boldsymbol{W}'\boldsymbol{W})^{-1}\hat{\boldsymbol{U}}\hat{\boldsymbol{\theta}}, \tag{13.13}$$

where we hope that an estimate $\hat{\boldsymbol{U}}$ of $\boldsymbol{U}$ is available from other data. In any particular case this could be calculated and examined.

We should notice that the result (13.13) is of the form we might expect. The effect of any errors is proportional to the regression coefficient, so that errors in a variable will not matter much if that variable is unimportant. It is also clear that the effect of errors can be very much worse than one might expect if the matrix $(\boldsymbol{W}'\boldsymbol{W})$ is nearly singular.

In order to state the working rules below we need some notation. Define

$$\boldsymbol{T} = \mathrm{diag}\{t_i\},$$

where t_i^2 is the ith diagonal element of $\boldsymbol{\Delta}'\boldsymbol{\Delta}/n$. For the case where the errors are random, t_i is the standard deviation of the errors in the ith row of $\boldsymbol{\Delta}$, and we have $t_i^2 = u_{ii}$.

13.3.1 Working rules

(1) Davies and Hutton (1975) give some working rules for checking on the effect of errors in explanatory variables. Firstly, we have seen that the effect of errors can be large if the matrix $\boldsymbol{W}$ is singular. The authors define 'distance from singularity' of a matrix $\boldsymbol{W}$ as the minimum of

$$\{\mathrm{tr}(\boldsymbol{D}'\boldsymbol{D})\}^{1/2},$$

where $(\boldsymbol{W} + \boldsymbol{T}\boldsymbol{D})$ is singular (that is, it has rank less than k). Davies and Hutton show that an estimate of this distance is given by

$$V = \{\textstyle\sum t_j^2 [\boldsymbol{W}'\boldsymbol{W}]_{ii}^{-1}\}^{-1/2}, \tag{13.14}$$

where

$$[\boldsymbol{W}'\boldsymbol{W}]_{jj}^{-1} \tag{13.15}$$

is the jth diagonal element of $\boldsymbol{W}'\boldsymbol{W}^{-1}$. If this quantity D is not larger than $\sqrt{(k_1 n)}$, where k_1 is the number of non-zero t_j or possibly $\sqrt{n}$ if n is large, some of the regression coefficients may have little meaning.

(2) If the test (1) is passed, we can perform an overall check of whether the biases resulting from errors are large. This is done by looking at the biases in the estimated regression coefficients, and comparing these with their estimated standard errors. By following this line of reasoning, Davies and Hutton suggest calculating

$$F = n^{1/2} \textstyle\sum |t_j \hat{\theta}_j| / \hat{\sigma}, \tag{13.16}$$

and if this is markedly less than 1 we conclude that biases are small with respect to the standard errors and can be ignored.

(3) The criterion given in (2) assumes that the errors are placed in the worst possible way. Therefore, if it is failed, further investigation

must be made. This is done by calculating

$$B = n\{\sum t_j^2 \widehat{\theta_j^2}\}/(\hat{\sigma}). \tag{13.17}$$

If this quantity is less than 1, the effect of errors is probably negligible, but if it is larger than 1, biases are likely to constitute a major part of the estimates of some regression parameters.

(4) It would be useful to examine the diagonal elements of

$$\boldsymbol{H} = \boldsymbol{W}(\boldsymbol{W}'\boldsymbol{W})^{-1}\boldsymbol{W}'.$$

to see if there are any points of high leverage. A moderate error in any such point could have a significant effect on the estimated regression coefficients, and yet go undetected by other residuals checks.

(5) Finally, it is possible to use equation (13.13) to estimate the biases in individual parameters.

13.3.2 Discussion

In applying these rules there are several quantities we need and, in particular, the e_j and the variance matrix of the errors, U. It is possible to obtain routine estimates of e_j from the data so that checks (1)–(4) can be performed automatically. The estimation of U is a different matter, although it will often be satisfactory to take this to be diagonal. In some cases it may be effective to calculate biases directly from (13.13) without doing the checks, and compare these biases with the estimated standard errors of the regression coefficients.

It will be noted that a vital part of this argument is to establish that the ordinary estimates of standard errors of regression coefficients are still valid, and for this we refer readers to Davies and Hutton (1975).

It should be stated that there has been insufficient published exploration of these formulae, and they could be sensitive to outliers, non-normality or other kinds of disturbance. For some work on this topic, see Hodges and Moore (1972), Swindel and Bower (1972), Davies and Hutton (1975), and Seber (1977).

13.4 Functional relationship model

In this section and the next one we deal with the functional relationship model suggested in Example 13.1, and a similar structural relationship model. Attempts to find satisfactory estimates of their parameters have revealed numerous difficulties with such models. Most of the discussion here will be for two observed variables; this case suffices to illustrate most of the problems.

It is instructive to examine an early attempt at obtaining maximum likelihood estimates and then see the many ways in which it can be criticized. The underlying relationship between 'true' values X and Y is assumed to have the form

$$Y = \alpha + \beta X, \tag{13.18}$$

where we observe pairs of variables

$$\begin{aligned} x &= X + \delta, \\ y &= Y + \varepsilon, \end{aligned} \tag{13.19}$$

to give data (x_i, y_i), $i = 1, 2, \ldots, n$. We wish to estimate the $n + 4$ parameters $\alpha, \beta, \sigma_\delta^2, \sigma_\varepsilon^2$ and $X_1, X_2, \ldots, X_n$ using n pairs of observations, so it is not surprising that we run into difficulties, since thc usual asymptotic properties of maximum likelihood do not hold in this situation.

The log-likelihood for the $2n$ observations is

$$L = -n\log(2\pi\sigma_\delta\sigma_\varepsilon) - \frac{1}{2\sigma_\delta^2}\sum_1^n (x_i - X_i)^2 - \frac{1}{2\sigma_\varepsilon^2}\sum_1^n (y_i - \alpha - \beta X_i)^2. \tag{13.20}$$

The following equations are then obtained by equating to zero the partial derivatives of L with respect to σ_δ, σ_ε, X_i, α and β respectively.

$$n\sigma_\delta^2 = \sum (x_i - X_i)^2, \tag{13.21}$$

$$n\sigma_\varepsilon^2 = \sum (y_i - \alpha - \beta X_i)^2, \tag{13.22}$$

$$(x_i - X_i)/\sigma_\delta^2 + \beta(y_i - \alpha - \beta X_i)/\sigma_\varepsilon^2 = 0, \qquad i = 1, 2, \ldots, n, \tag{13.23}$$

$$\sum (y_i - \alpha - \beta X_i) = 0, \tag{13.24}$$

$$\sum (y_i - \alpha - \beta X_i) X_i = 0. \tag{13.25}$$

Squaring in (13.23) and summing over i we obtain

$$\sum (x_i - \hat{X}_i)^2/\hat{\sigma}_\delta^4 = \hat{\beta}^2 \sum (y_i - \hat{\alpha} - \hat{\beta}\hat{X}_i)^2/\hat{\sigma}_\varepsilon^4.$$

Substituting (13.21) and (13.22) into this gives the remarkable result

$$\hat{\sigma}_\varepsilon^2 = \hat{\beta}^2 \hat{\sigma}_\delta^2. \tag{13.26}$$

(Since we have not used (13.24) and (13.25), a similar result would apply if α and β were known.) This relation between the slope of the functional relationship and the error variances in x and y was obtained by Lindley (1947), who noted that it was not satisfactory: there is no reason for supposing that such a relation holds, and a set of estimators satisfying it cannot be consistent.

It was thought that this result demonstrated a failure of the method of maximum likelihood. However, Solari (1969) has shown that equations (13.21) to (13.25) have two solutions and that the solution which gives the larger value of the likelihood corresponds not to a maximum but to a saddle point. Also the likelihood has essential singularities at all points at which either $\sigma_\delta = 0$ and $X_i = x_i$ $(i = 1, 2, \ldots, n)$ or $\sigma_\varepsilon = 0$ and $X_i = (y_i - \alpha)/\beta$ $(i = 1, 2, \ldots, n)$; in any neighbourhood of these points it may be made to approach any real value. Thus there is no maximum likelihood solution. (This type of problem is not confined to models such as the functional relationship in which the number of parameters increases with the number of observations; see Exercise 13.4.1.)

Copas (1972) points out that the singularities occur because the likelihood function given by (13.20) is based on an approximation which breaks down when either of the error variances tends to zero. It assumes that the observations x_i, y_i are recorded with complete accuracy, whereas any observation on a continuous variable is made with a grouping error; for example, a height recorded to the nearest centimetre as 179 cm represents a value in the range 178.5 cm to 179.5 cm. Suppose that the relative scale of variables x and y is chosen so that the values x_i and y_i are both recorded with a grouping interval of length h. The contribution to the likelihood of x_i is then equal to the probability associated with the interval $(x_i - \frac{1}{2}h,\ x_i + \frac{1}{2}h)$ under the model, that is

$$\Phi\{(x_i + \tfrac{1}{2}h - X_i)/\sigma_\delta\} - \Phi\{(x_i - \tfrac{1}{2}h - X_i)/\sigma_\delta\}, \qquad (13.27)$$

where Φ denotes the standard normal distribution function. If σ_δ is of the order of h or larger then (13.27) is well approximated by

$$h(2\pi)^{-1/2}\sigma_\delta^{-1}\exp\{-\tfrac{1}{2}(x_i - X_i)^2/\sigma_\delta^2\}.$$

With a similar approximation for the contribution from y_i, this leads to the likelihood in (13.20) (apart from a constant factor h^{2n}), but as either σ_δ or σ_ε tends to zero the approximation breaks down. (The same type of approximation is made when forming the likelihood for other models involving continuous variables, but the consequences are usually less serious.) Copas shows that when grouping is taken into account the likelihood is bounded and has no singularities. If the grouping interval is small then the maximum likelihood estimation of β is close to either the slope of the regression of y on x or the reciprocal of the slope of the regression of x on y according to whether $\sum(y_i - \bar{y})^2$ is smaller or larger than $\sum(x_i - \bar{x})^2$. In the first case, $\hat{\sigma}_\delta = 0$ and the $\hat{X}_i$ equal x_i approximately; in the second, $\hat{\sigma}_\varepsilon = 0$ and the $\hat{X}_i$ equal $(y_i - \hat{\alpha})/\hat{\beta}$ approximately. This solution, although it corresponds to the maximum of the likelihood rather than a saddle point, is still not satisfactory since it cannot be consistent if both error variances are positive. It also has the curious property that if we measured one variable to the nearest inch or pound, say, we could obtain estimates which suggest that all the error was in y, whereas if we measured the same variable to the nearest centimetre or kilogram it could appear that all the error was in x. It seems that, at least when the errors are normally distributed, it is difficult to distinguish when there is error in both variables rather than in only one using merely the n pairs of observations x_i, y_i.

A still more fundamental objection can be raised to these attempts to find estimates for the functional relationship model; this concerns the nature of the parameters $X_i, \ldots, X_n$, which appear in proportion to the number of observations. Nelder (1968) remarks 'It is not surprising that such parameters should lead to estimation difficulties, since they represent quantities which are not assumed constant over a defined population nor yet is any probability structure assigned to them; they exemplify, in fact, a non-statistical concept'. Such parameters are called 'incidental' and their presence leads to estimation difficulties in other statistical models. An example is the reconstruction of evolutionary trees, a problem even more slippery than the present one; see Felsenstein (1981).

As will become clear in the next section, the problems of estimation do not necessarily disappear even if we regard the X_i as random variables with a common distribution. In order to obtain reasonable estimates of the parameters in the model we require further information. This may, for example, take the form of knowledge of the error variances or their ratio, of how the X_i (not the x_i) are ordered, or of the values of a third variable which is correlated with X; consistent estimation is also possible if replicate observations can be made at common values of X or Y. In practice, we need to make use of whatever extra data are available or to design experiments, where possible, which provide suitable additional information, for example, by including replicate observations.

One type of additional information which can be used is a knowledge of the ratio of the error variances, say $\lambda = \sigma_\varepsilon^2/\sigma_\delta^2$. For example, Williams (1959, p. 200) describes an experiment designed to compare two measures of the impact strength of timber; pairs of specimens were taken from planks of wood and one was tested radially and the other tangentially to the growth rings. If we are willing to assume that the two 'true' readings for each plank are linearly related, then it is reasonable to assume that the variances of the corresponding errors are equal (so that $\lambda = 1$), since they arise from the same method of test. Another example is given in Exercise 13.4.4. Returning to equations (13.23) to (13.25), we replace σ_ε^2 by $\lambda\sigma_\delta^2$ in (13.23), sum over i and use (13.24) to obtain

$$\sum(x_i - \hat{X}_i) = 0. \tag{13.28}$$

with (13.24) this gives

$$\hat{\alpha} = \bar{y} - \hat{\beta}\bar{x}. \tag{13.29}$$

It is convenient to introduce the notations

$$x_i' = x_i - \bar{x}, \qquad y_i' = y_i - \bar{y}, \qquad \hat{X}_i' = \hat{X}_i - \bar{x};$$

using (13.28) and (13.29), (13.23) may be solved for $\hat{X}_i'$ to give

$$\hat{X}_i' = (\lambda x_i' + \hat{\beta} y_i')/(\lambda + \hat{\beta}^2). \tag{13.30}$$

In the same notation (13.25) becomes

$$\sum \hat{X}_i'(y_i - \hat{\beta}\hat{X}_i') = 0.$$

Combined with (13.30) this gives a quadratic equation in $\hat{\beta}$ which may be written as

$$\hat{\beta}^2 s_{xy} + \hat{\beta}(\lambda s_{xx} - s_{yy}) - \lambda s_{xy} = 0, \tag{13.31}$$

where

$$s_{xx} = \sum(x_i - \bar{x})^2/n, \qquad s_{xy} = \sum(x_i - \bar{x})(y_i - \bar{y})/n,$$

$$s_{yy} = \sum(y_i - \bar{y})^2/n. \tag{13.32}$$

The likelihood is maximized when $\hat{\beta}$ has the same sign as s_{xy}; this solution may be expressed as

$$\hat{\beta} = K + \sqrt{(K^2 + \lambda)}\,\mathrm{sgn}(s_{xy}), \tag{13.33}$$

where

$$K = \tfrac{1}{2}(s_{yy} - \lambda s_{xx})/s_{xy}.$$

This solution has the following properties.

(1) If X is observed without error, so that $\sigma_\delta = 0$ and $\lambda = \infty$, the estimate of β is the ordinary regression coefficient of y on x.
(2) If Y is observed without error, so that σ_ε and λ are zero, the estimate (13.33) is the reciprocal of the ordinary regression coefficient of x on y.
(3) If $\lambda = 1$ the line takes the direction of the first principal component.
(4) The estimator (13.33) is consistent for β if the sequence

$$\left\{ \sum_{i=1}^{n} X_i^2/n \right\}$$

has a finite positive limit as n tends to infinity.

The qualification on the behaviour of the X_i in (4) is required because the estimator is based on random variables whose distributions depend on the X_i. The presence of these parameters leads to a further difficulty; if σ_δ has also to be estimated from the data then its maximum likelihood estimator is not consistent. A consistent estimator can easily be found in this case, however.

In summary, the method of maximum likelihood does not lead to satisfactory estimators for the linear functional relationship model. Some progress can be made if further information is available, but even then the presence of incidental parameters can lead to difficulties because the standard asymptotic properties of maximum likelihood do not apply. This can be regarded as a fault of the model rather than of the method.

For access to some of the literature see Sprent (1969), Moran (1971), Kendall and Stuart (1979) Anderson (1984a).

Exercises 13.4

1. Observations $y_1, y_2, \ldots, y_n$ are assumed to form a random sample from a mixture of two univariate normal distributions. If ϕ denotes

the standard normal density function then the common mixture density has the form

$$f(y;\omega,\mu_1,\mu_2,\sigma_1,\sigma_2)=\omega\sigma_1^{-1}\phi\{(y-\mu_1)/\sigma_1\}$$
$$+(1-\omega)\sigma_2^{-1}\phi\{(y-\mu_2)/\sigma_2\}$$
$$(0<\omega<1,-\infty<\mu_1<\infty,-\infty<\mu_2<\infty,\sigma_1>0,\sigma_2>0).$$

Show that if grouping is ignored the likelhood function of the observations has essential singularities at all points at which for some i

$$\mu_1=y_i,\qquad \sigma_1=0 \quad \text{or} \quad \mu_2=y_i,\qquad \sigma_2=0.$$

(The likelihood may also have a number of local maxima.)

2. Show that the estimator (13.33) is consistent if the sequence of incidental parameters satisfies

$$n^{-1}\sum_{i=1}^{n}X_i\rightarrow M,\qquad n^{-1}\sum_{i=1}^{n}(X_i-M)^2\rightarrow V$$

as n tends to infinity. Show also that the maximum likelihood estimator of σ_δ^2 is given by

$$\tfrac{1}{2}(s_{yy}-2\hat{\beta}s_{xy}+\hat{\beta}^2s_{xx})/(\lambda+\hat{\beta}^2),$$

and that this estimator converges in probability to $\frac{1}{2}\sigma_\delta^2$.

3. A useful generalization of the case in which δ and ε are independent with variances in a known ratio is that in which the vector $[\delta,\varepsilon]$ has variance matrix of the form $\sigma^2\boldsymbol{G}$, where $\boldsymbol{G}$ is a known positive definite matrix with elements g_{11},g_{12},g_{22}. Show that the maximum likelihood estimate of β is the value minimizing the ratio

$$(\beta^2s_{xx}-2\beta s_{xy}+s_{yy})/(\beta^2g_{11}-2\beta g_{12}+g_{22}).$$

4. Suppose that $m>2$ simple linear regressions are fitted with the same response variable r and a common set of values for the explanatory variable z. The model for the ith regression is taken to be

$$E(r_{ij})=\mu_i+\lambda_iz_j,\qquad V(r_{ij})=\sigma^2\qquad (j=1,2,\ldots,n)$$

and the least squares estimators of λ_i and μ_i are denoted by l_i and m_i. Show that if the m regression lines are assumed to pass through a common unknown point with coordinates $(-\beta,\alpha)$ then the pairs of coefficients (λ_i,μ_i) satisfy a linear functional relationship, while the vector of estimators (l_i, m_i) has variance matrix of the form given in the last exercise. Hence show how to obtain point estimates of α and β. (This type of model with *concurrent* regressions is used in enzyme kinetics; see Crowder (1978).)

13.5 Structural relationship model

The difficulties with the functional relationship model arise in part at least, from the appearance of a new incidental parameter X_i with each pair of observations (x_i, y_i). The structural relationship model differs from the former model by including the assumption that the X_i form a random sample from some population. We might hope that this extra assumption would be sufficient to allow consistent estimation of at least the slope and intercept of the relationship, but this is not true in general.

We consider first the case in which X, δ and ε are independent and normal with distributions $N(\mu_X, \sigma_X^2)$, $N(0, \sigma_\delta^2)$ and $N(0, \sigma_\varepsilon^2)$. The ith vector of observations may be represented as

$$\begin{bmatrix} x_i \\ y_i \end{bmatrix} = \begin{bmatrix} X_i + \delta_i \\ \alpha + \beta X_i + \varepsilon_i \end{bmatrix}.$$

Using the additive property of normal random vectors, this is found to have the bivariate normal distribution $N_2(\mu, \Sigma)$ with

$$\mu = \begin{bmatrix} \mu_x \\ \alpha + \beta \mu_x \end{bmatrix}, \qquad \boldsymbol{\Sigma} = \begin{bmatrix} \sigma_x^2 + \sigma_\delta^2 & \beta\sigma_x^2 \\ \beta\sigma_x^2 & \beta^2\sigma_x^2 + \sigma_\varepsilon^2 \end{bmatrix}. \qquad (13.34)$$

The model has six parameters, $\alpha, \beta, \mu_x, \sigma_x, \sigma_\delta$ and σ_ε, but the bivariate normal distribution is defined by the five distinct elements of μ and $\boldsymbol{\Sigma}$. The parameters in the model are said to be not identifiable since different sets of values lead to the same joint distribution for the observed random variables. Thus without further information it is impossible to estimate these parameters; the same is true if one or more of the distributions is non-normal but only the first and second moments of the data are used. This result also shows that there can be no general method for consistently estimating β in the functional relationship model defined by (13.18) and (13.19), for any such method could not work if the X_i formed a random sample from a normal population. (This does not prevent proposals for such methods appearing from time to time!)

As with the functional relationship model, more progress can be made if extra information is available, for example, on the error variances. In particular, if the value λ of $\sigma_\varepsilon^2/\sigma_\delta^2$ can be specified then the maximum likelihood estimates of $\beta, \sigma_x, \sigma_\delta$ and σ_ε are obtained by equating the elements of $\boldsymbol{\Sigma}$ in (13.34) to the statistics s_{xx}, s_{yy} and s_{xy} defined in (13.32), which are the maximum likelihood estimates of the normal variances and covariance; the estimate of β turns out to be the same as (13.33). In contrast to the functional relationship model, the estimators of all the parameters are consistent because the standard asymptotic properties of the method are now applicable. Equating sample and population moments also gives the maximum likelihood estimates when one error variance is known, unless the equations lead to a negative value for one of the other variances, in which case the maximum occurs at a boundary point of the parameter space; see Kendall and Stuart (1979).

The problem of non-identifiability of the parameters in a structural relationship turns out to be confined to the case in which X, δ and ε are all normally distributed. Reiersøl (1950) shows that if δ and ε are normal then β is identifiable if and only if X is non-normal. However, this result does not solve the practical problem of estimation even when X is thought to be non-normal. Estimates may be found when the distribution of X has a particular parametric form – Chan (1982) suggests a method for the uniform distribution – but there appears to be no generally practicable method if the form is unspecified. Procedures based on moments higher than the second have been suggested, but are not guaranteed to work for an arbitrary non-normal distribution. Intuitively it would appear that good estimates would require a very non-normal distribution or a large sample size.

13.5.1 Discussion of functional and structural relationship models

The main conclusion from this section and the previous one is that the pairs of observations (x_i, y_i) do not provide enough information to estimate the parameters of a functional or structural relationship. Extra knowledge in some form is needed to obtain a solution. This may take various forms and a thorough discussion of these possibilities does not seem to be available.

One consequence of this conclusion is that it is not possible to correct a calculated regression coefficient for the effect of errors in the explanatory variable using only the values of the explanatory and response variables. A second concerns how the relationship between two variates should be described when they are to be treated on equal terms rather than as explanatory variable and response. Especially when the variables are highly correlated, there may be a temptation (for non-statisticians at least) to suppose that there is some 'true' relationship between the variables which is distinct from the two regressions. Methods for fitting a line representing this relationship are then proposed and used by workers in the relevant subject area; a favourite suggestion is a line through the mean vector with slope $\sqrt{(s_{yy}/s_{xx})}$, the geometric mean of the slope of the regression of y on x and the reciprocal of the slope for x on y. (See Jolicoeur (1975) for a comment on its use in fishery research.) It should be clear that any attempt to fit such a relationship is bound to fail in the absence of additional information.

Generalizations of the functional and structural relationship models considered here can be defined when more than two variables are observed. Methods of estimation (when they exist) are related to canonical analysis and factor analysis; see Sprent (1969), Theobald and Mallinson (1978) and Anderson (1984a and b). For access to the other literature, see the references mentioned at the end of Section 13.4. Non-linear functional and structural relationships are also of interest; see Exercise 13.5.3.

Exercises 13.5

1. Consider Exercise 13.2.1 again: what sort of additional information might be collected in order to investigate the relationship between the true values of pressure reduction and initial blood pressure?
2. Consider a structural relationship model in which there are three variables X, Y, Z which are measured with error and connected by two linear relationships. Thus observations are made on variables x, y, z, where

$$Y = \alpha_1 + \beta_1 X, \qquad Z = \alpha_2 + \beta_2 X,$$
$$x = X + \delta, \qquad y = Y + \varepsilon, \qquad z = Z + \zeta.$$

Show that this structure is identifiable if X and the error terms $\delta, \varepsilon, \zeta$ are independently and normally distributed with expectations $\mu_X, 0, 0, 0$ and variances $\sigma_X^2, \sigma_\delta^2, \sigma_\varepsilon^2, \sigma_\zeta^2$. Obtain the maximum likelihood estimates of the parameters of the model (assuming that all the variance estimates are positive). Is a model with only one linear relationship between X, Y, and Z identifiable?

3. Circular functional and structural relationships have been suggested as models for the positions of stones in megalithic rings. A functional relationship model for a circle with true centre at (ξ, η) (in Cartesian coordinates) and radius ρ assumes that the coordinates x_i, y_i of the centre of the ith stone are independent with distributions $N(\xi + \rho \cos \theta_i, \ \sigma^2)$, $N(\eta + \rho \sin \theta_i, \ \sigma^2)$. Treating $\theta_1, \theta_2, \ldots, \theta_n$ as separate parameters leads to maximum likelihood estimators which are not consistent. In the Ring of Brogar in Orkney, the angles made at the centre of the circle between the stones are very close to specified integer multiples of six degrees. If these angles are taken to be known exactly, then θ_i may be expressed as $\theta + t_i$ with the t_i known and θ unknown. Show that in this case maximum likelihood estimates of $\xi, \eta, \rho \cos \theta, \rho \sin \theta$ may be obtained by means of a multiple regression with $x_1, x_2, \ldots, x_n$, $y_1, y_2, \ldots, y_n$ as the responses (Chan, 1965; Berman, 1983).

13.6 Models with autocorrelated errors

Our regression model has the form

$$\mathbf{Y} = \boldsymbol{a}\boldsymbol{\theta} + \boldsymbol{\varepsilon}, \tag{13.35}$$

where $\mathbf{Y}$ is $(n \times 1)$, $\boldsymbol{a}$ is $(n \times k)$, and where $\boldsymbol{\varepsilon}$ is a random vector with

$$E(\boldsymbol{\varepsilon}) = 0, \qquad V(\boldsymbol{\varepsilon}) = \sigma_\varepsilon^2 \boldsymbol{V}. \tag{13.36}$$

Autocorrelation exists if the disturbance terms $\boldsymbol{\varepsilon}$ are correlated. This happens mostly with data which are taken serially in time, as in many econometric applications. Since $\boldsymbol{V}$ is usually unknown, and it involves $n(n-1)/2$ parameters, some kind of restrictions are necessary in order to obtain a solution. Often these restrictions take the form of assuming that the errors are observations on a particular stochastic process. Stationarity is an important assumption since it means that the distribution of $\boldsymbol{\varepsilon}$ does not depend on the observation number, t, and $\boldsymbol{V}$ has the form

$$\boldsymbol{V} = \begin{bmatrix} 1 & \rho_1 & \rho_2 & \cdots \\ \rho_1 & 1 & \rho_1 & \cdots \\ \rho_2 & \rho_1 & 1 & \\ \vdots & \vdots & \vdots & \end{bmatrix} \tag{13.37}$$

and there are just $(n-1)$ autocorrelation coefficients, $\rho_i, i = 1, 2, \ldots, (n-1)$. This is still too many coefficients and particular forms of stochastic process which have been considered are given below.

Often autocorrelation can arise because some important explanatory variables have been omitted—either because of ignorance or because of measurement difficulties. Frequently, the effect of omitted variables will not be immediate, but will be distributed over time, leading to autocorrelation in the disturbance terms. It is easy to mistake the 'omitted variable' case for genuine autocorrelation, and the effect of such an error is not known.

13.6.1 Some possible processes for regression disturbances

There are many possibilities, such as the following:

(1) An autoregressive process of order q, AR (q)

$$\varepsilon_t = \theta_1 \varepsilon_{t-1} + \theta_2 \varepsilon_{t-2} + \cdots + \theta_q \varepsilon_{t-q} + v_t. \tag{13.38}$$

(2) A moving average process of order q, MA (q)

$$\varepsilon_t = v_t + \alpha_1 t_{t-1} + \cdots + \alpha_q v_{t-q}. \tag{13.39}$$

(3) A combined moving average, autoregressive process of order (q, p), ARMA (q, p)

$$\varepsilon_t = \theta_1 \varepsilon_{t-1} + \cdots + \theta_q v_{t-q} + v_t + \alpha_1 v_{t-1} + \cdots + \alpha_p v_{t-p}. \tag{13.40}$$

In these models, θ_i and α_i are unknown parameters, and v_t are random disturbances such that

$$E(v_t) = 0, \qquad V(v_t) = \sigma_v^2, \qquad E(v_t v_s) = 0 \quad \text{if } t \neq s.$$

In the past the AR(1) process was often assumed, partly because of computational problems in fitting others. In any practical case, any of these could be considered now that computing difficulties have been removed. It is not easy, however, to set down a strategy for choosing a particular model, and little is known about the effect of using one model when another is true. It would be useful in practice to report results from two competing models.

We now consider examples of these models.

EXAMPLE 13.5

AR(1) process

$$\varepsilon_t = \theta \varepsilon_{t-1} + v_t \quad \text{leading to} \quad \sigma_\varepsilon^2 = \sigma_v^2/(1-\theta^2).$$

Autocorrelation function:

$$\rho_s = \theta^s. \tag{13.41}$$

EXAMPLE 13.6
MA(1) process

$$\varepsilon_t = v_t + \alpha v_{t-1} \quad \text{leading to} \quad \sigma_\varepsilon^2 = \sigma_v^2(1+\alpha^2).$$

Autocorrelation function:

$$\rho_s = \begin{cases} \alpha/(1+\alpha^2) & s=1, \\ 0, & s>1. \end{cases} \tag{13.42}$$

EXAMPLE 13.7
ARMA(1, 1) process

$$\varepsilon_t = \theta\varepsilon_{t-1} + v_t + \alpha v_{t-1}.$$

This leads to

$$\sigma_\varepsilon^2 = (1+\alpha^2+2\alpha\theta)\sigma_v^2/(1-\theta^2)$$

and the autocorrelation function is

$$\left.\begin{aligned} \rho_1 &= (1+\theta\alpha)(\theta+\alpha)/(1+\alpha^2+2\alpha\theta), \\ \rho_s &= \theta\rho_{s-1}, \qquad s>1, \end{aligned}\right\} \tag{13.43}$$

It is important to notice that ρ_s for AR processes dies out gradually, whereas for MA processes ρ_s becomes zero after a certain point.

These are not the only processes possible or even likely. In particular, alternatives to models with autocorrelated errors are models with lagged dependent variables. These latter models could be often more realistic but are not discussed here.

13.6.2 Fitting autocorrelated error models

There are two problems involved in fitting these models. The first is to choose a particular model to fit and the second is to choose the estimation method. The second of these problems is easier to deal with than the first.

If a particular model is chosen, OLS can be used and this will still give unbiased estimates of $\boldsymbol{\theta}$. The efficiency of OLS is often quite high, but some asymptotic results obtained by Watson and Hannan (1956) and Grenander and Rosenblatt (1957) imply that the loss in efficiency can be substantial. Further, the estimators of σ^2 and $V(\hat{\boldsymbol{\theta}})$ obtained by OLS are biased.

The principles behind possible estimation methods have been discussed briefly in Section 9.8. One possibility is to use generalized least squares for estimation. This leads to minimizing

$$(\mathbf{Y}-\boldsymbol{a\theta})'V^{-1}(\mathbf{Y}-\boldsymbol{a\theta}). \tag{13.44}$$

The suggested procedure is therefore to use OLS to get residuals, from which a particular process for the errors would be chosen and the parameters estimated. We can then minimize (13.44) for choice of $\boldsymbol{\theta}$. Alternatively, we could minimize (13.44) for both $\boldsymbol{\theta}$ and for estimates of the parameters in the model for the errors. Yet a further possibility, which is perhaps the best general procedure, is to use full maximum likelihood, but again, a particular model for error structure has to be assumed. These last two methods can lead to complex non-linear optimization problems. All of these methods have similar asymptotic properties.

In order to choose a particular model for the error structure, the best way to start is to use standard multiple regression to obtain residuals, and then study the autocorrelation function of the residuals. A fully worked-out methodology does not seem to be available, nor is it known what the errors are of using one model when another is true. For further details see Judge *et al.* (1980).

EXAMPLE 13.8

If the distribution of the errors is an AR(1) process then we have

$$V=\begin{bmatrix} 1 & \rho & \rho^2 & \cdots & \rho^{n-1} \\ \rho & 1 & \rho & \cdots & \rho^{n-2} \\ \cdots & \cdots & \cdots & \cdots & \cdots \\ \rho^{n-1} & \rho^{n-2} & & \cdots & 1 \end{bmatrix}.$$

For this matrix it can be shown that the inverse is

$$V^{-1}=\frac{1}{1-\rho^2}\begin{bmatrix} 1 & -\rho & 0 & \cdots & 0 & 0 \\ -\rho & 1+\rho^2 & -\rho & \cdots & 0 & 0 \\ 0 & -\rho & 1+\rho^2 & \cdots & 0 & 0 \\ \cdots & \cdots & \cdots & \cdots & \cdots & \cdots \\ 0 & 0 & 0 & \cdots & 1+\rho^2 & -\rho \\ 0 & 0 & 0 & \cdots & -\rho & 1 \end{bmatrix}.$$

We can use the transformation method of Section 9.8.2, and we find that

$$\boldsymbol{P}=\frac{1}{\sqrt{(1-\rho^2)}}\begin{bmatrix} \sqrt{(1-\rho^2)} & 0 & 0 & \cdots & 0 & 0 \\ -\rho & 1 & 0 & \cdots & 0 & 0 \\ 0 & -\rho & 1 & \cdots & 0 & 0 \\ \cdots & \cdots & \cdots & \cdots & \cdots & \cdots \\ 0 & & 0 & 0 & -\rho & 1 \end{bmatrix}.$$

Estimates can therefore be obtained by using GLS and this is done through use of (9.20) with $\boldsymbol{P}$ defined as above; see Section 9.8. If the parameter ρ is not known, but an AR (1) process for errors can be assumed, then the procedure is to apply OLS to obtain residuals, estimate ρ from these residuals, and then the above GLS method is applied.

Cook and Pocock (1983) fit an autocorrelated error model to data from the British Regional Heart Study. The response variable is the logarithm of the standardized mortality rate, and this is regressed on water hardness, percentage of days with rain, mean daily maximum temperature, percentage of manual workers and cars per 100 households for separate towns. It was thought that the correlation between towns would be a monotonic decreasing function of the distance between towns. Based on this, they adopted an *ad hoc* procedure for getting a model of $\boldsymbol{V}$. See their paper for details.

For further details of fitting autoregressive and moving average error models, see Judge *et al.* (1980).

13.7 Discussion

This chapter presents 'the tip of an iceberg' in discussing models somewhat similar to regression but which have a rather different structure for the error. The 'regression models of the second kind' case, introduced in Section 13.1, has not been discussed in detail and we refer readers to Nelder (1968).

A basic difficulty is that we are often in doubt as to which of these models best represents any particular situation. Further, we have, in general, little knowledge of the consequences of assuming one model when another is true, except that it is intuitively clear that the consequences could be quite serious.

All of this seems to indicate that there may well be a place for robust methods, which have good general properties, but which are not optimal for any one model. There is obviously a wide open field of research into these problems.

References

Anderson, T. W. (1984a) Estimating linear statistical relationships. *Ann. Statist.*, **12**, 1–45.

Anderson, T. W. (1984b) *An Introduction to Multivariate Statistical Analysis*, 2nd edn. Wiley, New York.

Berman, M. (1983) Estimating the parameters of a circle when angular differences are known. *Appl. Statist.*, **32**, 1–6.

Chan, N. N. (1965) On circular functional relationships. *J. Roy. Statist. Soc.* B, **27**, 45–56.

Chan, N. N. (1982) Linear structural relationships with unknown error variances. *Biometrika*, **69**, 277–279.

Cook, D. G. and Pocock, S. J. (1983) Multiple regression in geographical mortality studies, with allowance for spatially correlated errors. *Biometrics*, **39**, 361–371.

Copas, J. B. (1972) The likelihood surface in the linear functional relationship problem. *J. Roy. Statist. Soc.* B, **34**, 274–278.

Cox, D. R. (1968) Notes on some aspects of regression analysis. *J. Roy. Statist. Soc.* A, **131**, 265–279.

Crowder, M. J. (1978) On concurrent regression lines. *Appl. Statist.*, **27**, 310–318.

Davies, R. B. and Hutton, B. (1975) The effect of errors on the independent variables in linear regression. *Biometrika*, **62**, 383–391 (corrected **64**, 655).

Ehrenberg, A. S. C. (1968) The elements of lawlike relationships. *J. Roy. Statist. Soc* A, **31**, 280–302.

Felsenstein, J. (1981) Evolutionary trees from gene frequencies and quantitative characters: finding maximum likelihood estimates. *Evolution*, **35**, 1229–1242.

Grenander, U. and Rosenblatt, M. (1957) *Statistical Analysis of Stationary Time Series.* Wiley, New York.

Hald, A. (1952) *Statistical Theory with Engineering Applications.* Wiley, New York.

Hodges, S. D. and Moore, P. G. (1972) Data uncertainties and least squares regression. *Appl. Statist.*, **21**, 185–195.

Jolicoeur, P. (1975) Linear regressions in fishing research: some comments. *J. Fish. Res. Board Can.*, **30**, 409–434.

Judge, G. G., Griffith, W. E., Carter-Hill, R. and Lee, Tsoung-Chao (1980) *The Theory and Practice of Econometrics.* Wiley, New York.

Kendall, M. G. and Stuart, A. (1979) *The Advanced Theory of Statistics*, Vol. 2. 4th Edn, Griffin, London.

Lindley, D. V. (1947) Regression lines and the linear functional relationship. *J. Roy. Statist. Soc.* Suppl., **9**, 218–244.

Madansky, A. (1959) The fitting of straight lines when both variables are subject to error. *J. Amer. Statist. Assoc.*, **54**, 173–205.

Mayrovitz, H. N. and Roy, J. (1983) Microvascular blood flow: evidence indicating a cubic dependence on arteriolar diameter. *Amer. J. Physiol.*, **245**, H1031–H1038 (*Heart Circ. Physiol.*, **14**).

Moran, P. A. P. (1971) Estimating structural and functional relationships. *J. Multivar. Anal.*, **1**, 232–255.

Murray, C. D. (1926) The physiological principle of minimum work. 1: The vascular system and the cost of blood volume. *Proc. Nat. Acad. Sci.* USA, **12**, 207–213.

Nelder, J. A. (1968) Regression, model building and invariance. *J. Roy. Statist. Soc.* A, **131**, 303–315.

Reiersøl, O. (1950) Identifiability of a linear relation between variables which are subject to error. *Econometrica*, **18**, 375–389.

Seber, G. A. F. (1977) *Linear Regression Analysis.* Wiley, New York.

Solari, M. E. (1969) The 'maximum likelihood solution' of the problem of estimating a linear functional relationship. *J. Roy. Statist. Soc.* B, **31**, 372–375.

Sprent, P. (1969) *Models in Regression and Related Topics.* Methuen, London.

Swindel, B. F. and Bower, D. R. (1972) Rounding errors in independent variables in a general linear model. *Technometrics*, **14**, 215–218.

Theobald, C. M. and Mallinson, J. R. (1978) Comparative calibration, linear structural relationships and congeneric measurements. *Biometrics*, **34**, 39–45.

Watson, G. S. and Hannan, E. J. (1956) Serial correlation in regression analysis, II. *Biometrika*, **43**, 436–495.

Williams, E. J. (1959) *Regression Analysis.* Wiley, New York.

APPENDIX

Some test data sets

The data sets given below have been used in the literature in discussions related to the material presented in this volume. It is suggested that readers try out their own programs on these data sets. The references listed for each data set should only be consulted *after* attempts at analysis.

A.1 Longley's data

These data were used by Longley (1967) in testing the accuracy of regression programs. The objective of the analysis is to obtain an equation relating total derived employment to the other variables. The variables are:

Y: Total derived employment.
X_1: Gross National Product implicit price deflator (1954 = 100).
X_2: Gross National Product.
X_3: Unemployment.
X_4: Size of armed forces.
X_5: Non-institutional population 14 years of age and over.
X_6: Time.

Table A.1

X_1	X_2	X_3	X_4	X_5	X_6	Y
83.0	234 289	2 356	1 590	107 608	1947	60 323
88.5	259 426	2 325	1 456	108 632	1948	61 122
88.2	258 054	3 682	1 616	109 773	1949	60 171
89.5	284 599	3 351	1 650	110 929	1950	61 187
96.2	328 975	2 099	3 099	112 075	1951	63 221
98.1	346 999	1 932	3 594	113 270	1952	63 639
99.0	365 385	1 870	3 547	115 094	1953	64 989
100.0	363 112	3 578	3 350	116 219	1954	63 761
101.2	397 469	2 904	3 048	117 388	1955	66 019
104.6	419 180	2 822	2 857	118 734	1956	67 857
108.4	442 769	2 936	2 798	120 445	1957	68 169
110.8	444 546	4 681	2 637	121 950	1958	66 513
112.6	482 704	3 813	2 552	123 366	1959	68 655
114.2	502 601	3 931	2 514	125 368	1960	69 564
115.7	518 173	4 806	2 572	127 852	1961	69 331
116.9	554 894	4 007	2 827	130 081	1962	70 551

Longley, J. W. (1967) An appraisal of least squares programs for the electronic computer from the point of view of the user. *J. Amer. Statist. Assoc.* **62**, 819–841.

A.2 Black cherry tree data

These data were used as an example in the *Minitab Handbook*, Ryan *et al.* (1985). The objective of the analysis is to produce an equation for predicting the volume of black cherry trees from data on height and diameter at 4.5 feet above ground. A sample of trees was cut, and the measurement given below obtained. Volume is measured in cubic feet, height in feet and diameter in inches.

Table A.2

Diameter	*Height*	*Volume*
8.3	70	10.3
8.6	65	10.3
8.8	63	10.2
10.5	72	16.4
10.7	81	18.8
10.8	83	19.7
11.0	66	15.6
11.0	75	18.2
11.1	80	22.6
11.2	75	19.9
11.3	79	24.2
11.4	76	21.0
11.4	76	21.4
11.7	69	21.3
12.0	75	19.1
12.9	74	22.2
12.9	85	33.8
13.3	86	27.4
13.7	71	25.7
13.8	64	24.9
14.0	78	34.5
14.2	80	31.7
14.5	74	36.3
16.0	72	38.3
16.3	77	42.6
17.3	81	55.4
17.5	82	55.7
17.9	80	58.3
18.0	80	51.5
18.0	80	51.0
20.6	87	77.0

Atkinson, A. C. (1982) Regression diagnostics, transformations and constructed variables (with discussion). *J. Roy. Statist. Soc.* B, **44**, 1–36.
Ryan, B. F., Joiner, B. L. and Ryan, T. A., Jr (1985) *Minitab Handbook*. Duxbury Press, Boston.

A.3 Hocking and Pendleton's problem

These data were specially constructed by Hocking and Pendleton (1983). It is intended that a multiple regression analysis be performed for regressing Y on X_1, X_2 and X_3 (with a constant term).

Table A.3 (Reprinted from Hocking and Pendleton, 1983, p. 517, by courtesy of Marcel Dekker, Inc.)

	X_1	X_2	X_3	Y
1	12.980	0.317	9.998	57.702
2	14.295	2.028	6.776	59.296
3	15.531	5.305	2.947	56.166
4	15.133	4.738	4.201	55.767
5	15.342	7.038	2.053	51.722
6	17.149	5.982	−0.055	60.446
7	15.462	2.737	4.657	60.715
8	12.801	10.663	3.048	37.447
9	17.039	5.132	0.257	60.974
10	13.172	2.039	8.738	55.270
11	16.125	2.271	2.101	59.289
12	14.340	4.077	5.545	54.027
13	12.923	2.643	9.331	53.199
14	14.231	10.401	1.041	41.896
15	15.222	1.220	6.149	63.264
16	15.740	10.612	−1.691	45.798
17	14.958	4.815	4.111	58.699
18	14.125	3.153	8.453	50.086
19	16.391	9.698	−1.714	48.890
20	16.452	3.912	2.145	62.213
21	13.535	7.625	3.851	45.625
22	14.199	4.474	5.112	53.923
23	15.837	5.753	2.087	55.799
24	16.565	8.546	8.974	56.741
25	13.322	8.598	4.011	43.145
26	15.949	8.290	−0.248	50.706

Hocking, R. R. and Pendleton, O. J. (1983) The regression dilemma. *Commun. Statist. Theor. Methods*, **12** (5), 497–527, Marcel Dekker, Inc. NY.

A.4 Salinity forecasting

These data were obtained as part of a project to predict summer shrimp harvests from late-spring data, in North Carolina's Pimlico Sound. The particular data given below concern the prediction of salinity and this was a relatively small part of the substantive project. The data were collected during consecutive two-weekly periods in March to May for the years 1972–1977. The data recorded were 'Salinity', 'Sallag', which is the salinity lagged by two weeks, the fresh water flow 'Flow', the year, and the number of the two-weekly period for the year.

Table A.4

Observation	*Salinity*	*Sallag*	*Flow*	*Period*	*Year*
1	7.6	8.2	23.005	4	1972
2	7.7	7.6	23.873	5	
3	4.3	4.6	26.417	0	1973
4	5.9	4.3	24.868	1	
5	5.0	5.9	29.895	2	
6	6.5	5.0	24.200	3	
7	8.3	6.5	23.215	4	
8	8.2	8.3	21.862	5	
9	13.2	10.1	22.274	0	1974
10	12.6	13.2	23.830	1	
11	10.4	12.6	25.144	2	
12	10.8	10.4	22.430	3	
13	13.1	10.8	21.785	4	
14	12.3	13.1	22.380	5	
15	10.4	13.3	23.927	0	1975
16	10.5	10.4	33.443	1	
17	7.7	10.5	24.859	2	
18	9.5	7.7	22.686	3	
19	12.0	10.0	21.789	0	1976
20	12.6	12.0	22.041	1	
21	13.6	12.1	21.033	4	
22	14.1	13.6	21.005	5	
23	13.5	15.0	25.865	0	1977
24	11.5	13.5	26.290	1	
25	12.0	11.5	22.932	2	
26	13.0	12.0	21.313	3	
27	14.1	13.0	20.769	4	
28	15.1	14.1	21.393	5	

Carroll, R. J. and Ruppert, D. (1985) Transformations in regression: a robust analysis. *Technometrics*, **27**, 1–12.

Ruppert, D. and Carroll, R. J. (1980) Trimmed least squares estimation in the linear model. *J. Amer. Statist. Assoc.*, **75**, 828–838.

Author index

Reference lists are indicated by *italic page numbers*

Subject index

Figures and Tables are indicated by *italic page numbers.*